COMBINING PATTERN CLASSIFIERS

COMBINING PATTERN CLASSIFIERS

Methods and Algorithms

Second Edition

LUDMILA I. KUNCHEVA

Published by John Wiley & Sons, Inc., Hoboken, New Jersey.
Published simultaneously in Canada.

Library of Congress Cataloging-in-Publication Data

Kuncheva, Ludmila I. (Ludmila Ilieva), 1959–
  Combining pattern classifiers : methods and algorithms / Ludmila I. Kuncheva. – Second edition.
    pages cm
  Includes index.
  ISBN 978-1-118-31523-1 (hardback)
1. Pattern recognition systems.   2. Image processing–Digital techniques.   I. Title.
  TK7882.P3K83 2014
  006.4–dc23
                                                                    2014014214

Printed in the United States of America.

10 9 8 7 6 5 4 3 2 1

To Roumen, Diana and Kamelia

CONTENTS

2   Base Classifiers   49

PREFACE

Pattern recognition is everywhere. It is the technology behind automatically identifying fraudulent bank transactions, giving verbal instructions to your mobile phone, predicting oil deposit odds, or segmenting a brain tumour within a magnetic resonance image.

A decade has passed since the first edition of this book. Combining classifiers, also known as "classifier ensembles," has flourished into a prolific discipline. Viewed from the top, classifier ensembles reside at the intersection of engineering, computing, and mathematics. Zoomed in, classifier ensembles are fuelled by advances in pattern recognition, machine learning and data mining, among others. An ensemble aggregates the "opinions" of several pattern classifiers in the hope that the new opinion will be better than the individual ones. *Vox populi, vox Dei.*

The interest in classifier ensembles received a welcome boost due to the high-profile Netflix contest. The world's research creativeness was challenged using a difficult task and a substantial reward. The problem was to predict whether a person will enjoy a movie based on their past movie preferences. A Grand Prize of $1,000,000 was to be awarded to the team who first achieved a 10% improvement on the classification accuracy of the existing system Cinematch. The contest was launched in October 2006, and the prize was awarded in September 2009. The winning solution was nothing else but a rather fancy classifier ensemble.

What is wrong with the good old single classifiers? Jokingly, I often put up a slide in presentations, with a multiple-choice question. The question is "Why classifier ensembles?" and the three possible answers are:

(a) because we like to complicate entities beyond necessity (anti-Occam's razor);

(b) because we are lazy and stupid and cannot be bothered to design and train one single sophisticated classifier; and

(c) because democracy is so important to our society, it must be important to classification.

Funnily enough, the real answer hinges on choice (b). Of course, it is not a matter of laziness or stupidity, but the realization that a complex problem can be elegantly solved using simple and manageable tools. Recall the invention of the error back-propagation algorithm followed by the dramatic resurfacing of neural networks in the 1980s. Neural networks were proved to be universal approximators with unlimited flexibility. They could approximate any classification boundary in any number of dimensions. This capability, however, comes at a price. Large structures with a vast number of parameters have to be trained. The initial excitement cooled down as it transpired that massive structures cannot be easily trained with sufficient guarantees of good generalization performance. Until recently, a typical neural network classifier contained one hidden layer with a dozen neurons, sacrificing the so acclaimed flexibility but gaining credibility. Enter classifier ensembles! Ensembles of simple neural networks are among the most versatile and successful ensemble methods.

But the story does not end here. Recent studies have rekindled the excitement of using massive neural networks drawing upon hardware advances such as parallel computations using graphics processing units (GPU) [75]. The giant data sets necessary for training such structures are generated by small distortions of the available set. These conceptually different rival approaches to machine learning can be regarded as divide-and-conquer and brute force, respectively. It seems that the jury is still out about their relative merits. In this book we adopt the divide-and-conquer approach.

THE PLAYING FIELD

Writing the first edition of the book felt like the overwhelming task of bringing structure and organization to a hoarder's attic. The scenery has changed markedly since then. The series of workshops on Multiple Classifier Systems (MCS), run since 2000 by Fabio Roli and Josef Kittler [338], served as a beacon, inspiration, and guidance for experienced and new researchers alike. Excellent surveys shaped the field, among which are the works by Polikar [311], Brown [53], and Valentini and Re [397]. Better still, four recent texts together present accessible, in-depth, comprehensive, and exquisite coverage of the classifier ensemble area: Rokach [335], Zhou [439], Schapire and Freund [351], and Seni and Elder [355]. This gives me the comfort and luxury to be able to skim over topics which are discussed at length and in-depth elsewhere, and pick ones which I believe deserve more exposure or which I just find curious.

As in the first edition, I have no ambition to present an accurate snapshot of the state of the art. Instead, I have chosen to explain and illustrate some methods and algorithms, giving sufficient detail so that the reader can reproduce them in code.

Although I venture an opinion based on general consensus and examples in the text, this should not be regarded as a guide for preferring one method to another.

SOFTWARE

A rich set of classifier ensemble methods is implemented in WEKA[1] [167], a collection of machine learning algorithms for data-mining tasks. PRTools[2] is a MATLAB toolbox for pattern recognition developed by the Pattern Recognition Research Group of the TU Delft, The Netherlands, led by Professor R. P. W. (Bob) Duin. An industry-oriented spin-off toolbox, called "perClass"[3] was designed later. Classifier ensembles feature prominently in both packages.

PRTools and perClass are instruments for advanced MATLAB programmers and can also be used by practitioners after a short training. The recent edition of MATLAB Statistics toolbox (2013b) includes a classifier ensemble suite as well.

Snippets of MATLAB DIY (do-it-yourself) code for illustrating methodologies and concepts are given in the chapter appendices. MATLAB was seen as a suitable language for such illustrations because it often looks like executable pseudo-code. A programming language is like a living creature—it grows, develops, changes, and breeds. The code in the book is written by today's versions, styles, and conventions. It does not, by any means, measure up to the richness, elegance, and sophistication of PRTools and perClass. Aimed at simplicity, the code is not fool-proof nor is it optimized for time or other efficiency criteria. Its sole purpose is to enable the reader to grasp the ideas and run their own small-scale experiments.

STRUCTURE AND WHAT IS NEW IN THE SECOND EDITION

The book is organized as follows.

Chapter 1, Fundamentals, gives an introduction of the main concepts in pattern recognition, Bayes decision theory, and experimental comparison of classifiers. A new treatment of the classifier comparison issue is offered (after Demšar [89]). The discussion of bias and variance decomposition of the error which was given in a greater level of detail in Chapter 7 before (bagging and boosting) is now briefly introduced and illustrated in Chapter 1.

Chapter 2, Base Classifiers, contains methods and algorithms for designing the individual classifiers. In this edition, a special emphasis is put on the stability of the classifier models. To aid the discussions and illustrations throughout the book, a toy two-dimensional data set was created called the fish data. The Naïve Bayes classifier and the support vector machine classifier (SVM) are brought to the fore as they are often used in classifier ensembles. In the final section of this chapter, I introduce the triangle diagram that can enrich the analyses of pattern recognition methods.

---

[1]http://www.cs.waikato.ac.nz/ml/weka/
[2]http://prtools.org/
[3]http://perclass.com/index.php/html/

Chapter 3, Multiple Classifier Systems, discusses some general questions in combining classifiers. It has undergone a major makeover. The new final section, "Quo Vadis?," asks questions such as "Are we reinventing the wheel?" and "Has the progress thus far been illusory?" It also contains a bibliometric snapshot of the area of classifier ensembles as of January 4, 2013 using Thomson Reuters' Web of Knowledge (WoK).

Chapter 4, Combining Label Outputs, introduces a new theoretical framework which defines the optimality conditions of several fusion rules by progressively relaxing an assumption. The Behavior Knowledge Space method is trimmed down and illustrated better in this edition. The combination method based on singular value decomposition (SVD) has been dropped.

Chapter 5, Combining Continuous-Valued Outputs, summarizes classifier fusion methods such as simple and weighted average, decision templates and a classifier used as a combiner. The division of methods into class-conscious and class-independent in the first edition was regarded as surplus and was therefore abandoned.

Chapter 6, Ensemble Methods, grew out of the former Bagging and Boosting chapter. It now accommodates on an equal keel the reigning classics in classifier ensembles: bagging, random forest, AdaBoost and random subspace, as well as a couple of newcomers: rotation forest and random oracle. The Error Correcting Output Code (ECOC) ensemble method is included here, having been cast as "Miscellanea" in the first edition of the book. Based on the interest in this method, as well as its success, ECOC's rightful place is together with the classics.

Chapter 7, Classifier Selection, explains why this approach works and how classifier competence regions are estimated. The chapter contains new examples and illustrations.

Chapter 8, Diversity, gives a modern view on ensemble diversity, raising at the same time some old questions, which are still puzzling the researchers in spite of the remarkable progress made in the area. There is a frighteningly large number of possible "new" diversity measures, lurking as binary similarity and distance measures (take for example Choi et al.'s study [74] with 76, s-e-v-e-n-t-y s-i-x, such measures). And we have not even touched the continuous-valued outputs and the possible diversity measured from those. The message in this chapter is stronger now: we hardly need any more diversity measures; we need to pick a few and learn how to use them. In view of this, I have included a theoretical bound on the kappa-error diagram [243] which shows how much space is still there for new ensemble methods with engineered diversity.

Chapter 9, Ensemble Feature Selection, considers feature selection *by* the ensemble and *for* the ensemble. It was born from a section in the former Chapter 8, Miscellanea. The expansion was deemed necessary because of the surge of interest to ensemble feature selection from a variety of application areas, notably so from bioinformatics [346]. I have included a stability index between feature subsets or between feature rankings [236].

I picked a figure from each chapter to create a small graphical guide to the contents of the book as illustrated in Figure 1.

The former Theory chapter (Chapter 9) was dissolved; parts of it are now blended with the rest of the content of the book. Lengthier proofs are relegated to the respective

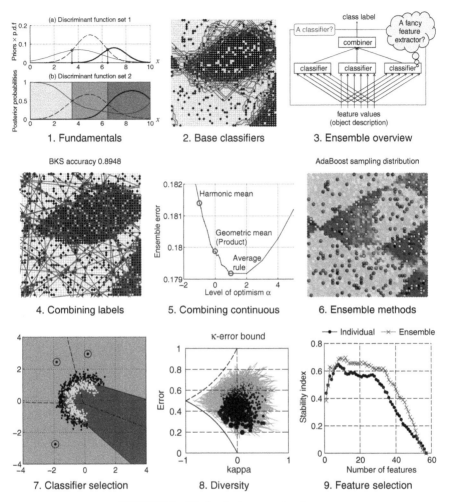

FIGURE 1   The book chapters at a glance.

chapter appendices. Some of the proofs and derivations were dropped altogether, for example, the theory behind the magic of AdaBoost. Plenty of literature sources can be consulted for the proofs and derivations left out.

The differences between the two editions reflect the fact that the classifier ensemble research has made a giant leap; some methods and techniques discussed in the first edition did not withstand the test of time, others were replaced with modern versions. The dramatic expansion of some sub-areas forced me, unfortunately, to drop topics such as cluster ensembles and stay away from topics such as classifier ensembles for: adaptive (on-line) learning, learning in the presence of concept drift, semi-supervised learning, active learning, handing imbalanced classes and missing values. Each of these sub-areas will likely see a bespoke monograph in a not so distant future. I look forward to that.

I am humbled by the enormous volume of literature on the subject, and the ingenious ideas and solutions within. My sincere apology to those authors, whose excellent research into classifier ensembles went without citation in this book because of lack of space or because of unawareness on my part.

WHO IS THIS BOOK FOR?

The book is suitable for postgraduate students and researchers in computing and engineering, as well as practitioners with some technical background. The assumed level of mathematics is minimal and includes a basic understanding of probabilities and simple linear algebra. Beginner's MATLAB programming knowledge would be beneficial but is not essential.

LUDMILA I. KUNCHEVA

Bangor, Gwynedd, UK
December 2013

ACKNOWLEDGEMENTS

I am most sincerely indebted to Gavin Brown, Juan Rodríguez, and Kami Kountcheva for scrutinizing the manuscript and returning to me their invaluable comments, suggestions, and corrections. Many heartfelt thanks go to my family and friends for their constant support and encouragement. Last but not least, thank you, my reader, for picking up this book.

LUDMILA I. KUNCHEVA

Bangor, Gwynedd, UK
December 2013

1

FUNDAMENTALS OF PATTERN RECOGNITION

1.1 BASIC CONCEPTS: CLASS, FEATURE, DATA SET

A wealth of literature in the 1960s and 1970s laid the grounds for modern pattern recognition [90,106,140,141,282,290,305,340,353,386]. Faced with the formidable challenges of real-life problems, elegant theories still coexist with ad hoc ideas, intuition, and guessing.

Pattern recognition is about assigning labels to objects. Objects are described by features, also called attributes. A classic example is recognition of handwritten digits for the purpose of automatic mail sorting. Figure 1.1 shows a small data sample. Each 15×15 image is one object. Its class label is the digit it represents, and the features can be extracted from the binary matrix of pixels.

1.1.1 Classes and Class Labels

Intuitively, a class contains similar objects, whereas objects from different classes are dissimilar. Some classes have a clear-cut meaning, and in the simplest case are mutually exclusive. For example, in signature verification, the signature is either genuine or forged. The true class is one of the two, regardless of what we might deduce from the observation of a particular signature. In other problems, classes might be difficult to define, for example, the classes of left-handed and right-handed people or ordered categories such as "low risk," "medium risk," and "high risk."

Combining Pattern Classifiers: Methods and Algorithms, Second Edition. Ludmila I. Kuncheva.
© 2014 John Wiley & Sons, Inc. Published 2014 by John Wiley & Sons, Inc.

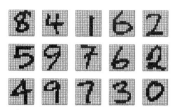

FIGURE 1.1   Example of images of handwritten digits.

We shall assume that there are c possible classes in the problem, labeled from ω_1 to ω_c, organized as a set of labels $\Omega = \{\omega_1, \dots, \omega_c\}$, and that each object belongs to one and only one class.

1.1.2   Features

Throughout this book we shall consider numerical features. Such are, for example, systolic blood pressure, the speed of the wind, a company's net profit in the past 12 months, the gray-level intensity of a pixel. Real-life problems are invariably more complex than that. Features can come in the forms of categories, structures, names, types of entities, hierarchies, so on. Such nonnumerical features can be transformed into numerical ones. For example, a feature "country of origin" can be encoded as a binary vector with number of elements equal to the number of possible countries where each bit corresponds to a country. The vector will contain 1 for a specified country and zeros elsewhere. In this way one feature gives rise to a collection of related numerical features. Alternatively, we can keep just the one feature where the categories are represented by different values. Depending on the classifier model we choose, the ordering of the categories and the scaling of the values may have a positive, negative, or neutral effect on the relevance of the feature. Sometimes the methodologies for quantifying features are highly subjective and heuristic. For example, sitting an exam is a methodology to quantify a student's learning progress. There are also unmeasurable features that we as humans can assess intuitively but can hardly explain. Examples of such features are sense of humor, intelligence, and beauty.

Once in a numerical format, the feature values for a given object are arranged as an n-dimensional vector $\mathbf{x} = [x_1, \dots, x_n]^T \in \mathbb{R}^n$. The real space $\mathbb{R}^n$ is called the *feature space*, each axis corresponding to a feature.

Sometimes an object can be represented by multiple, disjoint subsets of features. For example, in identity verification, three different sensing modalities can be used [207]: frontal face, face profile, and voice. Specific feature subsets are measured for each modality and then the feature vector is composed of three sub-vectors, $\mathbf{x} = [\mathbf{x}^{(1)}, \mathbf{x}^{(2)}, \mathbf{x}^{(3)}]^T$. We call this *distinct pattern representation* after Kittler et al. [207]. As we shall see later, an ensemble of classifiers can be built using distinct pattern representation, with one classifier on each feature subset.

1.1.3   Data Set

The information needed to design a classifier is usually in the form of a labeled *data set* $\mathbf{Z} = \{\mathbf{z}_1, \dots, \mathbf{z}_N\}$, $\mathbf{z}_j \in \mathbb{R}^n$. The class label of $\mathbf{z}_j$ is denoted by $y_j \in \Omega$, $j = 1, \dots, N$. A typical data set is organized as a matrix of N rows (objects, also called examples or instances) by n columns (features), with an extra column with the class labels

$$
\text{Data set} = \begin{bmatrix} z_{11}, & z_{12}, & \cdots & z_{1n} \\ z_{21}, & z_{22}, & \cdots & z_{2n} \\ \vdots & & & \\ z_{N1}, & z_{N2}, & \cdots & z_{Nn} \end{bmatrix} \quad \text{Labels} = \begin{bmatrix} y_1 \\ y_2 \\ \vdots \\ y_N \end{bmatrix}.
$$

Entry $z_{j,i}$ is the value of the i-th feature for the j-th object.

Example 1.1   A shape–color synthetic data set

Consider a data set with two classes, both containing a collection of the following objects: $\triangle$, $\square$, $\bigcirc$, $\blacktriangle$, $\blacksquare$, and $\bullet$. Figure 1.2 shows an example of such a data set. The collections of objects for the two classes are plotted next to one another. Class ω_1 is shaded. The features are only the shape and the color (black or white); the positioning of the objects within the two dimensions is not relevant. The data set contains 256 objects. Each object is labeled in its true class. We can code the color as 0 for white and 1 for black, and the shapes as triangle $= 1$, square $= 2$, and circle $= 3$.

FIGURE 1.2   A shape–color data set example. Class ω_1 is shaded.

Based on the two features, the classes are not completely separable. It can be observed that there are mostly circles in ω_1 and mostly squares in ω_2. Also, the proportion of black objects in class ω_2 is much larger. Thus, if we observe a color and a shape, we can make a decision about the class label. To evaluate the distribution of different objects in the two classes, we can count the number of appearances of each object. The distributions are as follows:

Object	△	□	○	▲	■	●
Class ω_1	9	22	72	1	4	20
Class ω_2	4	25	5	8	79	7
Decision	ω_1	ω_2	ω_1	ω_2	ω_2	ω_1

With the distributions obtained from the given data set, it makes sense to choose class ω_1 if we have a circle (of any color) or a white triangle. For all other possible combinations of values, we should choose label ω_2. Thus using only these two features for labeling, we will make 43 errors (16.8%).

A couple of questions spring to mind. First, if the objects are not discernible, how have they been labeled in the first place? Second, how far can we trust the estimated distributions to generalize over unseen data?

To answer the first question, we should be aware that the features supplied by the user are not expected to be perfect. Typically there is a way to determine the true class label, but the procedure may not be available, affordable, or possible at all. For example, certain medical conditions can be determined only *post mortem*. An early diagnosis inferred through pattern recognition may decide the outcome for the patient. As another example, consider classifying of expensive objects on a production line as good or defective. Suppose that an object has to be destroyed in order to determine the true label. It is desirable that the labeling is done using measurable features that do not require breaking of the object. Labeling may be too expensive, involving time and expertise which are not available. The problem then becomes a pattern recognition one, where we try to find the class label as correctly as possible from the available features.

Returning to the example in Figure 1.2, suppose that there is a third (unavailable) feature which could be, for example, the horizontal axis in the plot. This feature would have been used to label the data, but the quest is to find the best possible labeling method without it.

The second question "How far can we trust the estimated distributions to generalize over unseen data?" has inspired decades of research and will be considered later in this text.

▉ Example 1.2   The Iris data set

The Iris data set was collected by the American botanist Edgar Anderson and subsequently analyzed by the English geneticist and statistician Sir Ronald Aylmer Fisher in 1936 [127]. The Iris data set has become one of the iconic hallmarks of pattern

FIGURE 1.3   Iris flower specimen

recognition and has been used in thousands of publications over the years [39, 348]. This book would be incomplete without a mention of it.

The Iris data still serves as a prime example of a "well-behaved" data set. There are three balanced classes, each represented with a sample of 50 objects. The classes are species of the Iris flower (Figure 1.3): setosa, versicolor, and virginica. The four features describing an Iris flower are sepal length, sepal width, petal length, and petal width. The classes form neat elliptical clusters in the four-dimensional space. Scatter plots of the data in the spaces spanned by the six pairs of features are displayed in Figure 1.4. Class setosa is clearly distinguishable from the other two classes in all projections.

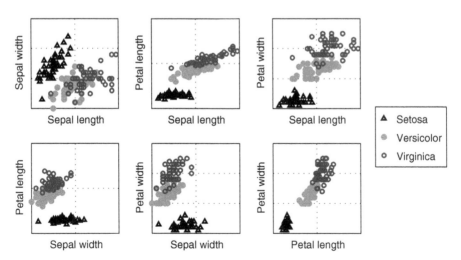

FIGURE 1.4   Scatter plot of the Iris data in the two-dimensional spaces spanned by the six pairs of features.

1.1.4 Generate Your Own Data

Trivial as it might be, sometimes you need a piece of code to generate your own data set with specified characteristics in order to test your own classification method.

1.1.4.1 The Normal Distribution   The normal distribution (or also Gaussian distribution) is widespread in nature and is one of the fundamental models in statistics. The one-dimensional normal distribution, denoted $N(\mu, \sigma^2)$, is characterized by mean $\mu \in \mathbb{R}$ and variance $\sigma^2 \in \mathbb{R}$. In n dimensions, the normal distribution is characterized by an n-dimensional vector of the mean, $\boldsymbol{\mu} \in \mathbb{R}^n$, and an $n \times n$ covariance matrix Σ. The notation for an n-dimensional normally distributed random variable is $\mathbf{x} \sim N(\boldsymbol{\mu}, \Sigma)$. The normal distribution is the most natural assumption reflecting the following situation: there is an "ideal prototype" ($\boldsymbol{\mu}$) and all the data are distorted versions of it. Small distortions are more likely to occur than large distortions, causing more objects to be located in the close vicinity of the ideal prototype than far away from it. The scatter of the points around the prototype $\boldsymbol{\mu}$ is associated with the covariance matrix Σ_i.

The probability density function (pdf) of $\mathbf{x} \sim N(\boldsymbol{\mu}, \Sigma)$ is

$$p(\mathbf{x}) = \frac{1}{(2\pi)^{\frac{n}{2}} \sqrt{|\Sigma|}} \exp\left\{ -\frac{1}{2}(\mathbf{x} - \boldsymbol{\mu})^T \Sigma^{-1}(\mathbf{x} - \boldsymbol{\mu}) \right\}, \tag{1.1}$$

where $|\Sigma|$ is the determinant of Σ. For the one-dimensional case, x and μ are scalars, and Σ reduces to the variance σ^2. Equation 1.1 simplifies to

$$p(x) = \frac{1}{\sqrt{2\pi}\,\sigma} \exp\left\{ -\frac{1}{2}\left(\frac{x - \mu}{\sigma}\right)^2 \right\}. \tag{1.2}$$

⬛ **Example 1.3   Cloud shapes and the corresponding covariance matrices**
Figure 1.5 shows four two-dimensional data sets generated from the normal distribution with different covariance matrices shown underneath.

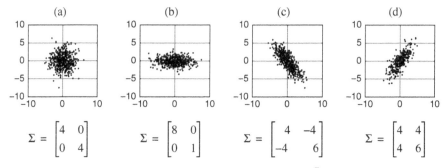

FIGURE 1.5   Normally distributed data sets with mean $[0, 0]^T$ and different covariance matrices shown underneath.

Figures 1.5a and 1.5b are generated with *independent* (noninteracting) features. Therefore, the data cloud is either spherical (Figure 1.5a), or stretched along one or more coordinate axes (Figure 1.5b). Notice that for these cases the off-diagonal entries of the covariance matrix are zeros. Figures 1.5c and 1.5d represent cases where the features are *dependent*. The data for this example was generated using the function samplegaussian in Appendix 1.A.1.

In the case of *independent features* we can decompose the n-dimensional pdf as a product of n one-dimensional pdfs. Let σ_k^2 be the diagonal entry of the covariance matrix Σ for the k-th feature, and μ_k be the k-th component of $\boldsymbol{\mu}$. Then

$$p(\mathbf{x}) = \frac{1}{(2\pi)^{\frac{n}{2}} \sqrt{|\Sigma|}} \exp\left\{ -\frac{1}{2}(\mathbf{x} - \boldsymbol{\mu})^T \Sigma^{-1}(\mathbf{x} - \boldsymbol{\mu}) \right\}$$

$$= \prod_{k=1}^{n} \left(\frac{1}{\sqrt{(2\pi)}\,\sigma_k} \exp\left\{ -\frac{1}{2}\left(\frac{x_k - \mu_k}{\sigma_k} \right)^2 \right\} \right). \tag{1.3}$$

The *cumulative distribution function* for a random variable $X \in \mathbb{R}$ with a normal distribution, $\Phi(z) = P(X \leq z)$, is available in tabulated form from most statistical textbooks.[1]

1.1.4.2 Noisy Geometric Figures

Sometimes it is useful to generate your own data set of a desired shape, prevalence of the classes, overlap, and so on. An example of a challenging classification problem with five Gaussian classes is shown in Figure 1.6 along with the MATLAB code that generates and plots the data.

One possible way to generate data with specific geometric shapes is detailed below. Suppose that each of the c classes is described by a shape, governed by parameter t.

```
1   x1 = samplegaussian(300,[0 0],[4 0;0 4]);
2   x2 = samplegaussian(300,[4 6],[8 0;0 1]);
3   x3 = samplegaussian(300,[-2 5],[4 -4;-4 6]);
4   x4 = samplegaussian(300,[3 -2],[4 4;4 6]);
5   x5 = samplegaussian(300,[7 0],[1 0;0 10]);
6   data = [x1;x2;x3;x4;x5];
7   t = ones(300,1);
8   labels = [t;2*t;3*t;4*t;5*t];
9   scatter(data(:,1),data(:,2),[],...
10      labels,'linewidth',2.5)
11  axis equal off
```

FIGURE 1.6   An example of five Gaussian classes generated using the samplegaussian function from Appendix 1.A.1.

---

[1] $\Phi(z)$ can be approximated with error at most 0.005 for $0 \leq z \leq 2.2$ as [150]

$$\Phi(z) = 0.5 + \frac{z(4.4 - z)}{10}.$$

The noise-free data is calculated from t, and then noise is added. Let t_i be the parameter for class ω_i, and $[a_i, b_i]$ be the interval for t_i describing the shape of the class. Denote by p_i the desired prevalence of class ω_i. Knowing that $p_1 + \cdots + p_c = 1$, we can calculate the approximate number of samples in a data set of N objects. Let N_i be the desired number of objects from class ω_i. The first step is to sample uniformly N_i values for t_i from the interval $[a_i, b_i]$. Subsequently, we find the coordinates $x_1, \ldots, x_n$ for each element of t_i. Finally, noise is added to all values. (We can use the `randn` MATLAB function for this purpose.) The noise could be scaled by multiplying the values by different constants for the different features. Alternatively, the noise could be scaled with the feature values or the values of t_i.

🖳 Example 1.4   Ellipses data set

The code for producing this data set is given in Appendix 1.A.1. We used the parametric equations for two-dimensional ellipses:

$$x(t) = x_c + a\cos(t)\cos(\phi) - b\sin(t)\sin(\phi),$$

$$y(t) = y_c + a\cos(t)\sin(\phi) - b\sin(t)\cos(\phi),$$

where (x_c, y_c) is the center of the ellipse, a and b are respectively the major and the minor semi-axes of the ellipse, and ϕ is the angle between the x-axis and the major axis. To traverse the whole ellipse, parameter t varies from 0 to 2π.

Figure 1.7a shows a data set where the random noise is the same across both features and all values of t. The classes have equal proportions, with 300 points from each class. Using a single ellipse with 1000 points, Figure 1.7b demonstrates the effect of scaling the noise with the parameter t. The MATLAB code is given in Appendix 1.A.1.

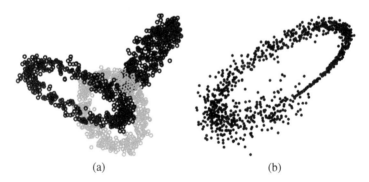

(a)	(b)

FIGURE 1.7   (a) The three-ellipse data set; (b) one ellipse with noise variance proportional to the parameter t.

1.1.4.3 *Rotated Checker Board Data.*   This is a two-dimensional data set which spans the unit square $[0, 1] \times [0, 1]$. The classes are placed as the light and the dark squares of a checker board and then the whole board is rotated at an angle α. A

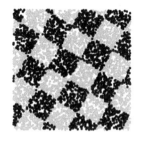 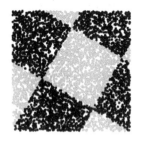

samplecb(5000,0.2,pi/6)   samplecb(5000,0.5,-pi/3)

$(a = 0.2, \alpha = \frac{\pi}{6})$ $(a = 0.5, \alpha = -\frac{\pi}{3})$

FIGURE 1.8   Rotated checker board data (100,000 points in each plot).

parameter a specifies the side of the individual square. For example, if $a = 0.5$, there will be four squares in total before the rotation. Figure 1.8 shows two data sets, each containing 5,000 points, generated with different input parameters. The MATLAB function samplecb(N,a,alpha) in Appendix 1.A.1 generates the data.

The properties which make this data set attractive for experimental purposes are:

- The two classes are perfectly separable.
- The classification regions for the same class are disjoint.
- The boundaries are not parallel to the coordinate axes.
- The classification performance will be highly dependent on the sample size.

1.2   CLASSIFIER, DISCRIMINANT FUNCTIONS, CLASSIFICATION REGIONS

A *classifier* is any function that will assign a class label to an object $\mathbf{x}$:

$$D : \mathbb{R}^n \to \Omega. \tag{1.4}$$

In the "canonical model of a classifier" [106], c *discriminant functions* are calculated

$$g_i : \mathbb{R}^n \to \mathbb{R}, \quad i = 1, \dots, c, \tag{1.5}$$

each one yielding a score for the respective class (Figure 1.9). The object $\mathbf{x} \in \mathbb{R}^n$ is labeled to the class with the highest score. This labeling choice is called the *maximum membership rule*. Ties are broken randomly, meaning that $\mathbf{x}$ is assigned randomly to one of the tied classes.

The discriminant functions partition the feature space $\mathbb{R}^n$ into c *decision regions* or *classification regions* denoted $\mathcal{R}_1, \dots, \mathcal{R}_c$:

$$\mathcal{R}_i = \left\{ \mathbf{x} \,\middle|\, \mathbf{x} \in \mathbb{R}^n, g_i(\mathbf{x}) = \max_{k=1,\dots,c} g_k(\mathbf{x}) \right\}, \quad i = 1, \dots, c. \tag{1.6}$$

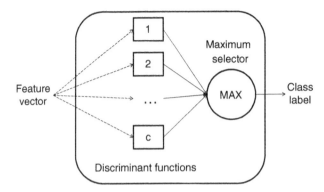

FIGURE 1.9 Canonical model of a classifier. An *n*-dimensional feature vector is passed through *c* discriminant functions, and the largest function output determines the class label.

The decision region for class ω_i is the set of points for which the *i*-th discriminant function has the highest score. According to the maximum membership rule, all points in decision region $\mathcal{R}_i$ are assigned to class ω_i. The decision regions are specified by the classifier D, or equivalently, by the discriminant functions G. The boundaries of the decision regions are called *classification boundaries* and contain the points for which the highest discriminant functions tie. A point on the boundary can be assigned to any of the bordering classes. If a decision region $\mathcal{R}_i$ contains data points from the labeled set **Z** with true class label ω_j, $j \neq i$, classes ω_i and ω_j are called *overlapping*. If the classes in **Z** can be separated completely by a hyperplane (a point in $\mathbb{R}$, a line in $\mathbb{R}^2$, a plane in $\mathbb{R}^3$), they are called *linearly separable*.

Note that overlapping classes in a given partition can be nonoverlapping if the space was partitioned in a different way. If there are no identical points with different class labels in the data set **Z**, we can *always* partition the feature space into pure classification regions. Generally, the smaller the overlapping, the better the classifier. Figure 1.10 shows an example of a two-dimensional data set and two sets of classification regions. Figure 1.10a shows the regions produced by the nearest neighbor classifier, where every point is labeled as its nearest neighbor. According to these boundaries and the plotted data, the classes are nonoverlapping. However, Figure 1.10b shows the optimal classification boundary and the optimal classification regions which guarantee the minimum possible error for unseen data generated from the same distributions. According to the optimal boundary, the classes are overlapping. This example shows that by striving to build boundaries that give a perfect split we may over-fit the training data.

Generally, *any* set of functions $g_1(\mathbf{x}), \dots, g_c(\mathbf{x})$ is a set of *discriminant functions*. It is another matter how successfully these discriminant functions separate the classes.

Let $G^* = \{g_1^*(\mathbf{x}), \dots, g_c^*(\mathbf{x})\}$ be a set of *optimal* (in some sense) discriminant functions. We can obtain infinitely many sets of *optimal* discriminant functions from G^* by applying a monotonic transformation $f(g_i^*(\mathbf{x}))$ that preserves the order of the function values for every $\mathbf{x} \in \mathbb{R}^n$. For example, $f(\zeta)$ can be a $\log(\zeta)$ or a^ζ, for $a > 1$.

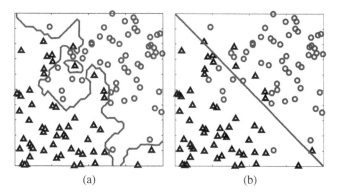

(a)   (b)

FIGURE 1.10 Classification regions obtained from two different classifiers: (a) the 1-nn boundary (nonoverlapping classes); (b) the optimal boundary (overlapping classes).

Applying the same f to all discriminant functions in G^*, we obtain an equivalent set of discriminant functions. Using the maximum membership rule, $\mathbf{x}$ will be labeled to the same class by any of the equivalent sets of discriminant functions.

1.3 CLASSIFICATION ERROR AND CLASSIFICATION ACCURACY

It is important to know how well our classifier performs. The *performance* of a classifier is a compound characteristic, whose most important component is the classification accuracy. If we were able to try the classifier on all possible input objects, we would know exactly how accurate it is. Unfortunately, this is hardly a possible scenario, so an estimate of the accuracy has to be used instead.

Classification error is a characteristic dual to the classification accuracy in that the two values sum up to 1

$$\text{Classification error} = 1 - \text{Classification accuracy}.$$

The quantity of interest is called the *generalization error*. This is the expected error of the trained classifier on unseen data drawn from the distribution of the problem.

1.3.1 Where Does the Error Come From? Bias and Variance

Why cannot we design the perfect classifier? Figure 1.11 shows a sketch of the possible sources of error. Suppose that we have chosen the classifier model. Even with a perfect training algorithm, our solution (marked as 1 in the figure) may be away from the best solution with this model (marked as 2). This *approximation error* comes from the fact that we have only a finite data set to train the classifier. Sometimes the training algorithm is not guaranteed to arrive at the optimal classifier with the given data. For example, the backpropagation training algorithm converges to a local

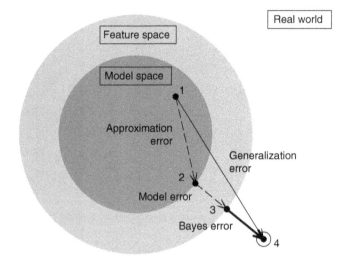

1: Our solution

2: Best possible solution with the chosen model

3: Best possible solution with the available features

4: The "real thing"

FIGURE 1.11   Composition of the generalization error.

minimum of the criterion function. If started from a different initialization point, the solution may be different. In addition to the approximation error, there may be a *model error*. Point 3 in the figure is the best possible solution in the given feature space. This point may not be achievable with the current classifier model. Finally, there is an irreducible part of the error, called *the Bayes error*. This error comes from insufficient representation. With the available features, two objects with the same feature values may have different class labels. Such a situation arose in Example 1.1.

Thus the true generalization error P_G of a classifier D trained on a given data set $\mathbf{Z}$ can be decomposed as

$$P_G(D, \mathbf{Z}) = P_A(\mathbf{Z}) + P_M + P_B, \tag{1.7}$$

where $P_A(\mathbf{Z})$ is the approximation error, P_M is the model error, and P_B is the Bayes error. The first term in the equation can be thought of as variance due to using different training data or non-deterministic training algorithms. The second term, P_M, can be taken as the bias of the model from the best possible solution.

The difference between bias and variance is explained in Figure 1.12. We can liken building the perfect classifier to shooting at a target. Suppose that our training algorithm generates different solutions owing to different data samples, different initialisations, or random branching of the training algorithm. If the solutions are

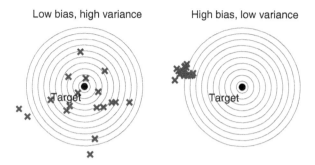

FIGURE 1.12   Bias and variance.

grouped together, variance is low. Then the distance to the target will be more due to the bias. Conversely, widely scattered solutions indicate large variance, and that can account for the distance between the shot and the target.

1.3.2   Estimation of the Error

Assume that a labeled data set $\mathbf{Z}_{ts}$ of size $N_{ts} \times n$ is available for testing the accuracy of our classifier, D. The most natural way to calculate an estimate of the error is to run D on all the objects in $\mathbf{Z}_{ts}$ and find the proportion of misclassified objects, called sometimes the *apparent error rate*

$$\hat{P}_D = \frac{N_{error}}{N_{ts}}. \tag{1.8}$$

Dual to this characteristic is the apparent classification accuracy which is calculated by $1 - \hat{P}_D$.

To look at the error from a probabilistic point of view, we can adopt the following model. The classifier commits an error with probability P_D on any object $\mathbf{x} \in \mathbb{R}^n$ (a wrong but useful assumption). Then the number of errors has a binomial distribution with parameters (P_D, N_{ts}). An estimate of P_D is $\hat{P}_D$. If N_{ts} and P_D satisfy the rule of thumb: $N_{ts} > 30$, $\hat{P}_D \times N_{ts} > 5$, and $(1 - \hat{P}_D) \times N_{ts} > 5$, the binomial distribution can be approximated by a normal distribution. The 95% confidence interval for the error is

$$\left[\hat{P}_D - 1.96 \sqrt{\frac{\hat{P}_D(1 - \hat{P}_D)}{N_{ts}}}, \quad \hat{P}_D + 1.96 \sqrt{\frac{\hat{P}_D(1 - \hat{P}_D)}{N_{ts}}} \right]. \tag{1.9}$$

By calculating the confidence interval we estimate how well *this* classifier (D) will fare on unseen data from the same problem. Ideally, we will have a large representative testing set, which will make the estimate precise.

1.3.3   Confusion Matrices and Loss Matrices

To find out how the errors are distributed across the classes we construct a *confusion matrix* using the testing data set, Z_{ts}. The entry a_{ij} of such a matrix denotes the number of elements from Z_{ts} whose true class is ω_i, and which are assigned by D to class ω_j. The estimate of the classification accuracy can be calculated as the trace of the matrix divided by the total sum of the entries. The additional information that the confusion matrix provides is *where* the misclassifications have occurred. This is important for problems with a large number of classes where a high off-diagonal entry of the matrix might indicate a difficult two-class problem that needs to be tackled separately.

▣ Example 1.5   Confusion matrix for the Letter data

The Letters data set, available from the UCI Machine Learning Repository Database, contains data extracted from 20,000 black-and-white images of capital English letters. Sixteen numerical features describe each image ($N = 20,000, c = 26, n = 16$). For the purpose of this illustration we used the hold-out method. The data set was randomly split into halves. One half was used for training a linear classifier, and the other half was used for testing. The labels of the testing data were matched to the labels obtained from the classifier, and the 26×26 confusion matrix was constructed. If the classifier was ideal, and all assigned and true labels were matched, the confusion matrix would be diagonal.

Table 1.1 shows the row in the confusion matrix corresponding to class "H." The entries show the number of times that true "H" is mistaken for the letter in the respective column. The boldface number is the diagonal entry showing how many times "H" has been correctly recognized. Thus, from the total of 350 examples of "H" in the testing set, only 159 have been labeled correctly by the classifier. Curiously, the largest number of mistakes, 33, are for the letter "O." Figure 1.13 visualizes the confusion matrix for the letter data set. Darker color signifies a higher value. The diagonal shows the darkest color, which indicates the high correct classification rate (over 69%). Three common misclassifications are indicated with arrows in the figure.

TABLE 1.1   The "H"-row in the Confusion Matrix for the Letter Data Set Obtained from a Linear Classifier Trained on 10,000 Points

"H" labeled as:	A	B	C	D	E	F	G	H	I	J	K	L	M
Times:	1	6	1	18	0	1	2	**159**	0	0	30	0	1

"H" labeled as:	N	O	P	Q	R	S	T	U	V	W	X	Y	Z
Times:	27	33	2	9	21	0	0	11	4	3	20	1	0

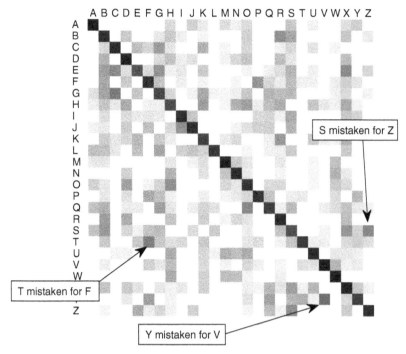

FIGURE 1.13   Graphical representation of the confusion matrix for the letter data set. Darker color signifies a higher value.

The errors in classification are not equally costly. To account for the different costs of mistakes, we introduce the *loss matrix*. We define a loss matrix with entries λ_{ij} denoting the loss incurred by assigning label ω_i, given that the true label of the object is ω_j. If the classifier is "unsure" about the label, it may refuse to make a decision. An extra class called "refuse-to-decide" can be added to the set of classes. Choosing the extra class should be less costly than choosing a wrong class. For a problem with c original classes and a refuse option, the loss matrix is of size $(c + 1) \times c$. Loss matrices are usually specified by the user. A zero–one loss matrix is defined as $\lambda_{ij} = 0$ for $i = j$ and $\lambda_{ij} = 1$ for $i \neq j$; that is, all errors are equally costly.

1.3.4   Training and Testing Protocols

The estimate $\hat{P}_D$ in Equation 1.8 is valid only for the given classifier D and the testing set from which it was calculated. It is possible to train a better classifier from different training data sampled from the distribution of the problem. What if we seek to answer the question "How well can this *classifier model* solve the problem?"

Suppose that we have a data set $\mathbf{Z}$ of size $N \times n$, containing n-dimensional feature vectors describing N objects. We would like to use as much as possible of the data to build the classifier (*training*), and also as much as possible unseen data to test its performance (*testing*). However, if we use all data for training and the same data for testing, we might *overtrain* the classifier. It could learn perfectly the available data but its performance on unseen data cannot be predicted. That is why it is important to have a separate data set on which to examine the final product. The most widely used training/testing protocols can be summarized as follows [216]:

- *Resubstitution*. Design classifier D on $\mathbf{Z}$ and test it on $\mathbf{Z}$. $\hat{P}_D$ is likely optimistically biased.
- *Hold-out*. Traditionally, split $\mathbf{Z}$ randomly into halves; use one half for training and the other half for calculating $\hat{P}_D$. Splits in other proportions are also used.
- *Repeated hold-out (Data shuffle)*. This is a version of the hold-out method where we do L random splits of $\mathbf{Z}$ into training and testing parts and average all L estimates of P_D calculated on the respective testing parts. The usual proportions are 90% for training and 10% for testing.
- *Cross-validation*. We choose an integer K (preferably a factor of N) and randomly divide $\mathbf{Z}$ into K subsets of size N/K. Then we use one subset to test the performance of D trained on the union of the remaining $K - 1$ subsets. This procedure is repeated K times choosing a different part for testing each time.
    To get the final value of $\hat{P}_D$ we average the K estimates.
    To reduce the effect of the single split into K folds, we can carry out repeated cross-validation. In an $M \times K$-fold cross validation, the data is split M times into K folds, and a cross-validation is performed on each such split. This procedure results in $M \times K$ estimates of $\hat{P}_D$, whose average produces the desired estimate. A 10×10-fold cross-validation is a typical choice of such a protocol.
- *Leave-one-out*. This is the cross-validation protocol where $K = N$, that is, one object is left aside, the classifier is trained on the remaining $N - 1$ objects, and the left out object is classified. $\hat{P}_D$ is the proportion of the N objects misclassified in their respective cross-validation fold.
- *Bootstrap*. This method is designed to correct for the optimistic bias of resubstitution. This is done by randomly sampling with replacement L sets of cardinality N from the original set $\mathbf{Z}$. Approximately 37% ($1/e$) of the data will not be chosen in a bootstrap replica. This part of the data is called the "out-of-bag" data. The classifier is built on the bootstrap replica and assessed on the out-of-bag data (testing data). L such classifiers are trained, and the error rates on the respective testing data are averaged. Sometimes the resubstitution and the out-of-bag error rates are taken together with different weights [216].

Hold-out, repeated hold-out and cross-validation can be carried out with *stratified sampling*. This means that the proportions of the classes are preserved as close as possible in all folds.

Pattern recognition has now outgrown the stage where the computation resource (or lack thereof) was the decisive factor as to which method to use. However, even with the modern computing technology, the problem has not disappeared. The ever growing sizes of the data sets collected in different fields of science and practice pose a new challenge. We are back to using the good old hold-out method, first because the others might be too time-consuming, and second, because the amount of data might be so excessive that small parts of it will suffice for training and testing. For example, consider a data set obtained from retail analysis, which involves hundreds of thousands of transactions. Using an estimate of the error over, say, 10,000 data points, can conveniently shrink the confidence interval and make the estimate sufficiently reliable.

It is now becoming common practice to use three instead of two data sets: one for training, one for *validation*, and one for testing. As before, the testing set remains unseen during the training process. The validation data set acts as pseudo-testing. We continue the training process until the performance improvement on the training set is no longer matched by a performance improvement on the validation set. At this point the training should be stopped so as to avoid overtraining. Not all data sets are large enough to allow for a validation part to be cut out. Many of the data sets from the UCI Machine Learning Repository Database[2] [22], often used as benchmarks in pattern recognition and machine learning, may be unsuitable for a three-way split into training/validation/testing. The reason is that the data subsets will be too small and the estimates of the error on these subsets would be unreliable. Then stopping the training at the point suggested by the validation set might be inadequate, the estimate of the testing accuracy might be inaccurate, and the classifier might be poor because of the insufficient training data.

When multiple training and testing sessions are carried out, there is the question of which of the classifiers built during this process we should use in the end. For example, in a 10-fold cross-validation, we build 10 different classifiers using different data subsets. The above methods are only meant to give us an estimate of the accuracy of a certain model built for the problem at hand. We rely on the assumption that the classification accuracy will change smoothly with the changes in the size of the training data [99]. Therefore, if we are happy with the accuracy and its variability across different training subsets, we should finally train a our chosen classifier on the whole data set.

1.3.5   Overtraining and Peeking

Testing should be done on previously *unseen* data. All parameters should be tuned on the training data. A common mistake in classification experiments is to select a feature set using the given data, and *then* run experiments with one of the above protocols to evaluate the accuracy of that set. This problem is widespread in bioinformatics and neurosciences, aptly termed "peeking" [308, 346, 348, 370]. Using the same data is likely to lead to an optimistic bias of the error.

▣   Example 1.6   Tuning a parameter on the testing set is wrong
Let $D(r)$ be a classifier with a parameter r such that varying r leads to different training accuracies. To demonstrate this effect, here we took a random training sample of

---

[2]http://www.ics.uci.edu/~mlearn/MLRepository.html

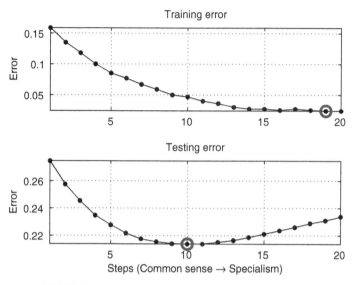

FIGURE 1.14   Example of overtraining: letter data set.

1000 objects from the letters data set. The remaining 19,000 objects were used for testing. A quadratic discriminant classifier (QDC) was used.[3] We vary a parameter r, $r \in [0, 1]$, called the regularization parameter, which determines to what extent we sacrifice adjustment to the given data in favor of robustness. For $r = 0$ there is no regularization; we have more accuracy on the training data and less certainty that the classifier will perform well on unseen data. For $r = 1$, the classifier might be less accurate on the training data but can be expected to perform at the same rate on unseen data. This dilemma can be translated into everyday language as "specific expertise" versus "common sense." If the classifier is trained to expertly recognize a certain data set, it might have this data-specific expertise and little common sense. This will show as high testing error. Conversely, if the classifier is trained to have good common sense, even if not overly successful on the training data, we might expect it to have common sense with any data set drawn from the same distribution.

In the experiment, r was decreased for 20 steps, starting with $r_0 = 0.4$ and taking r_{k+1} to be $0.8 \times r_k$. Figure 1.14 shows the training and the testing errors for the 20 steps.

This example is intended to demonstrate the overtraining phenomenon in the process of varying a parameter, therefore we will look at the tendencies in the error curves. While the training error decreases steadily with r, the testing error decreases to a certain point, and then increases again. This increase indicates overtraining, where the classifier becomes too much of a data-specific expert and loses common sense. A common mistake in this case is to declare that the QDC has a testing error of 21.37% (the minimum in the bottom plot). The mistake is in that the *testing* set was used to find the best value of r.

---

[3]Discussed in Chapter 2.

The problem of peeking, largely due to unawareness of its caveats, is alarmingly common in application studies on feature selection. In view of this, we discuss this issue further in Chapter 9.

1.4   EXPERIMENTAL COMPARISON OF CLASSIFIERS

There is no single "best" classifier. Classifiers applied to different problems and trained by different algorithms perform differently [107, 110, 173, 196]. Comparative studies are usually based on extensive experiments using a number of simulated and real data sets. When talking about experiment design, I cannot refrain from quoting again and again a masterpiece of advice by George Nagy titled *Candide's practical principles of experimental pattern recognition* [287] (Just a note—this is a joke! DO NOT DO THIS!)

- Comparison of classification accuracies. *Comparisons against algorithms proposed by others are distasteful and should be avoided. When this is not possible, the following Theorem of Ethical Data Selection may prove useful.*
- Theorem. *There exists a set of data for which a candidate algorithm is superior to any given rival algorithm. This set may be constructed by omitting from the test set any pattern which is misclassified by the candidate algorithm.*
- Replication of experiments. *Since pattern recognition is a mature discipline, the replication of experiments on new data by independent research groups, a fetish in the physical and biological sciences, is unnecessary. Concentrate instead on the accumulation of novel, universally applicable algorithms.*
- Casey's caution. *Do not ever make your experimental data available to others; someone may find an obvious solution that you missed.*

Albeit meant to be satirical, the above principles are surprisingly widespread and closely followed! Speaking seriously now, the rest of this section gives some practical tips and recommendations.

A point raised by Duin [110] is that the performance of a classifier depends upon the expertise and the *willingness* of the designer. There is not much to be done for classifiers with fixed structures and training procedures (called "automatic" classifiers in [110]). For classifiers with many training parameters however, we can make them work or fail. Keeping in mind that there are no rules defining a fair comparison of classifiers, here are a few (non-Candide's) guidelines:

1. Pick the training procedures in advance and keep them fixed during training. When publishing, give enough detail so that the experiment is reproducible by other researchers.
2. Compare modified versions of classifiers with the *original* (nonmodified) classifier. For example, a distance-based modification of k-nearest neighbors (k-nn) should be compared with the standard k-nn first, and then with other classifier models. If a slight modification of a certain model is being compared with

TABLE 1.2   The 2 × 2 Relationship Table with Counts

	D_2 correct (1)	D_2 wrong (0)
D_1 correct (1)	N_{11}	N_{10}
D_1 wrong (0)	N_{01}	N_{00}

Total, $N_{11} + N_{10} + N_{01} + N_{00} = N_{ts}$

a totally different classifier, then it is not clear who deserves the credit—the modification or the original model itself.

3. Make sure that all the information about the data is utilized by all classifiers to the largest extent possible. For example, a clever initialization of a method can make it favorite among a group of equivalent but randomly initialized methods.

4. Make sure that the testing set has not been seen at any stage of the training.

5. If possible, give also the complexity of the classifier: training and running times, memory requirements, and so on.

1.4.1   Two Trained Classifiers and a Fixed Testing Set

Suppose that we have two trained classifiers which have been run on the same testing data giving testing accuracies of 98% and 96%, respectively. Can we claim that the first classifier is significantly better than the second one?

McNemar test. The testing results for two classifiers D_1 and D_2 on a testing set with N_{ts} objects can be organized as shown in Table 1.2. We consider two output values: 0 for incorrect classification and 1 for correct classification. Thus N_{pq} is the number of objects in the testing set with output p from the first classifier and output q from the second classifier, $p, q \in \{0, 1\}$.

The null hypothesis H_0 is that there is no difference between the accuracies of the two classifiers. If the null hypothesis is correct, then the expected counts for both off-diagonal entries in Table 1.2 are $\frac{1}{2}(N_{01} + N_{10})$. The discrepancy between the expected and the observed counts is measured by the following statistic:

$$s = \frac{\left(|N_{01} - N_{10}| - 1\right)^2}{N_{01} + N_{10}}, \qquad (1.10)$$

which is approximately distributed as χ^2 with 1 degree of freedom. The "−1" in the numerator is a continuity correction [99]. The simplest way to carry out the test is to calculate s and compare it with the tabulated χ^2 value for, say, level of significance[4] $\alpha = 0.05$. If $s > 3.841$, we reject the null hypothesis and accept that the

---

[4]The *level of significance* of a statistical test is the probability of rejecting H_0 when it is true, in other words, the probability to "convict the innocent." This error is called *Type I error*. The alternative error, when we do not reject H_0 when it is in fact incorrect, is called *Type II error*. The corresponding name for it would be "free the guilty." Both errors are needed in order to characterize a statistical test. For example,

two classifiers have significantly different accuracies. A MATLAB function for this test, called mcnemar, is given in Appendix 1.A.2.

▣ Example 1.7   A comparison on the Iris data

We took the first two features of the Iris data (Example 1.2) and classes "versicolor" and "virginica." The data was split into 50% training and 50% testing parts. The testing data is plotted in Figure 1.15. The linear and the quadratic discriminant classifiers

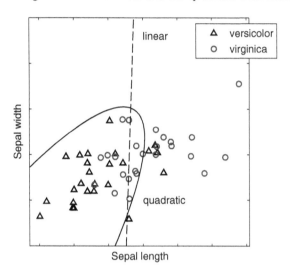

FIGURE 1.15   Testing data from the Iris data set and the decision boundaries of the linear and the quadratic discriminant classifiers.

(LDC and QDC, both detailed later) were trained on the training data. Their decision boundaries are plotted in Figure 1.15.

The confusion matrices of the two classifiers are as follows:

| | LDC | | | QDC | |
	Versicolor	Virginica		Versicolor	Virginica
Versicolor	20	5	Versicolor	20	5
Virginica	8	17	Virginica	14	11

Taking LDC to be classifier 1 and QDC, classifier 2, the values in Table 1.2 are as follows: $N_{11} = 31$, $N_{10} = 0$, $N_{01} = 6$, and $N_{00} = 13$. The difference is due to the six virginica objects in the "loop." These are correctly labeled by QDC and mislabeled by LDC. From Equation 1.10,

$$s = \frac{(|0 - 6| - 1)^2}{0 + 6} = \frac{25}{6} \approx 4.1667. \tag{1.11}$$

---

if we always accept H_0, there will be no Type I error at all. However, in this case the Type II error might be large. Ideally, both errors should be small.

Since the calculated s is greater than the tabulated value of 3.841, we reject the null hypothesis and accept that LDC and QDC are significantly different. Note that the Iris data is hardly large enough to be suitable for the hold-out protocol. It was used here for the purpose of the illustration only.

1.4.2   Two Classifier Models and a Single Data Set

Dietterich [99] details four important sources of variation that have to be taken into account when comparing classifier models.

1. *The choice of the testing set.* Different testing sets may rank differently classifiers which otherwise have the same accuracy across the whole population. Therefore, it is dangerous to draw conclusions from a single testing experiment, especially when the data size is small.

2. *The choice of the training set.* Some classifier models are called *unstable* [47] because small changes in the training set can cause substantial changes of the classifier trained on this set. Examples of unstable classifiers are decision tree classifiers and some neural networks.[5] Unstable classifiers are versatile models which are capable of adapting, so that most or all training examples are correctly classified. The instability of such classifiers is, in a way, the pay-off for their versatility. As we shall see later, unstable classifiers play a major role in classifier ensembles. Here we note that the variability with respect to the training data has to be accounted for.

3. *The internal randomness of the training algorithm.* Some training algorithms have a random component. This might be the initialization of the parameters of the classifier which are then fine-tuned (e.g., the backpropagation algorithm for training neural networks) or a stochastic procedure for tuning the classifier. Thus the trained classifier might be different for the same training set and even for the same initialization of the parameters.

4. *The random classification error.* Dietterich [99] considers the possibility of having mislabeled objects in the testing data as the fourth source of variability.

The above list suggests that multiple training and testing sets should be used, and multiple training runs should be carried out. Consider the task of comparing two classifier models (methods) over the *same data set* using one of the multi-test protocols. Let $E_{i,j}$ be the error of classifier i, $i \in \{1,2\}$, for the j-th testing set, $j = 1, \dots, K$. We can apply the traditional paired t-test for comparing the errors where $E_{1,j}$ are paired with $E_{2,j}$. However, this test may be inadequate because it does not take into account the fact that the training data used to build the classifier models are dependent [12, 89, 99, 286]. For example, in a K-fold cross-validation experiment, fold 1 will be used once for testing and $K - 1$ times for training the classifier. This may lead to overly liberal outcome of the statistical test, allowing the discovery of nonexisting significant differences between the two models. To avoid that, Nadeau and Bengio propose an amendment to the calculation of the *variance* of the error obtained as the average of T testing errors [286].

---

[5]All classifier models mentioned will be discussed later.

In the paired t-test, we calculate the differences $d_1, \ldots, d_T$, where $d_j = E_{1,j} - E_{2,j}$ for all j, and check the hypothesis that the mean of these differences is 0. Let σ_d be the empirical standard deviation of the differences. If the training data sets were independent, the standard deviation of the *mean* difference would be $\sigma_{d'} = \frac{\sigma_d}{\sqrt{T}}$. To account for the fact that the training data sets are not independent, we use instead

$$\sigma_{d'} = \sigma_d \sqrt{\frac{1}{T} + \frac{N_{\text{testing}}}{N_{\text{training}}}}, \tag{1.12}$$

where N_{training} and N_{testing} are the sizes of the training and the testing sets, respectively. For a K-fold cross-validation,

$$\sigma_{d'} = \sigma_d \sqrt{\frac{1}{K} + \frac{1}{K-1}} = \sigma_d \sqrt{\frac{2K-1}{K(K-1)}} . \tag{1.13}$$

This amendment holds for cross-validation, repeated cross-validation, data shuffle, and the bootstrap methods.

▣ Example 1.8   Correction of the variance for multiple testing sets

This example presents a Monte Carlo simulation to illustrate the need for the variance correction. Consider two Gaussian classes as shown in Figure 1.16. The classes have means $(-1, 0)$ and $(1, 0)$, and identity covariance matrices. We generated 200 data sets from this distribution; 20 points from class 1 and 20 points from class 2 in each data set. An example of such a set is circled in Figure 1.16. With each data set, we carried out 30 data shuffle runs by splitting the data into 90% training (36 data points) and 10% testing (4 data points). The LDC (detailed later) was trained on the training part and tested on the testing part.

FIGURE 1.16   Scatter plot of the two Gaussian classes. One of the 40-point data sets sampled from these classes is marked with circles.

Let $e_1, \ldots, e_{200}$ be the estimates of the classification error in the data shuffle experiment. Value e_i is the average of 30 testing errors with data set i (the data shuffle

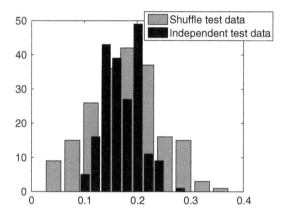

FIGURE 1.17   Histograms of the classification error for the two experiments.

protocol). Denote the mean and the standard deviation of these errors by μ_e and σ_e, respectively.

Consider now a matching experiment where we sample independently $200 \times 30 = 6000$ training sets of 36 objects and testing sets of 4 objects, storing the results as 200 batches of 30 runs. Denote the errors from the 200 batches by $q_1, \ldots, q_{200}$. Denote the mean and the standard deviation of these errors by μ_q and σ_q, respectively. Figure 1.17 shows the histograms for e and q. While the means of both errors are about 15.9% (the theoretical error), the spreads of the two histograms are different. In this example we obtained $\sigma_e = 0.0711$ and $\sigma_q = 0.0347$.

For each data shuffle experiment, we calculated not only the error e_i but also the standard deviation s_i. If the training and testing data were independently drawn, the standard error of the mean would be $\bar{s}_i = \frac{s_i}{\sqrt{30}}$. The average of $\bar{s}_i$ across i would be close to σ_e. However, this calculation gives a value of 0.0320, which is closer to σ_q than to σ_e, and does not properly reflect the larger spread seen in the histogram. Now we apply the correction and use $s_i^* = s_i \sqrt{\frac{1}{30} + \frac{1}{9}}$. The average of s_i^* across i is 0.0667, which is much closer to the observed σ_e.

Consider two models A and B, and T estimates of the classification error obtained through cross-validation or data shuffle. Denote these estimates by $a_1, a_2, \ldots, a_T$ and $b_1, b_2, \ldots, b_T$, respectively. The null hypothesis of the test, H_0, is that there is no difference between the mean errors of A and B for the given data set. The alternative hypothesis, H_1, is that there is difference. The step-by-step procedure for carrying out the amended paired t-test (two-tailed) is as follows:

1. Calculate the differences $d_i = a_i - b_i$, $i = 1, \ldots, T$. Calculate the mean and the standard deviation of d_i

$$m_\mathrm{d} = \frac{1}{T} \sum_{i=1}^{T} d_i, \qquad s_\mathrm{d} = \sqrt{\frac{1}{T-1} \sum_{i=1}^{T} (d_i - m_\mathrm{d})^2} \,.$$

2. Calculate the amended standard error of the mean

$$s'_d = s_d \sqrt{\frac{1}{T} + \frac{N_{testing}}{N_{training}}}.$$

3. Calculate the test statistic $t_d = \frac{m_d}{s'_d}$ and the degrees of freedom df $= T - 1$.

4. For a two-tailed t-test, find the p-value as

$$p = 2 \, F_t(-|t_d|, df),$$

where F_t is the Student's t cumulative distribution function.

If we set the alternative hypothesis H_1 to be "A has lower error than B" (one-tailed test), the p-value should be calculated as

$$p = F_t(t_d, df).$$

Comparing the obtained p-value with the chosen level of significance α, we reject H_0 if $p < \alpha$ and accept it otherwise. Function tvariance in Appendix 1.A.2 can be used for this calculation.

📖 Example 1.9   Paired t-test with corrected variance

Suppose that the values of the error (in %) in a 10-fold cross-validation experiment were as follows:

| Model A: | 7.4 | 18.1 | 13.7 | 17.5 | 13.0 | 12.5 | 8.9 | 12.1 | 12.4 | 7.4 |
| Model B: | 9.9 | 11.0 | 5.7 | 12.5 | 2.7 | 6.6 | 10.6 | 6.4 | 12.5 | 7.8 |

The mean of the difference between the errors of models A and B is $m = 3.73$ and the standard deviation is $s = 4.5090$. The standard error of the mean is therefore $\bar{s} = s/\sqrt{10} \approx 1.4259$. The p-value for the (traditional) two-tailed paired t-test ($10 - 1 = 9$ degrees of freedom) is

$$p = 2 \times F_t \left(-\frac{|m|}{\bar{s}}, df \right) = 2 \times F_t \left(-\frac{3.73}{1.4259}, 9 \right) \approx 0.0280 .$$

According to this test, at $\alpha = 0.05$, we can accept the alternative hypothesis that there is significant difference between the two classifier models. Knowing that the training and testing data were not independent, the amended standard deviation is $\bar{s}' = s \left(\sqrt{\frac{1}{10} + \frac{1}{9}} \right) \approx 2.0717$. The corrected p-value is

$$p' = 2 \times F_t \left(-\frac{|m|}{\bar{s}'}, df \right) = 2 \times F_t \left(-\frac{3.73}{2.0717}, 9 \right) \approx 0.1053.$$

This result does not give us ground to reject the null hypothesis and declare that there is difference between the two means.

The two examples highlight the importance of the variance correction when the training and testing data are dependent.

1.4.3   Two Classifier Models and Multiple Data Sets

Over the years, researchers have developed affinity for using extensively the UCI Machine Learning Repository [22] for drawing a sample of data sets and running comparative experiments on these [347]. We tend to over-tune our classification algorithms to these data sets and may ignore in the process data sets that present a real-life challenge [347]. If the claim is that algorithm A is better than algorithm B in general, then a large and diverse collection of data sets should be used.

The Wilcoxon signed rank test. Demšar proposes that the data sets chosen for the comparison of models A and B may be thought of as independent trials, but dissuades the reader from using a paired t-test [89]. The classification errors of different data sets are hardly commensurable. To bypass this problem, the Wilcoxon signed rank test was deemed more suitable. Let $a_1, a_2, \ldots, a_N$ and $b_1, b_2, \ldots, b_N$ in this context denote the error estimates of models A and B for the N *data sets* chosen for the experiment. These estimates can be obtained through any of the protocols, for example a 10-fold cross-validation. Again let $d_i = a_i - b_i$, $i = 1, \ldots, N$ be the differences of the errors. The Wilcoxon signed rank test does not take into account the exact value of d_i, only its relative magnitude. The null hypothesis of the test is that the data in vector d_i come from a continuous, symmetric distribution with zero median, against the alternative that the distribution does not have zero median. The test is applied in the following steps:[6]

1. Rank the absolute values of the distances $|d_i|$ so that the smallest distance receives rank 1 and the largest distance receives rank N. If there is a tie, all the ranks are shared so that the total sum stays $1 + 2 + \cdots + N$. For example, if there are four equal smallest distances, each will be assigned rank $(1 + 2 + 3 + 4)/4 = 2.5$. Thus each data set receives a rank r_i.
2. Split the ranks into positive and negative depending on the sign of d_i and calculate the sums:

$$R^+ = \sum_{d_i > 0} r_i + \frac{1}{2} \sum_{d_i = 0} r_i, \qquad R^- = \sum_{d_i < 0} r_i + \frac{1}{2} \sum_{d_i = 0} r_i.$$

3. Take as the test statistic $T = \min(R^+, R^-)$ and compare it with the critical value for the respective number of data sets N and the chosen level of significance. Table 1.A.1(a) in Appendix 1.A.2 gives the critical values for this test for $6 \le N \le 25$. For values $N > 25$, the following statistic is approximately normally distributed [89]:

$$z = \frac{T - \frac{1}{4}N(N + 1)}{\sqrt{\frac{1}{24}N(N + 1)(2N + 1)}}.$$

---

[6]Function `signrank` from the Statistics Toolbox of MATLAB can be used to calculate the p-value for this test.

The sign test. A simpler but less powerful alternative to this test is the sign test. This time we do not take into account the magnitude of the differences, only their sign. By doing so, we further avoid the problem of noncommensurable errors or differences thereof. For example, an error difference of 2% for a given data set may be more relevant than a difference of 4% on another data set. It is common practice to count WINS, DRAWS, and LOSSES, with or without statistical significance attached to these. The sign test is based on the intuition that if models A and B are equivalent, each one will score better than the other on approximately $N/2$ of the N data sets. Demšar [89] gives a useful table for checking the significance of the difference between models A and B tested on N data sets based on the sign test. We reproduce the table (with a small correction) as Table 1.A.1(b) and explain the calculation of the critical values in Appendix 1.A.2.

The table contains the required number of wins of A over B in order to reject H_0 and claim that model A is better than model B. The ties are split equally between A and B. For $N > 25$ data sets, we can use the normal approximation of the binomial distribution (mean $N/2$ and standard deviation $\sqrt{N}/2$). If the number of wins for A is greater than $N/2 + 1.96\sqrt{N}/2$, A is significantly better than B at $\alpha = 0.05$ (see [89] for more details).

1.4.4 Multiple Classifier Models and Multiple Data Sets

Demšar [89] recommends the Friedman test followed by the pairwise Nemenyi test for this task.

Friedman test with Iman and Davenport amendment. This is a nonparametric alternative of the analysis-of-variance (ANOVA) test. The classifier models are ranked on each of the N data sets. The best classifier receives rank 1 and the worst receives rank N. Tied ranks are shared equally as explained above. Let r_i^j be the rank of classifier model j on data set i, where $i = 1, \ldots, N$ and $j = 1, \ldots, M$. Let $R_j = \frac{1}{N} \sum_{i=1}^{N} r_i^j$ be the average rank of model j. The test statistic is

$$\chi_F^2 = \frac{12N}{M(M+1)} \left(\sum_{j=1}^{M} R_j^2 - \frac{M(M+1)^2}{4} \right). \tag{1.14}$$

The null hypothesis of the test H_0 is that all classifier models are equivalent. Under the null hypothesis, χ_F^2 follows the χ^2 distribution with $M - 1$ degrees of freedom (for $N > 10$ and $M > 5$ [89]).[7] Demšar advocates an amendment of the test statistic proposed by Iman and Davenport:

$$F_F = \frac{(N-1)\chi_F^2}{N(M-1) - \chi_F^2}, \tag{1.15}$$

---

[7]Function `friedman` from the Statistics Toolbox of MATLAB can be used to calculate the p-value for this test. Note that the MATLAB implementation contains an additional correction for tied ranks. This gives a slightly different test statistic compared to Equation 1.14 if there are tied ranks. See http://www.unistat.com/.

which follows the F-distribution with $(M - 1)$ and $(M - 1)(N - 1)$ degrees of freedom. The statistic's value is compared with the tabled critical values for the F-distribution (available in standard statistics textbooks), and if F_F is larger, we reject H_0 and accept that there is difference between the classifier models. Instead of using pre-tabulated critical values, a MATLAB function imandavenport for calculating the p-value of this test is given in Appendix 1.A.2.

▣ **Example 1.10   Comparison of 11 classifier models on 20 data sets**
This is a fictional example which demonstrates the calculation of the test statistic for the Friedman test and its modification. Table 1.3 displays the classification errors of the 11 classifier models for the 20 data sets. The data sets were arbitrarily named, just for fun, as the first 20 chemical elements of the periodic table. The "classification errors" were generated independently, as the absolute values of a normally distributed random variable with mean 0 and standard deviation 10, subsequently rounded to one decimal place. Thus, we do not expect to find significant differences between the models.

The corresponding ranks are shown in Table 1.4. Notice the shared ranks: both models C6 and C11 have the minimum error rate of 0.1 for data set "Nitrogen" and therefore equally share ranks 1 and 2, both models receiving rank 1.5. The average ranks are given in the bottom row. The Friedman statistic calculated as in

TABLE 1.3   Classification Errors of 11 Classifier Models on 20 Data Sets (A Fictional Example)

Data set	C1	C2	C3	C4	C5	C6	C7	C8	C9	C10	C11
Hydrogen	4.1	14.1	5.8	2.3	22.5	1.4	21.8	8.0	9.1	15.5	5.4
Helium	5.0	1.4	6.2	2.6	11.7	6.7	12.0	4.7	2.4	8.7	5.6
Lithium	0.8	12.9	5.4	9.0	12.0	8.5	3.7	0.5	6.5	1.5	19.8
Beryllium	1.6	4.9	4.8	21.6	6.6	0.6	24.1	3.5	12.7	3.9	5.4
Boron	5.3	6.2	1.1	0.9	3.3	2.9	7.4	12.6	5.5	13.2	1.4
Carbon	7.2	9.3	3.4	9.5	15.1	6.1	3.1	10.4	0.9	8.0	6.2
Nitrogen	8.5	2.8	9.3	11.7	8.9	0.1	5.4	4.3	3.5	1.4	0.1
Oxygen	8.0	2.0	0.1	17.4	9.6	4.1	1.2	0.8	9.9	4.2	11.1
Fluorine	7.3	1.4	11.4	3.7	4.9	9.4	9.6	17.9	4.3	8.2	1.9
Neon	16.9	8.8	4.3	11.9	4.4	8.4	2.5	8.0	10.0	8.5	11.2
Sodium	3.9	0.8	1.8	9.5	2.0	10.9	18.9	4.4	6.3	3.6	2.5
Magnesium	5.1	6.0	6.6	14.2	9.8	3.2	12.2	6.3	10.5	27.5	15.6
Aluminum	4.1	3.7	5.8	0.3	15.7	0.2	3.8	15.3	5.1	15.1	12.0
Silicon	10.7	8.4	16.2	2.6	3.7	11.6	0.5	27.3	3.3	4.3	2.4
Phosphorus	9.7	2.8	0.2	5.2	2.2	4.9	19.5	1.7	16.4	2.3	10.0
Sulfur	2.7	35.7	4.3	6.8	5.4	12.2	5.7	4.8	19.1	8.3	19.2
Chlorine	6.3	34.1	20.3	6.7	2.6	15.9	0.8	14.1	0.4	8.3	6.3
Argon	6.3	11.5	13.3	6.8	11.0	5.3	0.8	9.7	7.0	5.0	7.5
Potassium	0.8	7.9	3.2	3.1	5.5	2.6	7.9	2.9	1.7	23.2	2.1
Calcium	13.8	12.8	8.3	3.8	21.7	3.9	10.4	11.7	15.4	7.9	7.7

TABLE 1.4   Ranks of the 11 Classifier Models on the 20 Data Sets (Fictional Example)

Data set	C1	C2	C3	C4	C5	C6	C7	C8	C9	C10	C11
Hydrogen	3.0	8.0	5.0	2.0	11.0	1.0	10.0	6.0	7.0	9.0	4.0
Helium	5.0	1.0	7.0	3.0	10.0	8.0	11.0	4.0	2.0	9.0	6.0
Lithium	2.0	10.0	5.0	8.0	9.0	7.0	4.0	1.0	6.0	3.0	11.0
Beryllium	2.0	6.0	5.0	10.0	8.0	1.0	11.0	3.0	9.0	4.0	7.0
Boron	6.0	8.0	2.0	1.0	5.0	4.0	9.0	10.0	7.0	11.0	3.0
Carbon	6.0	8.0	3.0	9.0	11.0	4.0	2.0	10.0	1.0	7.0	5.0
Nitrogen	8.0	4.0	10.0	11.0	9.0	1.5	7.0	6.0	5.0	3.0	1.5
Oxygen	7.0	4.0	1.0	11.0	8.0	5.0	3.0	2.0	9.0	6.0	10.0
Fluorine	6.0	1.0	10.0	3.0	5.0	8.0	9.0	11.0	4.0	7.0	2.0
Neon	11.0	7.0	2.0	10.0	3.0	5.0	1.0	4.0	8.0	6.0	9.0
Sodium	6.0	1.0	2.0	9.0	3.0	10.0	11.0	7.0	8.0	5.0	4.0
Magnesium	2.0	3.0	5.0	9.0	6.0	1.0	8.0	4.0	7.0	11.0	10.0
Aluminum	5.0	3.0	7.0	2.0	11.0	1.0	4.0	10.0	6.0	9.0	8.0
Silicon	8.0	7.0	10.0	3.0	5.0	9.0	1.0	11.0	4.0	6.0	2.0
Phosphorus	8.0	5.0	1.0	7.0	3.0	6.0	11.0	2.0	10.0	4.0	9.0
Sulfur	1.0	11.0	2.0	6.0	4.0	8.0	5.0	3.0	9.0	7.0	10.0
Chlorine	4.5	11.0	10.0	6.0	3.0	9.0	2.0	8.0	1.0	7.0	4.5
Argon	4.0	10.0	11.0	5.0	9.0	3.0	1.0	8.0	6.0	2.0	7.0
Potassium	1.0	9.5	7.0	6.0	8.0	4.0	9.5	5.0	2.0	11.0	3.0
Calcium	9.0	8.0	5.0	1.0	11.0	2.0	6.0	7.0	10.0	4.0	3.0
R_j	5.22	6.28	5.50	6.10	7.10	4.88	6.28	6.10	6.05	6.55	5.95

Equation 1.14 is 6.9182. The p-value for the χ^2 distribution with $M - 1 = 10$ degrees of freedom is 0.7331. This value supports H_0: equal classifier models.

Applying the amendment from Equation 1.15, we arrive at $F_F = 0.6808$. The p-value of the F-test with $(M - 1)$ and $(M - 1)(N - 1)$ degrees of freedom is 0.7415, again supporting H_0.

The post-hoc test. If H_0 is rejected, Demšar [89] proposes the use of Nemenyi post-hoc test to find exactly where the differences are. All pairs of classifiers are examined. Two classifiers are declared different if their average ranks differ by more than a given critical value. For instance, for a pair of classifiers i and j, a test statistic is calculated using the average ranks R_i and R_j:

$$z = \frac{R_i - R_j}{\sqrt{\frac{M(M+1)}{6N}}}. \tag{1.16}$$

The number of pairwise comparisons $M(M - 1)/2$ determines the level of significance for this z-value. If the desired level of significance is α, the difference will be flagged as significant if the obtained p-value is smaller than $\frac{2\alpha}{M(M-1)}$. When a classifier model is singled out and compared with the remaining $M - 1$ models, the scaling constant is just $(M - 1)$ (Bonferroni-Dunn correction of the family-wise error). García and Herrera [147] explain in detail further step-wise procedures for post-hoc comparing of pairs of classifiers. The MATLAB code for both Nemenyi and Bonferroni-Dunn post-hoc tests is given in Appendix 1.A.2.

◼ Example 1.11 Post-hoc tests

The fictional comparison example was slightly modified. A constant of 0.8 was subtracted from the first column of Table 1.3, and all values in this column were multiplied by 0.5. This made classifier C1 better than all other classifier models. The ranks changed correspondingly, leading to the following average ranks:

	C1	C2	C3	C4	C5	C6	C7	C8	C9	C10	C11
R_j	3.10	6.47	5.65	6.35	7.25	5.28	6.42	6.30	6.20	6.80	6.17
Nemenyi					◼					◼	
Bonferroni	◼		◼	◼		◼	◼	◼	◼	◼	◼

The Friedman test statistic is 21.7273, giving a p-value of 0.0166. The Iman and Davenport amendment gives $F_F = 2.3157$ and a p-value of 0.0136. According to both tests, there is a difference among the 11 classifier models. The Nemenyi post-hoc test found significant differences at $\alpha < 0.05$ between C1 and C5 and also between C1 and C10 (two-tailed test). Nominating C1 as the classifier of interest, the Bonferroni–Dunn post-hoc test found C1 to be better (smaller error) than all classifiers except C3 and C6 (one-tailed test). The results from the post-hoc tests are shown underneath the average ranks above. A black square indicates that significant difference was found at $\alpha < 0.05$.

1.5 BAYES DECISION THEORY

1.5.1 Probabilistic Framework

Although many types of uncertainty exist, the probabilistic model fits surprisingly well in most pattern recognition problems. We assume that the class label ω is a random variable taking values in the set $\Omega = \{\omega_1, \dots, \omega_c\}$. The prior probabilities, $P(\omega_i)$, $i = 1, \dots, c$, constitute the probability mass function (pmf) of the variable ω:

$$0 \leq P(\omega_i) \leq 1, \quad \text{and} \quad \sum_{i=1}^{c} P(\omega_i) = 1. \tag{1.17}$$

We can construct a classifier based on this information only. To make the smallest possible number of mislabelings, we should always label an object with the class of the highest prior probability.

However, by measuring the relevant characteristics of the objects organized as the vector $\mathbf{x} \in \mathbb{R}^n$, we should be able to make a more accurate decision about this particular object. Assume that the objects from class ω_i are distributed in $\mathbb{R}^n$ according to the *class-conditional pdf* $p(\mathbf{x}|\omega_i)$, where $p(\mathbf{x}|\omega_i) \geq 0$, $\forall \mathbf{x} \in \mathbb{R}^n$, and

$$\int_{\mathbb{R}^n} p(\mathbf{x}|\omega_i) \, d\mathbf{x} = 1, \quad i = 1, \dots, c. \tag{1.18}$$

The likelihood of $\mathbf{x} \in \mathbb{R}^n$ is given by the *unconditional pdf*:

$$p(\mathbf{x}) = \sum_{i=1}^{c} P(\omega_i) \, p(\mathbf{x}|\omega_i). \tag{1.19}$$

Given the prior probabilities and the class-conditional pdfs, we can calculate the *posterior probability* that the true class label of the measured $\mathbf{x}$ is ω_i using the Bayes formula

$$P(\omega_i|\mathbf{x}) = \frac{P(\omega_i) \, p(\mathbf{x}|\omega_i)}{p(\mathbf{x})} = \frac{P(\omega_i) \, p(\mathbf{x}|\omega_i)}{\sum_{j=1}^{c} P(\omega_j) \, p(\mathbf{x}|\omega_j)}. \tag{1.20}$$

Equation 1.20 gives the probability mass function of the class label variable ω *for the observed* $\mathbf{x}$. The classification decision for that particular $\mathbf{x}$ should be made with respect to the posterior probability. Choosing the class with the highest posterior probability will lead to the smallest possible error when classifying any object with feature vector $\mathbf{x}$.

The probability model described above is valid for the discrete case as well. Let $\mathbf{x}$ be a discrete variable with possible values in $\mathbf{V} = \{\mathbf{v}_1, \dots, \mathbf{v}_s\}$. The only difference from the continuous-valued case is that instead of class-conditional pdf, we use *class-conditional pmf*, $P(\mathbf{x}|\omega_i)$, giving the probability that a particular value from $\mathbf{V}$ occurs if we draw at random an object from class ω_i. For all pmfs,

$$0 \leq P(\mathbf{x}|\omega_i) \leq 1, \quad \forall \mathbf{x} \in \mathbf{V}, \quad \text{and} \quad \sum_{j=1}^{s} P(\mathbf{v}_j|\omega_i) = 1. \tag{1.21}$$

1.5.2 Discriminant Functions and Decision Boundaries

The posterior probabilities can be used directly as the discriminant functions, that is,

$$g_i(\mathbf{x}) = P(\omega_i|\mathbf{x}), \quad i = 1, \dots, c. \tag{1.22}$$

Hence we can rewrite the maximum membership rule as

$$D(\mathbf{x}) = \omega_{i*} \in \Omega \quad \Longleftrightarrow \quad P(\omega_{i*}|\mathbf{x}) = \max_{i=1,\dots,c} \{P(\omega_i|\mathbf{x})\}. \tag{1.23}$$

In fact, a set of discriminant functions leading to the same classification regions would be

$$g_i(\mathbf{x}) = P(\omega_i) \, p(\mathbf{x}|\omega_i), \quad i = 1, \dots, c, \tag{1.24}$$

because the denominator of Equation 1.20 is the same for all i, and so will not change the ranking order of g_is. Another useful set of discriminant functions derived from the posterior probabilities is

$$g_i(\mathbf{x}) = \log(P(\omega_i)\, p(\mathbf{x}|\omega_i)), \qquad i = 1, \ldots, c. \tag{1.25}$$

▪ Example 1.12   Decision/classification boundaries

Let $x \in \mathbb{R}$. Figure 1.18 shows two sets of discriminant functions for three normally distributed classes with

$$
\begin{aligned}
P(\omega_1) &= 0.45, & p(x|\omega_1) &\sim N\left(4, (2.0)^2\right) \\
P(\omega_2) &= 0.35, & p(x|\omega_2) &\sim N\left(5, (1.2)^2\right) \\
P(\omega_3) &= 0.20, & p(x|\omega_3) &\sim N\left(7, (1.0)^2\right).
\end{aligned}
$$

Figure 1.18a depicts the first set of discriminant functions (Equation 1.24), obtained as $P(\omega_i)\, p(x|\omega_i)$, $i = 1, 2, 3$. The classification boundaries are marked with bullets on the x-axis. The posterior probabilities (Equation 1.22) are depicted as the second set of discriminant functions in Figure 1.18b. The classification regions specified by the boundaries are displayed with different shades of gray. Note that the same regions are found for both sets of discriminant functions.

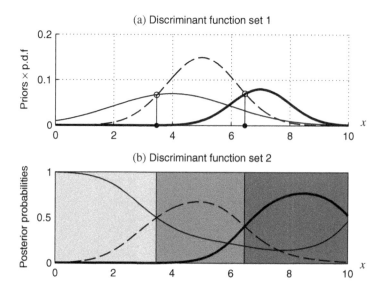

FIGURE 1.18   Plot of two equivalent sets of discriminant functions: (a) $P(\omega_1)p(x|\omega_1)$ (the thin line), $P(\omega_2)p(x|\omega_2)$ (the dashed line), and $P(\omega_3)p(x|\omega_3)$ (the thick line); (b) $P(\omega_1|x)$ (the thin line), $P(\omega_2|x)$ (the dashed line), and $P(\omega_3|x)$ (the thick line).

Sometimes more than two discriminant functions might tie at the boundaries. Ties are resolved randomly.

1.5.3   Bayes Error

Let D^* be a classifier which always assigns the class label with the largest posterior probability. Since for every $\mathbf{x}$ we can only be correct with probability

$$P(\omega_{i*}|\mathbf{x}) = \max_{i=1,\dots,c} \{P(\omega_i|\mathbf{x})\}, \tag{1.26}$$

there is some inevitable error. The overall probability of error of D^* is the sum of the errors of each individual $\mathbf{x}$ weighted by its likelihood value, $p(\mathbf{x})$, that is,

$$P_e(D^*) = \int_{\mathbb{R}^n} (1 - P(\omega_{i*}|\mathbf{x}))p(\mathbf{x})\,d\mathbf{x}. \tag{1.27}$$

It is convenient to split the integral into c integrals, one on each classification region. In this case, $\mathbf{x}$ will be given label ω_{i*} corresponding to the region's tag where $\mathbf{x}$ belongs. Then

$$P_e(D^*) = \sum_{i=1}^{c} \int_{\mathcal{R}_i^*} (1 - P(\omega_i|\mathbf{x}))p(\mathbf{x})\,d\mathbf{x}, \tag{1.28}$$

where $\mathcal{R}_i^*$ is the classification region for class ω_i, $\mathcal{R}_i^* \cap \mathcal{R}_j^* = \emptyset$ for any $j \neq i$ and $\cup_{i=1}^{c} \mathcal{R}_i^* = \mathbb{R}^n$. Substituting Equation 1.20 into Equation 1.28 and taking into account that $\sum_{i=1}^{c} \int_{\mathcal{R}_i^*} = \int_{\mathbb{R}^n}$,

$$P_e(D^*) = \sum_{i=1}^{c} \int_{\mathcal{R}_i^*} \left(1 - \frac{P(\omega_i)p(\mathbf{x}|\omega_i)}{p(\mathbf{x})}\right) p(\mathbf{x})\,d\mathbf{x} \tag{1.29}$$

$$= \int_{\mathbb{R}^n} p(\mathbf{x})\,d\mathbf{x} - \sum_{i=1}^{c} \int_{\mathcal{R}_i^*} P(\omega_i)p(\mathbf{x}|\omega_i)\,d\mathbf{x} \tag{1.30}$$

$$= 1 - \sum_{i=1}^{c} \int_{\mathcal{R}_i^*} P(\omega_i)p(\mathbf{x}|\omega_i)\,d\mathbf{x}. \tag{1.31}$$

Note that $P_e(D^*) = 1 - P_c(D^*)$, where $P_c(D^*)$ is the overall probability of correct classification of D^*, or the classification accuracy.

Consider a different classifier, D, which produces classification regions $\mathcal{R}_1, \dots, \mathcal{R}_c$, $\mathcal{R}_i \cap \mathcal{R}_j = \emptyset$ for any $j \neq i$ and $\cup_{i=1}^{c} \mathcal{R}_i = \mathbb{R}^n$. Regardless of the way the regions are formed, the error of D is

$$P_e(D) = \sum_{i=1}^{c} \int_{\mathcal{R}_i} (1 - P(\omega_i|\mathbf{x}))p(\mathbf{x})\,d\mathbf{x}. \tag{1.32}$$

The error of D^* is the smallest possible error, called the *Bayes error*. The example below illustrates this concept.

⬛ Example 1.13   Bayes error

Consider the simple case of $x \in \mathbb{R}$ and $\Omega = \{\omega_1, \omega_2\}$. Figure 1.19 displays the discriminant functions in the form $g_i(x) = P(\omega_i)p(x|\omega_i)$, $i = 1, 2$, $x \in [0, 10]$.

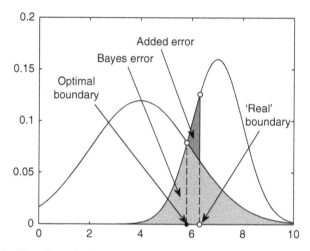

FIGURE 1.19   Plot of two discriminant functions $P(\omega_1)p(x|\omega_1)$ (left curve) and $P(\omega_2)p(x|\omega_2)$ (right curve) for $x \in [0, 10]$. The light-gray area corresponds to the Bayes error, incurred if the optimal decision boundary (denoted by •) is used. The dark-gray area corresponds to the additional error when another boundary (denoted by ○) is used.

For two classes,

$$P(\omega_1|x) = 1 - P(\omega_2|x), \tag{1.33}$$

and $P_e(D^*)$ in Equation 1.28 becomes

$$P_e(D^*) = \int_{\mathcal{R}_1^*} (1 - P(\omega_1|\mathbf{x}))p(\mathbf{x})\,d\mathbf{x} + \int_{\mathcal{R}_2^*} (1 - P(\omega_2|\mathbf{x}))p(\mathbf{x})\,d\mathbf{x} \tag{1.34}$$

$$= \int_{\mathcal{R}_1^*} P(\omega_2|\mathbf{x})p(\mathbf{x})\,d\mathbf{x} + \int_{\mathcal{R}_2^*} P(\omega_1|\mathbf{x})p(\mathbf{x})\,d\mathbf{x} \tag{1.35}$$

$$= \int_{\mathcal{R}_1^*} P(\omega_2)p(\mathbf{x}|\omega_2)\,d\mathbf{x} + \int_{\mathcal{R}_2^*} P(\omega_1)p(\mathbf{x}|\omega_1)\,d\mathbf{x}. \tag{1.36}$$

By design, the classification regions of D^* correspond to the true highest posterior probabilities. The bullet on the x-axis in Figure 1.19 splits $\mathbb{R}$ into $\mathcal{R}_1^*$ (to the left) and $\mathcal{R}_2^*$ (to the right). According to Equation 1.36, the Bayes error will be the area under $P(\omega_2)p(\mathbf{x}|\omega_2)$ in $\mathcal{R}_1^*$ plus the area under $P(\omega_1)p(\mathbf{x}|\omega_1)$ in $\mathcal{R}_2^*$. The total area corresponding to the Bayes error is marked in light gray. If the boundary is shifted to the left or right, additional error will be incurred. We can think of this boundary as coming from classifier D which is an imperfect approximation of D^*. The shifted boundary, depicted by an open circle, is called in this example the "real" boundary. Region $\mathcal{R}_1$ is therefore $\mathcal{R}_1^*$ extended to the right. The error calculated through Equation 1.36 is the area under $P(\omega_2)p(\mathbf{x}|\omega_2)$ in the whole of $\mathcal{R}_1$, and extra error will be incurred, measured by the area shaded in dark gray. Therefore, using the *true* posterior probabilities or an equivalent set of discriminant functions guarantees the smallest possible error rate, called the *Bayes error*.

Since the true probabilities are never available in practice, it is impossible to calculate the exact Bayes error or design the perfect Bayes classifier. Even if the probabilities were given, it will be difficult to find the classification regions in $\mathbb{R}^n$ and calculate the integrals. Therefore, we rely on estimates of the error as discussed in Section 1.3.

1.6   CLUSTERING AND FEATURE SELECTION

Pattern recognition developed historically as a union of three distinct but intrinsically related components: classification, clustering, and feature selection.

1.6.1   Clustering

Clustering aims to find groups in data. "Cluster" is an intuitive concept and does not have a mathematically rigorous definition. The members of one cluster should be similar to one another and dissimilar to the members of other clusters. A clustering algorithm operates on an unlabeled data set $\mathbf{Z}$ and produces a *partition* on it, denoted $P = (\mathbf{Z}^{(1)}, \ldots, \mathbf{Z}^{(c)})$, where $\mathbf{Z}^{(i)} \subseteq \mathbf{Z}$ and

$$\mathbf{Z}^{(i)} \cap \mathbf{Z}^{(j)} = \emptyset, \quad i, j = 1, \ldots, c, \quad i \neq j, \tag{1.37}$$

$$\bigcup_{i=1}^{c} \mathbf{Z}^{(i)} = \mathbf{Z}. \tag{1.38}$$

There is a vast amount of literature on clustering [18, 38, 126, 158, 195] looking for answers to the main questions, among which are:

- Is there really a structure in the data or are we imposing one by our clustering algorithms?
- How many clusters should we be looking for?
- How do we define similarity between objects in the feature space?
- How do we know whether our clustering results are good?

Two main groups of clustering algorithms are *hierarchical clustering* (agglomerative and divisive) and *nonhierarchical clustering*. The nearest neighbor (single

---

SINGLE LINKAGE CLUSTERING

1. Pick the number of clusters c and a similarity measure $S(a, b)$ between two objects a and b. Initialize the procedure by defining an individual cluster for each point in $\mathbf{Z}$.
2. Identify the two most similar clusters and join them as a new cluster, replacing the initial two clusters. The similarity between clusters A and B is measured as

$$\min_{a \in A, b \in B} S(a, b).$$

3. Repeat step 2 until c clusters are found.

FIGURE 1.20   The single linkage clustering algorithm.

---

c-MEANS CLUSTERING

1. Pick the number of clusters c and a similarity measure $S(a, b)$ between two objects a and b. Initialize the c cluster centers (i.e., by randomly selecting c points from $\mathbf{Z}$ to be these centers).
2. Label all points in $\mathbf{Z}$ with respect to their similarity to the cluster centers: each point is assigned to the cluster with the most similar center.
3. Calculate the new cluster centers using the points in the respective cluster.
4. Repeat steps 2 and 3 until no change in the centers occurs.

FIGURE 1.21   The c-means (k-means) clustering algorithm.

linkage) clustering algorithm shown in Figure 1.20 is an example of the hierarchical group whereas the c-means clustering algorithm (also called k-means) shown in Figure 1.21 is an example of the nonhierarchical group. Both algorithms are famous for their simplicity and elegance.[8]

■ Example 1.14   Clustering: there is no "best" algorithm
Consider a two-dimensional data set where 50 points are sampled from each of two normal distributions with means at (0,0) and (3,3), and identity covariance matrices (Figure 1.22a). The single linkage clustering algorithm is known for the "chain effect." An outlier would often present itself as a separate cluster, thereby preventing the algorithm from discovering meaningful balanced clusters. This is illustrated in Figure 1.22b where the two clusters found by the algorithm are plotted with different markers. The c-means algorithm, on the other hand, identifies the two clusters successfully (Figure 1.22c).

---

[8]Both single-linkage and c-means algorithms are available in many statistical software packages, including the Statistics Toolbox of MATLAB.

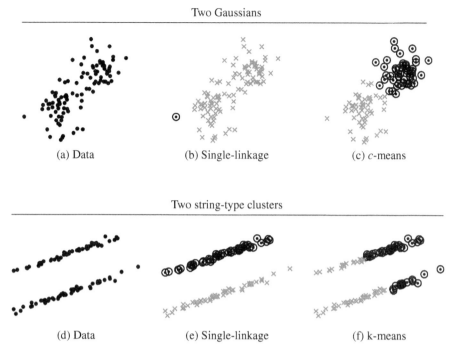

FIGURE 1.22 Examples of single linkage and k-means clustering on two synthetic data sets. The two clusters found by the algorithms are plotted with different markers: circles and gray crosses.

The second data set (Figure 1.22d) consists of two string-shaped clusters. This configuration is correctly identified by the single linkage but fools k-means into cutting both strings and finding nonexistent clusters.

Neither of the two algorithms is perfect, nor are the multitude of existing clustering algorithms. It may prove difficult to pick a suitable clustering algorithm for multi-dimensional data. Ensembles of "clusterers" are deemed to be more robust in that respect.

1.6.2   Feature Selection

Feature selection is the process of reducing the dimensionality of the feature space. Its aim is not only computational convenience but elimination of noise in the data so that it is easier to train an accurate and robust classifier. A myriad of insightful and comprehensive surveys, practitioners' guides, journal special issues, and conference tracks have been devoted to feature selection over the years [3, 42, 83, 164, 196, 214, 263, 284, 346]. Different methods and approaches have been recommended depending on the data types and sizes.

The two major questions that a feature selection method must address, separately or simultaneously, are:

1. Are the features evaluated individually? If not, how do we traverse the class of all subset-candidates?
2. What criterion do we apply to evaluate the merit of a given subset of features?

Consider for now question 1. The simplest way of selecting features is to rank them according to a certain test criterion and cut the list. Starting with key publications in the 1970s [77, 387], it is now well understood that features should be evaluated as a group rather than individually. By selecting the features individually, important dependencies may be overlooked. But evaluating *subsets* of features raises the question of computational complexity. If unlimited resources were available, exhaustive search could be carried out checking each and every possible subset. Sequential methods such as forward and backward selection, as well as floating search [315] have been found to be the best compromise between computation speed and accuracy. Figure 1.23 shows the sequential forward selection algorithm (SFS).

The output of SFS can be taken as the feature ranking determined by the order in which the features enter the set S in the algorithm.

The two basic approaches to question 2 are termed "wrapper" and "filter" [214]. The wrapper approach requires that a classifier model is chosen and trained on a given feature set. Its classification accuracy, evaluated on a validation set, is the measure of quality of that feature set. In the filter approach, some measure of separation between the classes in the space spanned by the feature set is used as a proxy for the classification accuracy. While the wrapper approach has been found to be generally more accurate, the filter approach is faster and easier to apply, which makes it a convenient compromise if a large number of feature subsets must be probed.

The MATLAB function sfs_filter(a,laba,d) in Appendix 1.A.3 carries out sequential forward selection of the features of data set a (columns of a) and returns the indices of the d features in order of selection. The criterion for evaluating the feature subset f is the Euclidean distance between the centroids of the classes in the space spanned by f but there are many alternatives offered by the pdist MATLAB function used within the code.

SEQUENTIAL FORWARD SELECTION (SFS)

1. Given is a feature set F. Choose a test criterion for a feature subset $f \subseteq F$ and a stopping criterion (e.g., a number of features $d \leq |F|$). Initialize the set of selected features $S = \emptyset$.
2. Taking all features in F that are not yet in S, add temporarily one feature at a time and measure the quality of the new set $S' = S \cup \{x\}, x \in F, x \notin S$.
3. Choose the feature with the highest criterion value and add it permanently to S.
4. Repeat steps 2 and 3 until the stopping criterion is met.

FIGURE 1.23   The sequential forward selection algorithm.

▣ Example 1.15   Feature selection: the peak effect

The data set "sonar" from the UCI ML Repository has 60 features and 2 classes. SFS was applied for feature selection with a filter approach. The quality of a feature subset was measured by the Euclidean distance between the two class centroids in the respective feature space (as in the MATLAB function `sfs_filter(a,laba,d)` in Appendix 1.A.3). One hundred runs were carried out where a randomly sampled half of the data set was used for training and the other half, for testing. The nearest neighbor classifier was applied for evaluating the selected feature subsets. Each split of the data produced a ranking of the 60 features. As an example, suppose that SFS on split j arranged the features as $\{32, 11, 6, 28, \dots\}$. For this split, feature 32 was the single best, $\{32, 11\}$ was the best pair containing feature 32, $\{32, 11, 6\}$ was the best set of three features containing features 32 and 11, and so on.

Consider plotting the classification accuracy, evaluated on the testing half of the data, when using feature sets $\{32\}$, $\{32, 11\}$, $\{32, 11, 6\}$, $\{32, 11, 6, 28\}$, and so on. It can be expected that the more features we include, the higher the accuracy will be, leading to the best accuracy with all features. The curves for the 100 data splits are shown in Figure 1.24 in gray. The average curve is depicted with a solid black line. The peak and the end points are marked and annotated. It can be seen that SFS reaches better accuracy with fewer features, called the "peak effect." For comparison, 100 random permutations of the features were generated, and an average curve was calculated in the same way as with the SFS rankings. As expected, the average random curve progresses gradually toward the same end point but without the peak effect.

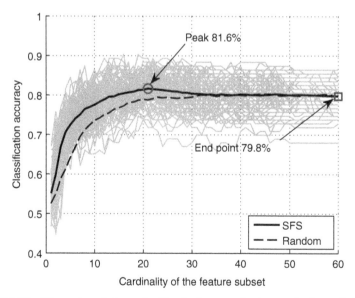

FIGURE 1.24   Illustration of the peak effect in feature selection on the "sonar" data, SFS filter and the nearest neighbor classifier.

This example demonstrates that feature selection could be beneficial not just for reducing computational complexity but also for increasing the classification accuracy. This problem is especially acute for very high-dimensional data where the number of features exceeds by orders of magnitude the number of samples, the so-called "wide" data sets.

The long-lasting and proliferate research on feature selection has delivered a refined collection of excellent feature selection algorithms such as the floating search [315], fast correlation-based filters (FCBF) [428], RELIEF [205], and SVM-RFE [165]. Yet, with the new challenges posed by the very high dimensionality of modern data, there is room for development. Saeys et al. [346] highlight the potential of ensemble feature selection methods to improve the stability and accuracy of the individual methods. We will touch upon feature selection *for ensembles* and *by ensembles* further in the book.

1.7   CHALLENGES OF REAL-LIFE DATA

Finally, pattern recognition branched out tremendously in the past couple of decades, taking what were curious little niches in the past into powerful independent research streams in their own right. Real-life data pose challenges such as

- *Unbalanced classes.* Many times the class of interest is like a "needle in a haystack." An example is detecting a face in an image. Suppose that the gray image has 500 rows and 600 columns of pixels, and a face is expected to be within a 50-by-50 square of pixels. Then there are $(500 - 49) \times (600 - 49) = 248,501$ candidate squares. If the image is a photograph of a person, the class "face" will contain a handful of objects (squares containing predominantly the face), and class "nonface" will contain all remaining squares. Thus class "face" will be a minute fraction of the data.

- *Uncertain labels.* Sometimes the labels of the objects cannot be assigned precisely. Take, for example, emotion recognition. Affective computing is gaining importance in psychological research, entertainment, and gaming industries. However, it is hardly possible to pinpoint and label the experienced emotion.

- *Massive volumes.* Computational costs, algorithmic tractability, and statistical validity of the results are only a few of the problems with very high-dimensional data and massive sample sizes.

- *Nonstationary distributions.* The data set collected at a certain time may become obsolete if the circumstances or the problem characteristic change. Adaptive classification is needed for such cases.

Standard and custom-tailored classifier ensemble methods are quickly turning into one of the most favorite tools in all these areas.

APPENDIX

1.A.1   DATA GENERATION

```
1  %--------------------------------------------------------------------%
2  function x = samplegaussian(N,mu,Sigma)
3  mu = mu(:); R = chol(Sigma);
4  for i = 1:N
5      x(i,:) = mu' + (R'*randn(size(mu)))';
6  end
7  %--------------------------------------------------------------------%
```

```
1  %--------------------------------------------------------------------%
2  % Subplot (a) ---
3  % Introduce the ellipse function
4  elx = @(t,xc,a,b,phi) xc+a*cos(t)*cos(phi)-b*sin(t)*sin(phi) ;
5  ely = @(t,yc,a,b,phi) yc+a*cos(t)*sin(phi)+b*sin(t)*cos(phi) ;
6
7  % Calculate the  ellipse equations
8  N = 500;
9  t = rand(1,N)*2*pi; % sample random points from the figure
10 el1x = elx(t,-6,2,6,-2); el1y = ely(t,0,2,6,-2);  % ellipse 1
11 el2x = elx(t,-2,4,3,-1); el2y = ely(t,-2,4,3,-1); % ellipse 2
12 el3x = elx(t,2,4,1,0.9); el3y = ely(t,4,4,1,0.9); % ellipse 3
13
14 % Add noise
15 edata = [el1x(:), el1y(:); el2x(:), el2y(:); el3x(:), el3y(:)];
16 edata = edata + randn(size(edata))*0.5;
17 w = ones(numel(el1x),1);
18 elabels = [w;w*2;w*3];
19
20 % Plot the data
21 figure, hold on
22 scatter(edata(:,1),edata(:,2),[],elabels,'linewidth',2.5)
23 axis equal off
24
25 % Subplot (b) ---
26 t = rand(1,1000)*2*pi; % sample random points from the figure
27 el1x = elx(t,0,3,9,-1); el1y = ely(t,0,3,9,-1);
28 el1x = el1x + randn(size(el1x)).*t*.2;
29 el1y = el1y + randn(size(el1y)).*t*.2;
30 figure
31 plot(el1x,el1y,'k.','markersize',15);
32 axis equal off
33 %--------------------------------------------------------------------%
```

```
1  %-------------------------------------------------------------------%
2  function [d, labd] = samplecb(N,a,alpha)
3  d = rand(N,2);
4  d_transformed = [d(:,1)*cos(alpha)-d(:,2)*sin(alpha),...
5      d(:,1)*sin(alpha)+d(:,2)*cos(alpha)];
6  s = ceil(d_transformed(:,1)/a)+floor(d_transformed(:,2)/a);
7  labd = 2 - mod(s,2);
8  %-------------------------------------------------------------------%
```

1.A.2   COMPARISON OF CLASSIFIERS

1.A.2.1   MATLAB Functions for Comparing Classifiers

The output of all hypothesis-testing functions is in the form $[H,p]$, where H is 0 if the null hypothesis is accepted, and 1 if the null hypothesis is rejected at significance level 0.05. The output p is the test p-value.

```
1  %-------------------------------------------------------------------%
2  function [H,p] = mcnemar(labels1, labels2, true_labels)
3  % --- McNemar test for two classifiers
4  % Needs Statistics Toolbox
5  % (all labels are integers 1,2,...)
6  v1 = labels1(:) == true_labels(:);
7  v2 = labels2(:) == true_labels(:);
8  t2(1,1) = sum(~v1&~v2);t2(1,2) = sum(~v1&v2);
9  t2(2,1) = sum(v1&~v2);t2(2,2) = sum(v1&v2);
10 % the two-way table [N00,N01;N10,N11]
11 % calculate the test statistic
12 if any([t2(1,2),t2(2,1)])
13     if t2(1,2) + t2(2,1) > 25
14         x2 = (abs(t2(1,2)-t2(2,1))-1)^2/(t2(1,2)+t2(2,1));
15         % find the p-value
16         p = 1 - chi2cdf(x2,1);
17     else % exact test using binomial distribution
18         % t2(1,2) is compared to a binomial distribution
19         % with size parameter equal to t2(1,2) + t2(2,1)
20         % and "probability of success" = 0.5,
21         p=binocdf(min(t2(1,2),t2(2,1)),t2(1,2)+t2(2,1),0.5)...
22             + 1 - binocdf(max(t2(1,2),t2(2,1))-1,t2(1,2)+...
                    t2(2,1),0.5);
23     end
24 else % identical classifiers
25     p = 1;
26 end
```

```
27  % calculate the hypothesis outcome at significance level 0.05
28  % H = 0 if the null hypothesis holds; H = 1 otherwise.
29  H = p < 0.05;
30  %-------------------------------------------------------------------%
```

```
 1  %-------------------------------------------------------------------%
 2  function [H,p] = tvariance(x,y,ts_tr_ratio)
 3  % --- paired t-test with corrected variance
 4  % Needs Statistics Toolbox
 5  d = x - y;
 6  md = mean(d); stdd = std(d);
 7  se_corrected = stdd * sqrt(1/numel(x) + ts_tr_ratio);
 8  t = md / se_corrected; % the test statistic
 9  % two-tailed test
10  p = 2 * tcdf(-abs(t),numel(x)-1);
11  % calculate the hypothesis outcome at significance level 0.05
12  % H = 0 if the null hypothesis holds; H = 1 otherwise.
13  H = p < 0.05;
14  %-------------------------------------------------------------------%
```

```
 1  %-------------------------------------------------------------------%
 2  function [H,p] = imandavenport(a)
 3  % --- Iman and Davenport test for N classifiers on M data sets
 4  % Needs Statistics Toolbox
 5  % a_ji is the error of model j on data set i
 6  % N rows, M columns
 7  [N,M] = size(a);
 8
 9  r = ranks(a')'; R = mean(r);
10  x2F =12*N/(M*(M+1))*(sum(R.^2) - M*(M+1)^2/4);
11
12  % ===
13  % MATLAB Stats Toolbox variant with additional correction
14  % for tied ranks:
15  % [~,t] = friedman(a,1,'off');
16  % x2F = t{2,5} % Friedman chi^2 statistic
17  % ===
18
19  FF = (N-1) * x2F / (N*(M-1) - x2F); % amended
20  p = 1 - fcdf(FF,(M-1),(M-1)*(N-1)); % p-value from the F-distribution
21  % calculate the hypothesis outcome at significance level 0.05
22  % H = 0 if the null hypothesis holds; H = 1 otherwise.
23  H = p < 0.05;
24  end
25
26  function ran = ranks(a)
```

```
27  [maxr,maxcol] = size(a);
28  ran = zeros(size(a));
29  for i = 1:maxcol
30      [~, rr] = sort(a(:,i)); [~, b2] = sort(rr);
31      for j = 1 : maxr % check for ties
32          inr = a(:,i) == a(j,i);
33          b2(inr) = mean(b2(inr));
34      end
35      ran(:,i) = b2;
36  end
37  end
38  %----------------------------------------------------------------------%
```

```
 1  %----------------------------------------------------------------------%
 2  function [H,p] = nemenyiposthoc(a)
 3  % --- Nemenyi post-hoc test
 4  % Needs Statistics Toolbox & function RANKS
 5  % a_ji is the error of model j on data set i
 6  % N rows, M columns
 7  [N,M] = size(a);
 8  r = ranks(a')'; R = mean(r);
 9  const = M * (M-1) / 2;
10  for i = 1:M-1
11      for j = i+1:M
12          z = (R(i)-R(j))/sqrt(M*(M+1)/(6*N));
13          p(i,j) = 2*normcdf(-abs(z)); % two-tailed test
14          p(j,i) = p(i,j);
15      end
16      p(i,i) = 1;
17  end
18  p(M,M) = 1;
19  p = min(1,p*const);
20
21  % calculate the hypothesis outcome at significance level 0.05
22  % H = 0 if the null hypothesis holds; H = 1 otherwise.
23  H = p < 0.05;
24  end
25  %----------------------------------------------------------------------%
```

```
 1  %----------------------------------------------------------------------%
 2  function [H,p] = bonferroniposthoc(a)
 3  % --- Bonferroni-Dunn post-hoc test
 4  % Needs Statistics Toolbox & function RANKS
 5  % a_ji is the error of model j on data set i
```

```
6   % N rows, M columns
7   % The classifier of interest is in column 1 of a.
8   % The output contains M-1 results from the
9   % comparisons of columns 2:M with column 1
10
11  [N,M] = size(a);
12
13  r = ranks(a')'; R = mean(r);
14  const = M - 1;
15
16  for i = 2:M
17      z = (R(1)-R(i))/sqrt(M*(M+1)/(6*N));
18      % p(i-1) = 2*normcdf(-abs(z)); % two-tailed test
19      p(i-1) = normcdf(z); % one-tailed test
20  end
21  p = min(1,p*const);
22
23  % calculate the hypothesis outcome at significance level 0.05
24  % H = 0 if the null hypothesis holds; H = 1 otherwise.
25  H = p < 0.05;
26  end
27  %-----------------------------------------------------------------%
```

1.A.2.2   Critical Values for Wilcoxon and Sign Test

Table 1.A.1 shows the critical values for comparing two classifier models on N data sets. Sub-table (a) gives the values for the Wilcoxon signed rank test (two-tailed), and sub-table (b), the values for the sign test (one-tailed). For sub-table (b), classifier model A is better than B if it wins on w_α or more data sets. Here we explain the calculation of the critical values for the sign test.

Suppose that in comparing classifier models A and B, both were tested on N data sets. Model A was found to be better on K out of the N sets. Can we claim that A is better than B? The null hypothesis of our test, H_0, is that there is no difference between A and B. Then the probability that A wins over B on a randomly chosen data set is $1/2$. The number of data sets where A wins over B in N attempts follows a binomial distribution with parameters N and $1/2$. Under the null hypothesis, the probability that A wins on K or fewer data sets is

$$F_b(K,N,0.5) = 0.5^N \sum_{i=0}^{K} \frac{N!}{i!(N-i)!}, \qquad (1.A.1)$$

where F_b is the cumulative distribution function of the binomial distribution. The alternative hypothesis, H_1, is that A is better than B (one-tailed test). To reject the

TABLE 1.A.1   Table of the Critical Values for Comparing Two Classifier Models on N Data Sets. (a) Wilcoxon Signed Rank Test (Two Tailed), (b) Sign Test (One Tailed). For Sub-table (b), Classifier Model A Is Better Than B if It Wins on w_α or More Data Sets

| | (a) | | | | (b) | |
| | α | | | | α | |
N	0.1	0.05	0.01	N	$w_{0.10}$	$w_{0.05}$
				5	5	5
6	0	–	–	6	6	6
7	2	0	–	7	6	7
8	4	2	0	8	7	7
9	6	3	2	9	7	8
10	8	5	3	10	8	9
11	11	7	5	11	9	9
12	14	10	7	12	9	10
13	17	13	10	13	10	10
14	21	16	13	14	10	11
15	25	20	16	15	11	12
16	30	24	20	16	12	12
17	35	28	23	17	12	13
18	40	33	28	18	13	13
19	46	38	32	19	13	14
20	52	43	38	20	14	15
21	59	49	43	21	14	15
22	66	56	49	22	15	16
23	73	62	55	23	16	16
24	81	69	61	24	16	17
25	89	77	68	25	17	18

null hypothesis and accept H_1 at level of significance α, K must be large enough. For example, let $N = 5$. Then $F_b(K, N, 0.5)$ is

K	0	1	2	3	4	5
$F_b(K, 5, 0.5)$	0.0313	0.1875	0.5000	0.8125	0.9688	1.0000

Because of the discrete nature of the problem, we cannot achieve the desired level of significance exactly. If the null hypothesis is correct, the probability of observing K or more wins (A better than B) is

$$1 - F_b(K - 1, 5, 0.5). \tag{1.A.2}$$

We must set the critical value of K so that this probability is smaller or equal to α. Any number of wins greater than or equal to this critical value will allow us to reject the null hypothesis at the desired level of significance α or better. Let $\alpha = 0.05$. We have

$$1 - F_b(4, 5, 0.5) = 0.0313 \qquad (1.A.3)$$

and

$$1 - F_b(3, 5, 0.5) = 0.1875. \qquad (1.A.4)$$

Then the critical value K^* is obtained from $K^* - 1 = 4$, hence $K^* = 5$.

Therefore, to construct the table with the critical values, we find

$$K' = \arg \min_{0 \leq K \leq N-1} \left\{ F_b(K, N, 0.5) \geq 1 - \alpha \right\}, \qquad (1.A.5)$$

and set $K^* = K' + 1$ as the critical value.

1.A.3   FEATURE SELECTION

```
1   %----------------------------------------------------------------------%
2   function S = sfs_filter(a,laba,d)
3   % --- Sequential Forward Selection - filter approach
4   % a - data set
5   % laba - labels 1,2,3,...,
6   % d - desired number of features
7   % S - indices of the selected features
8   %      (in order of selection)
9
10  c = max(laba); % number of classes
11  n = size(a,2); % number of features
12  F = ones(1,n); % features to choose from
13  S = []; % chosen subset (empty)
14
15  % calculate class means
16  x = zeros(c,n);
17  for k = 1:c
18      x(k,:) = mean(a(laba == k,:),1);
19  end
20
21  for i = 1:d
22      Remaining = find(F); % features not selected yet
23      for j = 1:numel(Remaining)
24          Sdash = S;
25          % temporarily add one feature
```

```
26              Sdash = [Sdash Remaining(j)];
27              % calculate the criterion
28              crit(j) = mean(pdist(x(:,Sdash)));
29          end
30          % choose the best feature to add
31          [~,best] = max(crit);
32          S = [S Remaining(best)]; % add the best feature
33          F(Remaining(best)) = 0; % remove from F
34      end
35  %------------------------------------------------------------------%
```

2

BASE CLASSIFIERS

The classifiers whose decisions are combined to form the ensemble are called "base classifiers." This chapter details some of the most popular base classifier models.

2.1 LINEAR AND QUADRATIC CLASSIFIERS

2.1.1 Linear Discriminant Classifier

Linear and quadratic classifiers are named after the type of discriminant functions they use. Let $\mathbf{x} \in \mathbb{R}^n$ be the object to classify in one of c classes. Let $\mathbf{w}_i \in \mathbb{R}^n$ be a vector with coefficients and w_{i0} be a constant free term. A linear classifier is any set of c linear functions, one for each class, $g_i : \mathbb{R}^n \to \mathbb{R}$, $i = 1, \ldots, c$,

$$g_i(\mathbf{x}) = w_{i0} + \mathbf{w}_i^T \mathbf{x}. \tag{2.1}$$

The tag of the largest $g_i(\mathbf{x})$ determines the class label.

2.1.1.1 Training Linear Discriminant Classifier. A straightforward way to train a linear discriminant classifier (LDC) is shown in Figure 2.1 and detailed below:

1. Estimate the prior probabilities for the classes. Let N_i be the number of objects in the data set $\mathbf{Z}$ from class ω_i, $i = 1, \ldots c$, and $y_j \in \Omega$ be the class label of $\mathbf{z}_j \in \mathbf{Z}$. Then

$$\hat{P}(\omega_i) = \frac{1}{N_i}, \quad i = 1, \ldots, c. \tag{2.2}$$

Combining Pattern Classifiers: Methods and Algorithms, Second Edition. Ludmila I. Kuncheva.
© 2014 John Wiley & Sons, Inc. Published 2014 by John Wiley & Sons, Inc.

LINEAR DISCRIMINANT CLASSIFIER (LDC)

Training

1. Given is a labeled data set **Z**.
2. Estimate the prior probabilities for the classes $\hat{P}(\omega_i)$ as in Equation 2.2.
3. Calculate estimates of the class means $\hat{\mu}_i$ from the data as in Equation 2.3.
4. Calculate the estimates of the covariance matrices for the classes, $\hat{\Sigma}_i$, using Equation 2.4.
5. Calculate the common covariance matrix $\hat{\Sigma}$ from Equation 2.5.
6. Calculate the coefficients and the free terms of the c discriminant functions using Equation 2.6.
7. Return $\mathbf{w}_i$ and w_{i0} for $i = 1, \ldots, c$.

Operation

1. To classify an object **x**, calculate the discriminant functions $g_i(\mathbf{x})$, $i = 1, \ldots, c$, using Equation 2.1.
2. Assign to **x** the class label with the maximum $g_i(\mathbf{x})$.

FIGURE 2.1   Training and operation of the linear discriminant classifier.

2. Calculate estimates of the class means from the data:

$$\hat{\mu}_i = \frac{1}{N_i} \sum_{y_j=\omega_i} \mathbf{z}_j. \tag{2.3}$$

3. Calculate the estimates of the covariance matrices for the classes, by[1]

$$\hat{\Sigma}_i = \frac{1}{N_i} \sum_{y_j=\omega_i} (\mathbf{z}_j - \hat{\mu}_i)(\mathbf{z}_j - \hat{\mu}_i)^T. \tag{2.4}$$

4. Calculate the common covariance matrix for LDC as the weighted average of the class-conditional covariance matrices:

$$\hat{\Sigma} = \frac{1}{N} \sum_{i=1}^{c} N_i \hat{\Sigma}_i. \tag{2.5}$$

5. Calculate the coefficients and the free terms of the c discriminant functions:

$$\mathbf{w}_i = \hat{\Sigma}^{-1} \hat{\mu}_i, \qquad w_{i0} = \log(\hat{P}(\omega_i)) - \frac{1}{2} \hat{\mu}_i^T \hat{\Sigma}^{-1} \hat{\mu}_i, \qquad i = 1, \ldots, c. \tag{2.6}$$

---

[1]We use the maximum likelihood estimate of the covariance matrices and note that this estimate is biased. For an unbiased estimate take $\hat{\Sigma}_i = \frac{1}{N_i-1} \sum_{y_j=\omega_i} (\mathbf{z}_j - \hat{\mu}_i)(\mathbf{z}_j - \hat{\mu}_i)^T$.

Training of linear classifiers has been rigorously studied in the early pattern recognition literature [106], dating back to the Fisher's linear discriminant, 1936 [127]. In reality, neither are the classes normally distributed nor are the true values of μ_i and Σ_i known. Nonetheless, LDC has been praised for its robustness, simplicity, and accuracy [178].

Note that small changes in the training data are not going to affect dramatically the estimates of the means and the covariance matrix, hence the discriminant functions are likely to be *stable*. This makes LDC an unlikely choice as the base classifier in classifier ensembles.

2.1.1.2 Regularization of LDC

Problems may arise if the common covariance matrix $\hat{\Sigma}$ is close to singular (ill-posed or poorly posed problem). A way of regularizing the estimates is to add a term to $\hat{\Sigma}_i$ which will shrink it toward a multiple of the identity matrix

$$\hat{\Sigma}_i(r) = (1 - r)\hat{\Sigma}_i + \frac{r}{n} \, tr\left(\hat{\Sigma}_i\right) \, I, \tag{2.7}$$

where $tr(\cdot)$ denotes the trace of the matrix, n is the dimensionality of the feature space, and I is the identity matrix of size $n \times n$. This estimate has the effect of equalizing the eigenvalues of $\hat{\Sigma}_i$ which counters the bias inherent in sample-based estimates of the eigenvalues [137]. The parameter $r \in [0, 1]$ determines to what extent we want to equalize the eigenvalues. For $r = 0$, there is no regularization and for $r = 1$, $\hat{\Sigma}_i$ is a diagonal matrix with eigenvalues equal to the averaged eigenvalues of the sample-based estimate of the covariance matrix.

2.1.1.3 Optimality of LDC.

When is LDC equivalent to the Bayes classifier? As discussed in the previous chapter, any set of discriminant functions obtained by a monotonic transformation from the posterior probabilities $P(\omega_i|\mathbf{x})$ constitute an optimal set in terms of minimum error. Let us form such a set by taking

$$g_i(\mathbf{x}) = \log\left(P(\omega_i)\, p(\mathbf{x}|\omega_i)\right), \quad i = 1, \ldots, c, \tag{2.8}$$

where $P(\omega_i)$ is the prior probability for class ω_i and $p(\mathbf{x}|\omega_i)$ is the class-conditional probability density function (pdf). Suppose that all classes are normally distributed with means μ_i and covariance matrices Σ_i, that is, $p(\mathbf{x}|\omega_i) \sim N\left(\mu_i, \Sigma_i\right)$, $i = 1, \ldots, c$. Then Equation 2.8 takes the form

$$g_i(\mathbf{x}) = \log(P(\omega_i)) + \log\left(\frac{1}{(2\pi)^{\frac{n}{2}} \sqrt{|\Sigma_i|}} \exp\left\{-\frac{1}{2}(\mathbf{x} - \mu_i)^T \Sigma_i^{-1}(\mathbf{x} - \mu_i)\right\}\right) \tag{2.9}$$

$$= \log(P(\omega_i)) - \frac{n}{2}\log(2\pi) - \frac{1}{2}\log(|\Sigma_i|) - \frac{1}{2}(\mathbf{x} - \mu_i)^T \Sigma_i^{-1}(\mathbf{x} - \mu_i). \tag{2.10}$$

Assume that all class-covariance matrices are the same, $\Sigma_i = \Sigma$ for all $i = 1, \ldots, c$. Opening the parentheses in the last term of Equation 2.10 and discarding all terms that do not depend on ω_i, we obtain a new set of discriminant functions

$$g_i(\mathbf{x}) = \log(P(\omega_i)) - \frac{1}{2}\boldsymbol{\mu}_i^T \Sigma^{-1} \boldsymbol{\mu}_i + \boldsymbol{\mu}_i^T \Sigma^{-1} \mathbf{x} = w_{i0} + \mathbf{w}_i^T \mathbf{x}, \qquad (2.11)$$

where $w_{i0} \in \mathbb{R}$ and $\mathbf{w}_i \in \mathbb{R}^n$ are respectively the free term and the coefficients in Equation 2.6. As we started with the optimal discriminant functions, and derived the LDC coefficients, LDC is equivalent to the Bayes classifier (guaranteeing minimum error) under the assumption that the classes are normally distributed and the covariance matrices are the same.

2.1.2   Nearest Mean Classifier

The nearest mean classifier (NMC) is the simplest variant of LDC where the object is assigned to the class with the nearest mean.

As the convention is that larger values of the discriminant function indicate larger preference for the respective class, we can take as discriminant functions the negative squared Euclidean distance to the class means:

$$g_i(\mathbf{x}) = -(\boldsymbol{\mu}_i - \mathbf{x})^T (\boldsymbol{\mu}_i - \mathbf{x}) \qquad (2.12)$$
$$= -\boldsymbol{\mu}_i^T \boldsymbol{\mu}_i + 2\boldsymbol{\mu}_i^T \mathbf{x} - \mathbf{x}^T \mathbf{x}. \qquad (2.13)$$

The quadratic term $\mathbf{x}^T \mathbf{x}$ in this equation does not depend on the class label, and therefore can be dropped, giving a set of new discriminant functions that are linear on $\mathbf{x}$:

$$g_i(\mathbf{x}) = -\boldsymbol{\mu}_i^T \boldsymbol{\mu}_i + 2\boldsymbol{\mu}_i^T \mathbf{x} = w_{i0} + \mathbf{w}_i^T \mathbf{x}. \qquad (2.14)$$

This classifier is identical to LDC when the covariance matrices for all classes are the identity matrices, and all prior probabilities are equal. Therefore, NMC inherits the LDC optimality for equiprobable classes following a normal distribution with identity covariance matrices.

2.1.3   Quadratic Discriminant Classifier

The quadratic discriminant classifier (QDC) is defined by a set of quadratic discriminant functions

$$g_i(\mathbf{x}) = w_{i0} + \mathbf{w}_i^T \mathbf{x} + \mathbf{x}^T W_i \mathbf{x}, \qquad \mathbf{x}, \mathbf{w}_i \in \mathbb{R}^n, \quad w_{i0} \in \mathbb{R}, \qquad (2.15)$$

where W_i is an $n \times n$ matrix. The QDC training follows steps 1–3 of the LDC training but skips step 4. Finally, the free terms, the coefficients, and the class-specific matrices are calculated as follows:

$$w_{i0} = \log(\hat{P}(\omega_i)) - \frac{1}{2}\hat{\mu}_i^T\hat{\Sigma}_i^{-1}\hat{\mu}_i - \frac{1}{2}\log(|\hat{\Sigma}_i|), \qquad (2.16)$$

$$\mathbf{w}_i = \hat{\Sigma}_i^{-1}\hat{\mu}_i, \qquad (2.17)$$

and

$$W_i = -\frac{1}{2}\hat{\Sigma}_i^{-1}. \qquad (2.18)$$

QDC extends the optimality of LDC to the case where the covariance matrices of the classes are not identical, that is, $p(\mathbf{x}|\omega_i) \sim N(\mu_i, \Sigma_i)$. The set of optimal discriminant functions is obtained from Equation 2.10 by discarding all terms that do not depend on the class label ω_i.

QDC is even more prone to problems with singular covariance matrices than LDC. Regularized versions of QDC have been proposed [137], where the covariance matrices are gradually driven to the common covariance matrix used in LDC. The process is again governed by a parameter that determines the extent of the regularization.

LDC and QDC are parametric classifiers because their training involves estimating the parameters of the assumed normal distributions.

LDC has a number of rivals which also produce a linear boundary between the classes, notable examples of which are the support vector machine (SVM) classifier with a linear kernel, the logistic classifier, and the Rosenblatt's perceptron [107, 179]. Their training is different and so are their optimality conditions. Nonetheless, LDC is an "evergreen" in pattern recognition and still enjoys widespread accolades.

2.1.4   Stability of LDC and QDC

LDC and QDC are usually seen as unsuitable for combining into an ensemble because of their stability. Recall Figure 1.12 from Chapter 1 illustrating bias and variance of the error. LDC and QDC are unlikely to change dramatically with small changes in the data set. If they happen to not be sufficiently adequate for the problem at hand, LDC and QDC will exhibit high bias (far from the true solution) and low variance (stable boundaries).

■ **Example 2.1   High bias and low variance of LDC and QDC**
The data set plotted in Figure 2.2a, named "fish data," consists of the 2500 nodes of a square two-dimensional grid $[0, 1] \times [0, 1]$. The class labels are indicated with black and gray.[2] A random 10 percent of the labels are flipped to create an imperfect

---

[2]A MATLAB function for generating the fish data is given in Appendix 2.A.1.

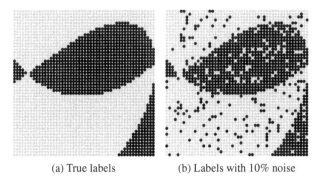

(a) True labels   (b) Labels with 10% noise

FIGURE 2.2   A synthetic data set "fish," 2500 points on a regular grid; class prevalence: 64% (gray) and 36% (black).

version of the data, shown in Figure 2.2b. The true labels against which classification error is measured are the ones in Figure 2.2a. The true labels are assigned as:

$$z_1 = x^3 - 2xy + 1.6y^2 < 0.4$$
$$z_2 = -x^3 + 2y \sin(x) + y < 0.7$$
$$\text{label}(x, y) = \text{xor}(z_1, z_2). \tag{2.19}$$

Twenty different boundaries are obtained by training the respective classifier with a randomly sampled 25% of the data points. The data and the boundaries are plotted in Figure 2.3.

The example demonstrates that there will hardly be much benefit from using LDC/QDC as base classifiers in an ensemble. The ensemble will be able to reduce the variance of the solution but will not improve on the bias. On the other hand, imagine linear classifiers of similar accuracy as the ones in Figure 2.3a, but much more scattered, as in subplot Figure 2.3c. If suitably combined, such classifiers have a much better chance of producing a reasonable ensemble classification boundary. The example hints at a very important issue in classifier ensembles—the *diversity* in the ensemble, discussed later in the book.

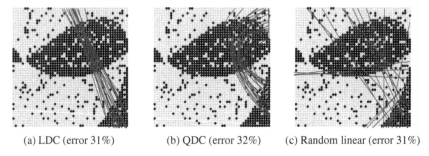

(a) LDC (error 31%)   (b) QDC (error 32%)   (c) Random linear (error 31%)

FIGURE 2.3   Decision boundaries of LDC, QDC, and a random linear classifier.

2.2   DECISION TREE CLASSIFIERS

2.2.1   Basics and Terminology

Seni and Elder [355] point out that at the KDNuggets poll in 2007, decision tree classifiers scored a clear top position as the "most frequently used" data mining method. In the same poll, neural networks and the SVM classifier were ranked respectively 8th and 9th, with various other classifier ensemble methods further down behind. In a similar poll in 2011, decision trees still reigned supreme, with SVM climbing to the 7th place, and ensemble methods to the 9th place, leaving neural networks in 11th place. Even though these polls were not taken across an unbiased and representative canvas of researchers, their results faithfully reflect the importance of decision trees.

What is so good about decision tree classifiers?

1. They can handle irrelevant and redundant variables. Each split uses a single best variable, hence irrelevant variables may never be picked.
2. Continuous-valued, discrete, and categorical variables can be handled together; there is no need to convert one type into another.
3. Scaling of the variables does not matter. Since each feature is handled separately to find a bespoke threshold, there is no need to normalize or re-scale the data to fit into a given interval. A distance is not trivial to formulate when the objects are described by categorical or mixed-type features. Decision trees have the advantage of bypassing this problem and are therefore *nonmetric methods* for classification [107].
4. If all the objects are distinguishable, that is, there are no identical elements of Z with different class labels, then we can build a tree classifier with zero resubstitution error. This fact places tree classifiers in the *unstable* group: capable of memorizing the training data so that small alterations of the data might lead to a differently structured tree classifier. As we shall see later, instability can be an advantage rather than a drawback for classifier ensembles.
5. Tree classifiers are intuitive because the decision process can be traced as a sequence of simple choices. Tree structures can capture a knowledge base in a hierarchical arrangement; most pronounced examples of which are botany, zoology, and medical diagnosis.
6. Training is reasonably fast while operation can be extremely fast.

Tree classifiers are usually described in graph terminology. A classification tree consists of a root, intermediate nodes (optional), and leaves. The root and the inter-mediate nodes branch the decision process, while the leaves assign the labels. An object that is to be labeled travels along a path in the tree and reaches a leaf where it receives its label. An example is given below.

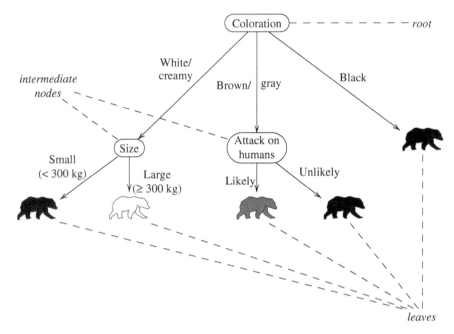

FIGURE 2.4   An illustration of a decision tree and related terminology.

Example 2.2   Terminology of tree classifiers

Shown in Figure 2.4 is a decision tree for distinguishing between three types of adult bears.[3] The three classes are

Ursus maritimus   *Ursus americanus*   *Ursus arctos*
(Polar Bear)   (American Black Bear)   (Grizzly Bear)

The features (attributes) are coloration, size, and attack on humans. (Curiously, almost white individuals of American Black Bear could be found in the north-western region of North America).

A special case of the decision tree classifier is the one-node tree [5], aptly called a "decision stump." The decision stump selects only one feature and makes a binary

---

[3]Based upon information found at: http://en.wikipedia.org/wiki/List_of_bears and http://en.wikipedia.org/wiki/American_black_bear

split on it. Such simple classifiers have been found to perform surprisingly well on an overwhelming part of the commonly used data sets [172, 185].

Usually one feature is used at each nonterminal node (monothetic trees). Subsets of features can be used at a node and the branching decisions can be made by a formula based on these features. The reasons to stick with single features are rather of psychological and engineering nature. If several features are involved in a split, the interpretation of the final decision as a chain of simple decisions might be jeopardized.

2.2.2 Training of Decision Tree Classifiers

Decision tree training lends itself to an elegant recursive tree-growing algorithm as shown in Figure 2.5 [51]. Starting at the root, decide whether the data is pure enough to warrant termination of the training. If yes, make a leaf and add it to the tree. If not, find the feature with the maximum discrimination ability. Split the data into left and right children nodes according to the best threshold for that feature. Repeat the procedure for the left and then for the right child, taking forward the respective portions of the data reaching that node.

A termination criterion could be, for example, that all objects be labeled correctly. Having constructed a perfect tree, we have to *prune* it to counter overtraining. This process is called *post-pruning*. Alternatively, we may use some measurable objective function to decide when to stop splitting, called also *pre-pruning*.

To automate the tree construction, it is reasonable to choose *binary trees*, which means that each nonterminal node has exactly two children nodes. For any node

---

DECISION TREE CLASSIFIER

Training

1. Given is a labeled data set $\mathbf{Z}$, a feature evaluation criterion and a leaf creating criterion.
2. Start with an empty tree and consider the root node.
3. If the leaf-creating criterion is satisfied, create a leaf and assign to it the label most represented in the data that arrived at that node.
4. Otherwise, add an intermediate node to the tree. Use the feature evaluation criterion to find the best feature and the respective threshold. Use these to split the data and send the parts to the Left and Right children nodes.
5. Repeat from step 3 for both children nodes until the whole of $\mathbf{Z}$ is distributed into leaf nodes.
6. Return the tree.

Operation

1. To label a new data point $\mathbf{x}$, start at the root of the tree and follow the path according to the feature values of $\mathbf{x}$.
2. Assign to $\mathbf{x}$ the label of the leaf it finally arrives at.

FIGURE 2.5 Training and operation of a decision tree classifier.

with multiple answers, there are equivalent representations with binary nodes. For continuous-valued features, the question for a binary split is usually of the form "is $x \leq x_s$?" where x_s is the node's threshold. For ordinal categorical features with M successive categories, there are $M - 1$ possible splits. For nominal features with M possible values, there are $2^{M-1} - 1$ possible splits (the number of pairs of nonempty subsets). Some tree construction algorithms (C4.5 and J48 in WEKA) generate M children nodes instead of making a binary split.

2.2.3 Selection of the Feature for a Node

The compound objective in designing a tree classifier involves accuracy and simplicity. The construction of the tree splits a given training set hierarchically until either all the objects within the same sub-region have the same class label, or the sub-region is *pure enough*.

Consider a c-class problem with label set $\Omega = \{\omega_1, \ldots, \omega_c\}$. Let P_j be the probability for class ω_j at a certain node t of a decision tree. We can estimate these probabilities as the proportion of points from the respective class within the data set that reached node t. The *impurity* of the distribution of the class labels at t can be measured in different ways.

Entropy-based measure of impurity

$$i(t) = - \sum_{j=1}^{c} P_j \log P_j, \tag{2.20}$$

where $0 \times \log 0 = 0$. For a pure region (only one class label), impurity takes its minimum value, $i(t) = 0$. The most impure situation is when the classes have uniform distribution. In this case, impurity is maximum, $i(t) = \log c$.

Gini impurity

$$i(t) = 1 - \sum_{j=1}^{c} P_j^2. \tag{2.21}$$

Again, for a pure region, $i(t) = 0$. The highest impurity is $i(t) = \frac{c-1}{c}$, in the case of uniform distribution of the class labels. The Gini index can be thought of as the expected classification error incurred if a class label was drawn randomly from the distribution of the labels at t.

Misclassification impurity

$$i(t) = 1 - \max_{j=1}^{c} \{P_j\}. \tag{2.22}$$

Of the three indices, misclassification impurity is most related to the classification accuracy. It gives the expected error if the node was replaced by a leaf and the chosen label was the one corresponding to the largest P_j.

Assume that we split t into two children nodes t_0 and t_1 based on a binary feature X. The gain in splitting t is in the drop of impurity on average, denoted $\Delta i(t)$:

$$\Delta i(t) = i(t) - P(X = 0) \times i(t_0) - P(X = 1) \times i(t_1)$$
$$= i(t) - P(X = 0) \times i(t_0) - (1 - P(X = 0)) \times i(t_1), \qquad (2.23)$$

where $P(\zeta)$ is the probability of event ζ.

If the features we are using are binary, then the task of selecting the best feature for node t is easy: try each one in turn and pick the feature with the highest $\Delta i(t)$. However, for features with multiple categories and for continuous-valued features, we have to find first the optimal threshold to split t into left (t_L) and right (t_R).

Example 2.3   Calculation of impurity indices

Figure 2.6 shows the three indices calculated for the fish data set (Figure 2.2b). Fifty split points were checked on each axis and the indices Δi were plotted as functions of the split point. The functions for feature X are projected underneath the data sub-plot, and the functions for feature Y are projected on the top left sub-plot.

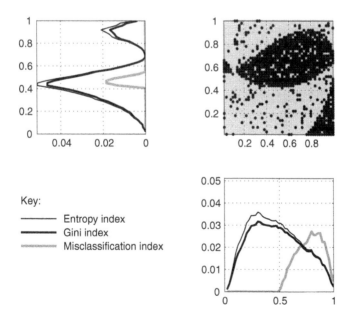

FIGURE 2.6   Three indices of impurity used in growing decision trees: entropy-based, Gini, and misclassification, calculated for 50 split points for each of X and Y for the fish data set.

TABLE 2.1 **Maximal Values of Δi and the Respective Split Points for the Three Impurity Indices**

	Entropy		Gini		Misclassification	
	X	Y	X	Y	X	Y
Δi	0.0359	**0.0511**	0.0317	**0.0463**	**0.0272**	0.0184
x_s, y_s	0.3061	0.4286	0.3061	0.4490	0.7959	0.4490

Table 2.1 shows the values of Δi for the three indices and the suggested split points. For example, if we use Gini for building the tree, we should prefer the split on Y, point $y_s = 0.4490$, because this gives the highest value of Δi. On the other hand, if we used the misclassification index, we should split on X at $x_s = 0.7959$.

Gini index is often used in tree construction [107, 330] because it can distinguish between choices for which the misclassification index gives the same value. The choice of impurity index does not seem to be of vital importance for the quality of the trained tree classifier compared to the importance of the stopping criterion and the pruning method [107].

Note that the choice of the best feature for the current node is a typical example of a *greedy algorithm*. The optimality of the local choices does not guarantee that the constructed tree will be globally optimal.

2.2.4 Stopping Criterion

The tree construction could be carried on until there are no further impure nodes. If the training data contains no objects with the same feature values and different class labels, then a perfect tree can be constructed. However, such a perfect result could be due to overtraining, and so the tree classifier would not be of much value. One solution to this problem is to stop the training before reaching pure nodes. How do we decide where to stop? If the splitting is stopped too early, we might leave the classifier under-trained. Below we summarize the options suggested in [107].

Validation set. Put part of the training data aside as a validation set. After each split, check the classification accuracy of the tree on the validation set. When the error begins to increase, stop splitting.

Impurity reduction threshold. Set a small impurity reduction threshold β. When the greatest possible reduction of impurity at node t is $\Delta i(t) \le \beta$, stop splitting and label this node as the majority of the points. This approach does not exactly answer the question "where to stop?" because the stopping is determined by the choice of β, and this choice lies again with the designer.

Number/percentage threshold. We may decide not to split a node if it contains less than k points or less than l percent of the total number of points in the training set. The argument is that continuing the splitting beyond this point will most probably lead to overtraining.

Complexity penalty. Penalize the complexity of the tree by using a criterion such as

$$\text{minimize} \quad \alpha \times \text{size} + \sum_{\text{leaves } t} i(t), \tag{2.24}$$

where "size" can be the total number of nodes, branches or leaves of the tree, and α is a positive constant. Again, the problem here is choosing the value of α so that the "right" balance between complexity and accuracy is achieved.

Hypothesis testing. A hypothesis can be tested to decide whether a further split is beneficial or not. Consider a two-class problem. Assume that there are n data points at node t, n_1 of which have labels ω_1 and n_2 have labels ω_2. Assume that the best feature for the node has been found to be X, and the best split of X is at some point x_s. The left and the right children nodes obtain n_L and n_R number of points, respectively. Denote by n_{L1} the number of points from ω_1 in the left child node, and by n_{L2} the number of points from ω_2 in the left child node. If the distributions of class labels at the children nodes are identical to that at the parent node, then no purity is gained and the split is meaningless. To test the equivalence of the distributions at the children and the parent's node, calculate

$$\chi_L^2 = \frac{(n \times n_{L1} - n_L \times n_1)^2}{n \times n_L \times n_1} + \frac{(n \times n_{L2} - n_L \times n_2)^2}{n \times n_L \times n_2} \tag{2.25}$$

and

$$\chi_R^2 = \frac{(n \times n_{R1} - n_R \times n_1)^2}{n \times n_R \times n_1} + \frac{(n \times n_{R2} - n_R \times n_2)^2}{n \times n_R \times n_2}. \tag{2.26}$$

We can take the average $\chi^2 = \frac{1}{2}\left(\chi_L^2 + \chi_R^2\right)$, which, after simple algebraic transformations is

$$\chi^2 = \frac{1}{2}\left(\chi_L^2 + \chi_R^2\right) = \frac{n}{2n_R}\chi_L^2 = \frac{n}{2n_L}\chi_R^2. \tag{2.27}$$

If χ^2 is greater than the tabular value with the specified level of significance and one degree of freedom, then the split is worthwhile. If the calculated value is smaller than the tabulated value, we should stop splitting and should label the node as ω_1 if $n_1 > n_2$, and ω_2 otherwise. The lower the level of significance, the smaller the tree would be because greater differences in the distributions will be required to permit splitting.

For c classes, the degrees of freedom will be $c - 1$, and the χ^2 statistic should be calculated as the average of χ_L^2 and χ_R^2, where

$$\chi_L^2 = \sum_{i=1}^{c} \frac{(n \times n_{Li} - n_L \times n_i)^2}{n \times n_L \times n_i} \tag{2.28}$$

and

$$\chi_R^2 = \sum_{i=1}^{c} \frac{(n \times n_{Ri} - n_R \times n_i)^2}{n \times n_R \times n_i}.$$

Note that the χ^2 statistic (Equation 2.25) or (Equation 2.28) can be used for a direct comparison with a threshold set by the user. If the calculated value is greater than the threshold, we should split the node, otherwise we label it as a leaf. Appropriate values of the threshold vary between 0 (full tree) and 10 (heavy pre-pruning).

◾ Example 2.4   Step-by-step decision tree construction

Consider a data set with two classes as shown in Figure 2.7a. The training starts by examining the leaf-creating criterion at the root. We have chosen the χ^2 criterion and a threshold of 5. At the root node $\chi^2 = 23.52 > 5$, therefore we split the node. The best feature-threshold pair was identified using the Gini criterion: feature x_1 with threshold 0.37.

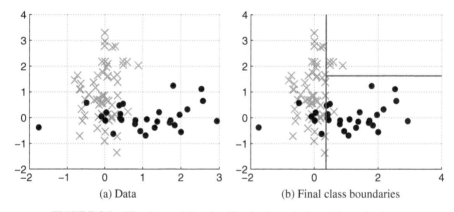

(a) Data   (b) Final class boundaries

FIGURE 2.7   The data and the classification boundaries of the trained tree.

As shown in Figure 2.8, the data is split into left and right parts according to the vertical line in the top graph. The left part goes to the left child node where $\chi^2 = 4.66 < 5$, which indicates that a leaf should be formed. The class label of the leaf is the predominant label in the data that has arrived at that leaf, hence class "cross."

The χ^2 criterion at the right child node indicates that a new split must be carried out. This time the best feature-threshold pair was feature x_2, threshold 1.63. Both children nodes satisfy the leaf criterion, which terminates the training process. The final classification boundaries of the decision tree classifier are shown in Figure 2.7b.

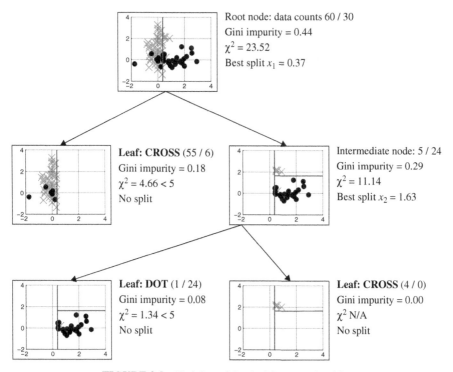

FIGURE 2.8   Training of the decision tree classifier.

A DIY MATLAB function for constructing a decision tree classifier is given in Appendix 2.A.2. The code implements the Gini index for the splits and the χ^2 criterion for pre-pruning. The code is meant only for illustration purposes and may be quite slow on large data sets.

2.2.5   Pruning of the Decision Tree

Sometimes early stopping can be too short-sighted and prevent further beneficial splits. This phenomenon is called the *horizon effect* [107]. To counter the horizon effect, we can grow the full tree and then prune it to a smaller size. The pruning seeks a balance between the increase of the training error and the decrease of the size of the tree. Downsizing the tree will hopefully reduce overtraining. There are different criteria and methods to prune a tree summarized by Esposito et al. [124].

As an example, consider *Reduced Error Pruning (REP)*. This method is perceived as the simplest pruning method. It uses an additional training set, called the "pruning set," unseen during the growing stage. Starting at the bottom (leaves) and working our way up to the top (the root), a simple error check is calculated for all nonleaf nodes. We replace the node with a leaf and label it to the majority class, then calculate the error of the new tree *on the pruning set*. If this error is smaller than the error of the whole tree on the pruning set, we replace the node with the leaf. Otherwise, we keep the subtree. This bottom-up procedure guarantees that the minimum tree that is optimal with respect to the pruning set error will be found. Also, REP has low complexity because each internal node is visited just once. A drawback of REP is that it has a tendency toward over-pruning [124].

2.2.6   C4.5 and ID3

The methodology for tree construction explained so far is within the CART framework [51]. The two main alternatives for designing trees are the ID3 algorithm and the C4.5 algorithm.

The ID3 algorithm is due to Quinlan [320], a third algorithm from a series of *interactive dichotomizer* algorithms. It is designed for nominal data, so all continuous-valued variables are first discretized and the categories are treated as unordered. The main characteristic of ID3 is that each node has as many children nodes as the number of categories of the (nominal) feature at that node. Since a feature is completely "used up" on a single node, the tree can only be grown up to maximum n layers, where n is the total number of features. Pruning of the resultant tree can be done as well. To select a feature for a node, we should use a modified version of the impurity criteria because the formulas introduced above inherently favor multiple splits over two-way splits without a good reason. Instead of the absolute impurity reduction Δi, we should use a scaled version thereof, called the *gain ratio impurity*. In an M-way split, let P_i be the proportion of objects going into the i-th child node. Then

$$\Delta i_M = \frac{\Delta i}{-\sum_{i=1}^{M} P_i \log P_i}. \tag{2.29}$$

The main problem with ID3 when continuous-valued variables are concerned is the way these are discretized into categories. There is an inevitable loss of information involved in the process. Moreover, a careless formulation of the categories might "kill" an otherwise important variable.

The C4.5 algorithm avoids the discretization problem by using the continuous-valued variables as in CART, and the nominal variables as in ID3. Duda et al. [107] note that C4.5 is the currently most popular tree construction algorithm among the machine learning researchers.

2.2.7   Instability of Decision Trees

Decision tree classifiers are unstable in that small alteration of the training data may lead to different splits and hence differently shaped classification regions.

■ **Example 2.5**   We built 20 decision trees for the fish data (Figure 2.2b) on randomly sampled 25% of the data. Figure 2.9a shows the classification regions

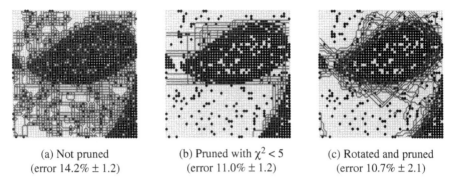

| (a) Not pruned | (b) Pruned with $\chi^2 < 5$ | (c) Rotated and pruned |
| (error 14.2% ± 1.2) | (error 11.0% ± 1.2) | (error 10.7% ± 2.1) |

FIGURE 2.9   Decision boundaries of the decision tree classifier.

when the trees are grown without any pruning. The noise points which happen to fall in the training set of a given tree appeared as little "islands" in the decision region of the opposite class. This explains the web of boundary lines in Figure 2.9a. The average error rate of the 20 trees is 14.2%. The pruned trees show low bias (decision regions follow the true classification regions) and a relatively good diversity of the boundaries as shown in Figure 2.9b. This makes the decision tree classifier a good candidate for a base ensemble classifier. Finally, to demonstrate the further potential of the decision tree classifier, we created 20 more trees, but this time each sampled training data set was rotated at a random angle between 0 and 2π radians. By design, the classification boundaries of the standard decision trees are composed of segments parallel to the coordinate axes. Therefore, by rotating the data we force boundaries at different angles (Figure 2.9c). The average classification error with the rotated data was lower than the error with the original data while the standard deviation was higher. Low classification error and high diversity are an ideal combination for classifier ensembles. A highly successful ensemble model called "Rotation Forest," which will be detailed later in the book, is based on the idea of combining decision trees built on rotated data.

2.2.8 Random Trees

Breiman proposed a random tree classifier, which underpins one of the most successful ensemble models to date—the random forest [50]. The difference from the standard tree training is that, at each node, the feature to split upon is chosen from a random subset of the original features. The cardinality of this subset, M, is a parameter of the training algorithm; as a rule of thumb, M is often taken to be the square root of the total number of features. Breiman argues that the algorithm is not overly sensitive to M. Random trees are important in the context of classifier ensembles. A small decline in the classification accuracy is traded to gain some increase of the diversity between the base classifiers, thereby opening the perspective for a better ensemble. The diversity will come from the randomized splits. The random trees are grown without pre- or post-pruning, which contributes to their diversity.

2.3   THE NAÏVE BAYES CLASSIFIER

Naïve Bayes or also "Idiot's Bayes" [172] is a simple and often surprisingly accurate classification technique [139, 212, 213, 218, 251, 253, 413]. Consider an object represented by a feature vector $\mathbf{x} = [x_1, \ldots, x_n]^T$ that is to be assigned to one of c predefined classes, $\omega_1, \ldots, \omega_c$. Minimum classification error is guaranteed if the class with the largest posterior probability, $P(\omega_i \mid \mathbf{x})$, is chosen. To calculate posterior probabilities, the Bayes formula is used with estimates of the prior probabilities, $P(\omega_i)$, and the class-conditional pdf, $p(\mathbf{x} \mid \omega_i)$:

$$P(\omega_i \mid \mathbf{x}) = \frac{P(\omega_i)p(\mathbf{x} \mid \omega_i)}{\sum_{j=1}^{c} P(\omega_j)p(\mathbf{x} \mid \omega_j)}, \quad i = 1, \ldots, c. \tag{2.30}$$

Obtaining an accurate estimate of the joined pdf is difficult, especially if the dimensionality of the feature space, n, is large. The "naïvety" of the Naïve Bayes model comes from the fact that the features are assumed to be conditionally independent. In this case the joined pdf for a given class is the product of the marginal pdfs:

$$p(\mathbf{x} \mid \omega_i) = \prod_{j=1}^{n} p(x_j \mid \omega_i), \quad i = 1, \ldots, c. \tag{2.31}$$

Accurate estimates of the marginal pdfs can be obtained from much smaller amounts of data compared to these for the joint pdf. This makes the Naïve Bayes classifier (NB) so attractively simple. The assumption of conditional independence among features may look too restrictive. Nonetheless NB has demonstrated robust and accurate performance across various domains, often reported as "surprisingly" accurate, even where the assumption is clearly false [172].

The feature's pdfs can be estimated in different ways. For a categorical feature with a small set of possible values we can estimate the conditional frequencies of all categories. For continuous-valued features, we can use a parametric or a nonparametric estimate of $p(\mathbf{x}|\omega_i)$. The pdf estimates are subsequently plugged in Equation 2.31 and the posterior probabilities are calculated through Equation 2.30. The class with the largest posterior probability is assigned to $\mathbf{x}$. One possible variant of the parametric approach is to assume normal distribution for each feature and each class, and estimate the respective mean and standard deviation. For the nonparametric approach, each feature is discretized, and the pdf is estimated for all discrete values. This can be done, for example, by a histogram with K equally spaced bins. A generic version of the NB training and classification algorithms is shown in Figure 2.10.

NAÏVE BAYES CLASSIFIER (NB)

Training

1. Given is a labeled data set $\mathbf{Z}$ (n features and c classes) and a pdf approximation algorithm for a single feature.
2. Estimate the prior probabilities for the classes, $\hat{P}(\omega_i), i = 1, \ldots, c$.
3. For each class and each feature, approximate the class-conditional pdf, $\hat{p}(x_j|\omega_i)$, $i = 1, \ldots, c, j = 1, \ldots, n$.
4. Return $\hat{P}(\omega_i)$ and $\hat{p}(x_j|\omega_i)$.

Operation

1. To label a new data point $\mathbf{x}$, calculate $\hat{p}(x_j|\omega_i)$ for the feature values in $\mathbf{x}$, $i = 1, \ldots, c$, $j = 1, \ldots, n$.
2. Calculate the discriminant functions

$$g_i(\mathbf{x}) = \hat{P}(\omega_i) \prod_{j=1}^{n} \hat{p}(x_j|\omega_i), \quad i = 1, \ldots, c.$$

3. Assign to $\mathbf{x}$ the class label with the maximum $g_i(\mathbf{x})$.

FIGURE 2.10   Training and operation of the Naïve Bayes classifier.

■ Example 2.6   Classification boundaries of the Naïve Bayes classifier (NB)
Consider the fish data set shown in Figure 2.2b. Twenty random subsamples were taken, each containing a quarter of the points on the grid. Figure 2.11a shows the classification boundaries for the parametric Naïve Bayes classifier (normal distribution for each feature and each class). Figure 2.11b shows the boundaries where each of the 20 samples was rotated at a random angle between 0 and 2π radians. The plots show that NB is reasonably sensitive to alterations of the data such as sub-sampling and rotation, and has a slightly better accuracy than LDC and QDC. The MATLAB code for the NB training and classification is given in Appendix 2.A.2.2.

   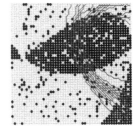

(a) Original (error 28.0% ±1.1)   (b) Rotated (error 27.8% ± 1.5)

FIGURE 2.11   Decision boundaries of the parametric Naïve Bayes classifier (normal distribution) for the fish data with 10% label noise.

2.4   NEURAL NETWORKS

Artificial Neural Networks (ANNs or simply NNs) originated from the idea to model mathematically human intellectual abilities by biologically plausible engineering designs. Meant to be massive, parallel computational schemes resembling a real brain, NNs evolved to become a valuable classification tool with a significant influence on pattern recognition theory and practice. Neural networks are often used as base classifiers in multiple classifier systems [360]. Similarly to tree classifiers, NNs are unstable, in that small changes in the training data might lead to a large change in the classifier, both in its structure and parameters.

Literature on NNs is excessive and continuously growing [40, 180, 330], discussing NNs at various theoretical and algorithmic depth.

Consider an n-dimensional pattern recognition problem with c classes. The NN obtains a feature vector $\mathbf{x} = [x_1, \dots, x_n]^T \in \mathbb{R}^n$ at its input and produces values for the c discriminant functions $g_1(\mathbf{x}), \dots, g_c(\mathbf{x})$ at its output. Typically NNs are trained to minimize the squared error on a labeled training set $\mathbf{Z} = \{\mathbf{z}_1, \dots, \mathbf{z}_N\}, \mathbf{z}_j \in \mathbb{R}^n$, and $y_j \in \Omega$:

$$
E = \frac{1}{2} \sum_{j=1}^{N} \sum_{i=1}^{c} \left(g_i(\mathbf{z}_j) - \mathcal{I}(\omega_i, y_j) \right)^2,
\tag{2.32}
$$

where $\mathcal{I}(\omega_i, y_j)$ is an indicator function taking value 1 if the label of $\mathbf{z}_j$, y_j is ω_i, and 0 otherwise. It has been shown that the set of discriminant functions obtained by minimizing Equation 2.32 approach the posterior probabilities for the classes for data size $N \to \infty$ [328, 341, 406], that is,

$$
\lim_{N \to \infty} g_i(\mathbf{x}) = P(\omega_i | \mathbf{x}), \ \mathbf{x} \in \mathbb{R}^n.
\tag{2.33}
$$

This result was brought to light in connection with NNs but, in fact, it holds for any classifier which can approximate an arbitrary discriminant function with a specified precision. This *universal approximation* property has been proven for the two important NN models: the *Multi-Layered Perceptron* (MLP) and the *Radial Basis Function* (RBF) networks (for summaries of the literature and proofs refer to [40] and [330]). Various NN training protocols and algorithms have been developed, and these have been the key to the success of NN classifiers.

2.4.1   Neurons

The processing units in the human brain are neurons of different specialization and functioning. The earliest models of neurons, including the one proposed by McCulloch and Pitts [269] and Fukushima's cognitron [142], reprinted in the collection [17], were more similar to the biological neuron than later models. For example, they incorporated both activating and veto-type inhibitory inputs. To avoid confusion, artificial neurons are often given other names: "nodes" [313], "units" [40, 330],

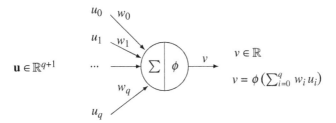

FIGURE 2.12 The NN processing unit.

"neurodes" [268]. Simple models will need a large structure for the whole system to work well (e.g., the weightless neural networks [7]) while for more complex models of neurons a few units will suffice. In both cases proper algorithms are needed to train the structure and parameters of the NN. Complex models without good training algorithms are not of much use. The basic scheme of a processing node is shown in Figure 2.12.

Let $\mathbf{u} = [u_0, \ldots, u_q]^T \in \mathbb{R}^{q+1}$ be the input vector to the node and $v \in \mathbb{R}$ be its output. We call $\mathbf{w} = [w_0, \ldots, w_q]^T \in \mathbb{R}^{q+1}$ a vector of *synaptic weights*. The processing element implements the function

$$v = \phi(\xi); \quad \xi = \sum_{i=0}^{q} w_i u_i, \tag{2.34}$$

where $\phi : \mathbb{R} \to \mathbb{R}$ is the *activation function* and ξ is the *net sum*. Typical choices for ϕ are

- The Heaviside (threshold) function:

$$\phi(\xi) = \begin{cases} 1, & \text{if } \xi \geq 0, \\ 0, & \text{otherwise.} \end{cases} \tag{2.35}$$

- The sigmoid function:

$$\phi(\xi) = \frac{1}{1 + \exp(-\xi)}. \tag{2.36}$$

- The identity function:

$$\phi(\xi) = \xi. \tag{2.37}$$

The three activation functions are shown in Figure 2.13.

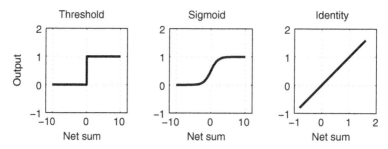

FIGURE 2.13   Threshold, sigmoid, and identity activation functions.

The sigmoid activation function is the most widely used one because:

- It can model both linear and threshold functions to a desirable precision.
- The sigmoid function is differentiable, which is important for the NN training algorithms. Moreover, the derivative on ξ has the simple form $\phi'(\xi) = \phi(\xi)(1 - \phi(\xi))$.

The weight "$-w_0$" is used as a *bias* and the corresponding input value u_0 is set to 1. Equation (2.34) can be rewritten as

$$v = \phi\left(\zeta - (-w_0)\right) = \phi\left(\sum_{i=1}^{q} w_i u_i - (-w_0)\right),\qquad(2.38)$$

where ζ is now the weighted sum of the inputs from 1 to q. Geometrically, the equation

$$\sum_{i=1}^{q} w_i u_i - (-w_0) = 0\qquad(2.39)$$

defines a hyperplane in $\mathbb{R}^q$. Therefore a node with a threshold activation function (2.35) responds with value +1 to all inputs $[u_1, \ldots, u_q]^T$ on the one side of the hyperplane, and value 0 on the other side.

2.4.2   Rosenblatt's Perceptron

An important model of a neuron was defined by Rosenblatt [340]. It is called *perceptron* and is famous for its training algorithm. The perceptron is implemented as Equation 2.34 with a threshold activation function

$$\phi(\xi) = \begin{cases} 1, & \text{if } \xi \geq 0, \\ -1, & \text{otherwise.} \end{cases}\qquad(2.40)$$

This one-neuron classifier separates two classes in $\mathbb{R}^n$ by the linear discriminant function defined by $\xi = 0$. The vectors from one of the classes get an output value of $+1$, and from the other class, -1. The algorithm starts with random initial weights $\mathbf{w}$ and proceeds by modifying the weights as each object from $\mathbf{Z}$ is submitted. The weight modification takes place *only if* the current object $\mathbf{z}_j$ is misclassified (appears on the "wrong" side of the hyperplane). The weights are corrected by

$$\mathbf{w} \leftarrow \mathbf{w} - v\eta\mathbf{z}_j, \tag{2.41}$$

where v is the output of the node for $\mathbf{z}_j$ and η is a parameter specifying the *learning rate*. Beside its simplicity, the perceptron training has the following interesting properties:

- If the two classes are *linearly separable* in $\mathbb{R}^n$, the algorithm *always converges in a finite number of steps* to a linear discriminant function that gives no resubstitution errors on $\mathbf{Z}$, for any η. (This is called "the perceptron convergence theorem.")
- If the two classes are not linearly separable in $\mathbb{R}^n$, the algorithm will loop infinitely through $\mathbf{Z}$ and never converge. Moreover, there is no guarantee that if we terminate the procedure at some stage, the resultant linear function is the one with the smallest possible misclassification count on $\mathbf{Z}$.

2.4.3   Multi-Layer Perceptron

By connecting perceptrons we can design an NN structure called the *Multi-Layer Perceptron* (MLP). MLP is a *feed-forward* structure because the output of the input layer and all intermediate layers is submitted only to the higher layer. The generic model of a feed-forward NN classifier is shown in Figure 2.14.

Here "layer" means a layer of neurons. The feature vector $\mathbf{x}$ is submitted to an input layer, and at the output layer there are c discriminant functions $g_1(\mathbf{x}), \dots, g_c(\mathbf{x})$. The number of hidden layers and the number of neurons at each hidden layer are not limited. The most common default conventions are:

- The activation function at the input layer is the identity function (2.37).
- There are no lateral connections between the nodes at the same layer (feed-forward structure).
- Nonadjacent layers are not connected directly.
- All nodes at all hidden layers have the same activation function ϕ.

This model is not as constrained as it might look. It has been proven that an MLP with a single hidden layer and threshold nodes can approximate *any* function with a specified precision [40, 330, 349]. However, the proofs did not offer a feasible training algorithm for the MLP. The conception and subsequent refinement of the *error backpropagation* training algorithm in the 1980 heralded a new era in the NN field, which continues to this day.

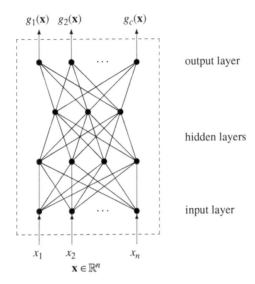

$g_1(\mathbf{x})$ $g_2(\mathbf{x})$ $g_c(\mathbf{x})$

output layer

hidden layers

input layer

x_1 x_2 x_n

$\mathbf{x} \in \mathbb{R}^n$

FIGURE 2.14   A generic model of an MLP classifier.

The error backpropagation training of an MLP updates the network weights iteratively until the output error drops below a given threshold or until the limit of the number of iterations is reached. It is a gradient descend (greedy) algorithm, which converges to a set of weights corresponding to a local minimum of the output error. Different random initializations may lead to different minima. The algorithm goes through the following steps.

1. Choose the MLP structure: number of hidden layers, number of nodes at each layer, and a differentiable activation function. Pick the learning rate $\eta > 0$, the error goal $\epsilon > 0$, and T, the number of training epochs.[4]
2. Initialize the training procedure by picking small random values for all weights (including biases) of the NN.
3. Set the initial error at $E = \infty$, the epoch counter at $t = 1$, and the object counter at $j = 1$.
4. While $(E > \epsilon$ and $t \leq T)$ do
   (a) Submit $\mathbf{z}_j$ as the next training example.
   (b) *Forward propagation.* Calculate the output of the NN with the current weights.
   (c) *Backward propagation.* Calculate the error term at each node at the output layer. Calculate recursively all error terms at the nodes of the hidden layers.*
   (d) For each hidden and each output node update the weights using the learning rate η to scale the update.*

---

[4] An epoch is a pass of the whole data set, object by object, through the training algorithm.

(e) Calculate E using the current weights and Equation 2.32.

(f) If $j = N$ (a whole pass through $\mathbf{Z}$—an epoch—is completed), then set $t = t + 1$ and $j = 0$. Else, set $j = j + 1$.

5. End of the while loop.

Marked with * are steps for which the details are omitted. The full algorithmic detail can be recovered from the MATLAB code given in the Appendix 2.A.2.3. The MLP trained through this code has a single hidden layer (the most popular choice) with a number of nodes given as an input parameter.

There are two ways to implement the training procedure: batch and online. The version explained above is the online version where the updates of the weights take place after the submission of each object. In the batch version, the errors are accumulated and the updating is done only after all of $\mathbf{Z}$ is seen, that is, there is one update after each epoch.

■ **Example 2.7   Classification boundaries of the MLP classifier**
Figure 2.15 shows the classification boundaries for the fish data of 20 MLP classifiers run with different random initializations. The figure shows why the MLP classifier is one of the preferred base models in classifier ensembles: the boundaries are fairly precise and diverse at the same time.

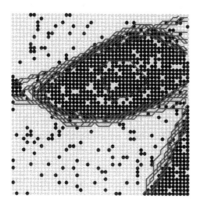

FIGURE 2.15   Decision boundaries of the MLP classifier with eight hidden neurons (average training error $17.45\% \pm 6.67$) for the fish data with 10% label noise.

2.5   SUPPORT VECTOR MACHINES

2.5.1   Why Would It Work?

Ever since its conception, the SVM classifier has been a prominent landmark in statistical learning theory [61]. The success of SVM can be attributed to two ideas: (1) a transformation of the original space into a very high-dimensional new space and (2) identifying a "large margin" linear boundary in the new space.

📖 Example 2.8   Linear discrimination in a higher-dimensional space

To explain why the first idea works, consider the one-dimensional two-class data set shown with two types of markers on the x-axis in Figure 2.16.

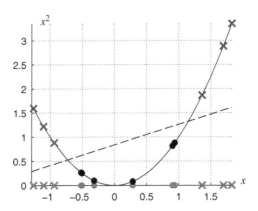

FIGURE 2.16   One-dimensional data set, which becomes linearly separable in the space (x, x^2).

The two classes are not linearly separable in the one-dimensional space x. However, by adding a second dimension, x^2, the classes become linearly separable, as shown in the figure. The linear classification boundary in $\mathbb{R}^2$ translates into a nonlinear boundary in the original one-dimensional space.

2.5.2   Classification Margins

The second idea is related to the concept of a *margin*. We can define a margin for an object $\mathbf{x}$ as the *signed distance* from $\mathbf{x}$ to the classification boundary. If $\mathbf{x}$ is in its correct classification region, the margin is positive, otherwise the margin is negative.

Distance from a point to a hyperplane. Let $\mathbf{w}^T\mathbf{x} - w_0 = 0$ be the equation of a hyperplane in $\mathbb{R}^n$ and $A \in \mathbb{R}^n$ be the point of interest. Denote by B be the point on the hyperplane such that $\vec{AB}$ is orthogonal to the hyperplane. With no loss of generality, Figure 2.17 illustrates the argument in two dimension.

Since $\mathbf{w}$ is a normal vector to the line (hyperplane in the general case), vector $\vec{AB}$ is collinear to $\mathbf{w}$, that is, $\vec{AB} = \vec{A} - \vec{B} = k\mathbf{w}$, where $\vec{A}$ and $\vec{B}$ are the position vectors of points A and B. Then

$$\mathbf{w}^T(\vec{A} - \vec{B}) = k\mathbf{w}^T\mathbf{w}. \tag{2.42}$$

Since B lies on the line, $\mathbf{w}^T\vec{B} - w_0 = 0$. Then

$$\mathbf{w}^T\vec{A} + w_0 - \mathbf{w}^T\vec{B} + w_0 = k\mathbf{w}^T\mathbf{w} \tag{2.43}$$

$$\mathbf{w}^T\vec{A} + w_0 = k\mathbf{w}^T\mathbf{w} \tag{2.44}$$

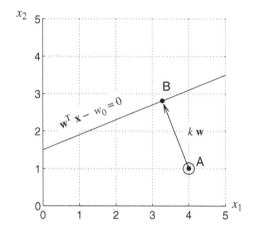

FIGURE 2.17   Illustration of the distance calculation from A to the line.

hence

$$k = \frac{\mathbf{w}^T \vec{A} + w_0}{\mathbf{w}^T \mathbf{w}} = \frac{\mathbf{w}^T \vec{A} + w_0}{||\mathbf{w}||^2}. \tag{2.45}$$

The distance between A and the line is the magnitude of $\vec{AB}$:

$$||\vec{AB}|| = k||\mathbf{w}|| = \frac{\mathbf{w}^T \vec{A} + w_0}{||\mathbf{w}||}. \tag{2.46}$$

The SVM classifier is based on the theory of Structural Risk Minimisation (SRM) [403]. According to SRM, maximising the margins of a classifier on the training data may lead to a lower generalization error.

■ Example 2.9   Classification margins
To illustrate classification margins, assume that the task is to design a classifier for the two-dimensional data set shown in Figure 2.18.

The cumulative distributions of the classification margins for the two lines are shown in Figure 2.19. Ideally, all margins should be large and positive, so the cumulative distribution is a singleton with value 1 (100%) at the maximum possible margin. The distribution for the better boundary line (B) is shifted to the right, indicating that there is a larger proportion of large positive margins for line B than for line A. The point on the vertical line at margin 0 is the proportion of negative margins, which is exactly the training error. When the training error is zero, we should aspire to build a classifier whose margin distribution is shifted to the right as much as possible.

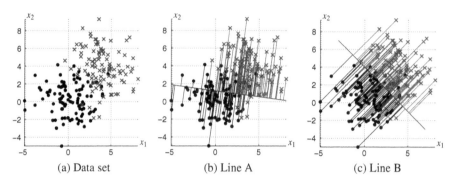

(a) Data set   (b) Line A   (c) Line B

FIGURE 2.18   Data sets; two classification boundaries and the respective classification margins.

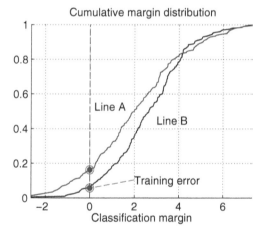

FIGURE 2.19   Cumulative distribution of the classification margins for the two boundary lines in Figure 2.18.

2.5.3   Optimal Linear Boundary

Let us start with two linearly separable classes. Any boundary that separates the classes will give zero training error. However, the boundary can be improved further by trying to increase the margins. In addition to ensuring zero training error, the SVM classifier is built in such a way that the linear boundary is as far as possible from the nearest data points.

Consider a data set $\mathbf{Z} = \{\mathbf{z}_1, \ldots, \mathbf{z}_N\}$, $\mathbf{z}_i \in \mathbb{R}^n$, labeled in two classes using $y \in \{-1, 1\}$ as the class labels. Suppose that the classes are linearly separable by a boundary with equation $\mathbf{w}^T\mathbf{x} - w_0 = 0$. The class labels are assigned by the sign of the left-hand side of the equation. Because of the linear separability, there exists a nonnegative constant m such that

$$y_i(\mathbf{w}^T \mathbf{z}_i - w_0) \geq m, \quad \forall\, i = 1, \dots, N. \tag{2.47}$$

Dividing both sides by $||\mathbf{w}||$, the left-hand side becomes the (signed) margin of object $\mathbf{z}_i$:

$$y_i \left(\frac{\mathbf{w}^T}{||\mathbf{w}||} \mathbf{z}_i - \frac{w_0}{||\mathbf{w}||} \right) \geq \frac{m}{||\mathbf{w}||}. \tag{2.48}$$

Absorbing $||\mathbf{w}||$ into new constants, equation (2.48) can be rewritten as

$$y_i(\beta^T \mathbf{z}_i - \beta_0) \geq M. \tag{2.49}$$

Thus SVM looks for a maximum margin classifier by solving the following optimization problem [61, 179]:

$$\max_{\beta, \beta_0, ||\beta||=1} M$$
$$\text{subject to } y_i(\beta^T \mathbf{z}_i - \beta_0) \geq M, \quad i = 1, \dots, N. \tag{2.50}$$

▆ Example 2.10 Support vectors

This example shows the trained SVM classifier for a simple data set (Figure 2.20). The points which are at the exact minimum distance corresponding to the minimum margin are called the *support vectors*. These points are marked in the figure with circles.

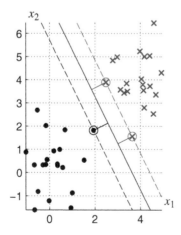

FIGURE 2.20   An example of the SVM classification boundary and the support vectors.

2.5.4 Parameters and Classification Boundaries of SVM

The optimization problem can be posed and solved for nonseparable classes too. In this case, SVM needs a parameter C which measures what penalty is assigned to errors [33].

To project the original space into a new high-dimensional space, SVM uses the so-called "kernel trick." This means that there is no need for the original space to be explicitly transformed into the new space. Instead, for a given data set $\mathbf{Z} = \{\mathbf{z}_1, \dots, \mathbf{z}_N\}$, the SVM optimisation procedure seeks to find the coefficients α_i and w_0 in the following expression:

$$f(\mathbf{x}) = \sum_{i=1}^{N} \alpha_i K(\mathbf{x}, \mathbf{z}_i) + w_0, \tag{2.51}$$

where $K(.,.)$ is a kernel function. The training of the SVM classifier results in all α_i being zero, except for the coefficients multiplying the support vectors. Popular choices for kernel functions are:

- The polynomial kernel family with parameter d:

$$K(\mathbf{x}, \mathbf{y}) = (1 - \mathbf{x}^T \mathbf{y})^d, \tag{2.52}$$

  the most used of which is the linear kernel ($d = 1$).
- The radial-basis kernel:

$$K(\mathbf{x}, \mathbf{y}) = \exp(-\gamma ||\mathbf{x} - \mathbf{y}||^2). \tag{2.53}$$

  Interpreting γ as $\frac{1}{2\sigma^2}$, this model is also called the Gaussian kernel.
- The neural network kernel

$$K(\mathbf{x}, \mathbf{y}) = \tanh(\kappa_1 (\mathbf{x}^T \mathbf{y}) + \kappa_2), \tag{2.54}$$

where κ_1 and κ_2 are adjustable parameters.

Notice that the calculation of the optimal boundary is carried out in the high-dimensional space. Depending on the chosen parameters of SVM, a linear boundary in this space can be projected as a nonlinear boundary in the original space. If training errors are heavily penalized, the classifier may end up overtrained, with jagged classification boundary. The following example illustrates the nonlinear boundaries obtained though the Gaussian kernel for different values of the penalizing parameter C.

■ **Example 2.11 SVM boundaries and the effect of C**
Figure 2.21 shows the classification boundaries of the SVM classifier for the fish data set. Twenty sets of 500 points were randomly sampled from the grid data and an SVM classifier with the Gaussian kernel was trained on each set.

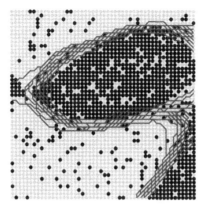

FIGURE 2.21   Decision boundaries of the SVM classifier with $C = 1$ (average error over the whole data set $17.01\% \pm 0.84$).

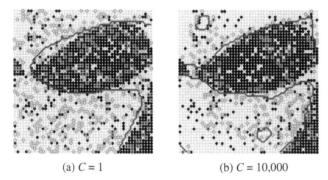

(a) $C = 1$ (b) $C = 10,000$

FIGURE 2.22   Decision boundaries of the SVM classifier: (a) $C = 1$; 479 support vectors, training error rate 16.40%; (b) $C = 10,000$; 319 support vectors, training error 12.00%.

Figure 2.22 shows the SVM classification boundaries for two values of the penalizing parameter C and the Gaussian kernel. The data for the training were again a random sample of 500 points from the grid. It can be seen that for the smaller value C (Figure 2.22a), the boundaries are smoother and the training error rate is higher than for the larger C. However, penalizing the errors too strongly pushes the SVM classifier into learning the noise in the data, which can be seen as the little "islands" in Figure 2.22b. The support vectors are also marked in the figure.

Tuning of the SVM parameters is usually done by cross-validation on the training part of the data [67]. Depending on the choices of a kernel and its parameters, the classification boundaries could be flexible and fairly accurate, which makes the SVM classifier a good ensemble candidate.

2.6   THE *k*-NEAREST NEIGHBOR CLASSIFIER (*k*-nn)

The *k*-nearest neighbor classifier (*k*-nn) is one of the most theoretically elegant and simple classification techniques [81, 107, 141]. Let $\mathbf{V} = \{\mathbf{v}_1, \ldots, \mathbf{v}_M\}$ be a labeled *reference set* containing M points in $\mathbb{R}^n$, referred to as *prototypes*. Each prototype is labeled in one of the c classes. Usually, $\mathbf{V}$ is the whole of $\mathbf{Z}$. To classify an input $\mathbf{x}$, the k nearest prototypes are retrieved from $\mathbf{V}$ together with their class labels. The input $\mathbf{x}$ is labeled to the most represented class within this set of neighbors.

The error rate of the *k*-nn classifier, $P_{\text{k-nn}}$ satisfies

$$\lim_{M,k\to\infty,\frac{k}{M}\to 0} P_{\text{k-nn}} = P_{\text{B}}, \tag{2.55}$$

where P_{B} is the Bayes error rate. When $k = 1$, the *k*-nn classifier becomes the famous *nearest neighbor classifier*, denoted 1-nn. For 1-nn, when $N \to \infty$, the error rate $P_{\text{1-nn}}$ is bounded from above by twice the Bayes error rate [107]:

$$P_{\text{1-nn}} \leq 2P_{\text{B}}. \tag{2.56}$$

The classification regions of the 1-nn rule are unions of *Voronoi diagrams*. The Voronoi cell V for $\mathbf{z}_j \in \mathbf{Z}$ is defined as the set of points in $\mathbb{R}^n$ whose nearest neighbor from $\mathbf{Z}$ is $\mathbf{z}_j$:

$$V(\mathbf{z}_j) = \left\{ \mathbf{x} \,\middle|\, \mathbf{x} \in \mathbb{R}^n, d(\mathbf{z}_j, \mathbf{x}) = \min_{\mathbf{z}_k \in \mathbf{Z}} \ d(\mathbf{z}_k, \mathbf{x}) \right\}, \tag{2.57}$$

where $d(\cdot, \cdot)$ is a distance in $\mathbb{R}^n$. A typical choice is the Euclidean distance.

▣ Example 2.12   Voronoi cells
MATLAB code for 1-nn is given in Appendix 2.A.2.4. Using this code and the fish data set with 20% label noise (Figure 2.23a), we ran 5000 iterations. In each iteration we chose randomly seven data points as the reference set. The error on the training

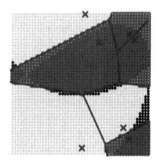

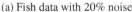
(a) Fish data with 20% noise   (b) Voronoi cells

FIGURE 2.23   The fish data with 20% label noise and the Voronoi cells for the reference set of seven prototypes found through random search.

(noisy) set was estimated. The best set of prototypes among the 500 selections is displayed with cross markers, and their Voronoi cells are outlined in Figure 2.23b. The error rate with respect to the noise-free data set is 12.44%.

Some of the drawbacks of k-nn are that it is computationally expensive, adversely sensitive to noisy and redundant features and the choice of the similarity function. Besides, k-nn it does not have a natural mechanism for handling missing values.

Diversity can be induced into the k-nn classifier by using: different subsets of features, different distance metrics in $\mathbb{R}^n$, different values of k, and edited versions of $\mathbf{Z}$ as the reference set $\mathbf{V}$. The classifier is sensitive to the choice of features on which the distance is calculated. This suggests that feature sub-sampling should be considered when building ensembles of k-nn classifiers. Taking different subsets of the data will not alter the k-nn classifier dramatically, making it a stable classifier, and unsuitable for ensembles. The example below demonstrates the use of k-nn in an ensemble setting based on random data editing.

▣ Example 2.13   Edited 1-nn in an ensemble setting

Again we used the fish data set with 20% label noise and ran a Monte Carlo experiment with 5000 iterations. In each iteration we chose randomly seven data points as the reference set. The error on the training (noisy) set was estimated. At the end of the run, we chose the best 25 prototype sets, with the smallest training error. Figure 2.24 shows the following boundaries overlaid on the noisy training set: (a) the 25 1-nn classifiers each using a reference set of seven prototypes; (b) the 1-nn classifier using the pooled 25 prototype sets together; (c) the majority vote ensemble classifier across the 25 individual 1-nn; and (d) the 1-nn classifier on the result of another Monte Carlo experiment with 5000 iterations looking for a reference set with $25 \times 7 = 175$ prototypes.

The example shows that 1-nn can be used in ensemble context from at least two different perspectives: aggregating the decision of classifiers with heavily pruned reference sets and pooling sets of prototypes found through a random search or a similar algorithm.

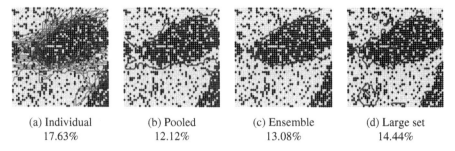

| (a) Individual | (b) Pooled | (c) Ensemble | (d) Large set |
| 17.63% | 12.12% | 13.08% | 14.44% |

FIGURE 2.24   Boundaries of 1-nn classifiers. The values underneath are the error rates with respect to the noise-free data.

2.7 FINAL REMARKS

How do we choose a particular classifier model? In spite of this being the most important question in pattern recognition and machine learning, we are still looking for answers. The technical question of how we compare N classifiers on M data sets was discussed in the previous chapter. However, we cannot generalize without limit the findings of any such experiment, regardless of how extensive that may be. Duin [110] points out that a classifier preferred by the author of a study can be shown to dominate the off-the-shelf models due to preferential parameter tuning or data set selection. This may well happen by luck. Over the years, pattern recognition has witnessed the rise and demise of many classifier models. Some of them have withstood the test of time, for example, the simple decision tree classifier and the nearest neighbor classifier. Why are there so many models around?

The No Free Lunch Theorem partly answers this question. Cited after [107], this theorem states that

> *"On the criterion of generalisation performance, there are no context- or problem-independent reasons to favour one learning or classification method to another."*

This seems to be throwing a spanner in the wheel! Why bother proposing new models or perfecting the old ones? The theorem merely reinforces the message that although there is general wisdom about classifiers, for example, that ensembles outperform single classifiers, this is not guaranteed for all data sets or the data set at hand.

2.7.1 Simple or Complex Models?

It is intuitively clear that simple models or stable classifiers are less likely to be overtrained than more sophisticated models. However, simple models might not be versatile enough to fit complex classification boundaries. More complex models such as neural networks have a better flexibility but require more system resources and are prone to overtraining. What do "simple" and "complex" mean in this context? The main aspects of complexity can be summarized as [11]:

- Training time and training complexity.
- Memory requirements (e.g., the number of the parameters of the classifier that are needed for its operation).
- Running complexity.

An often cited postulate attributed to William of Ockham (c. 1285–1349), termed *Occam's razor*, says that *"Entities should not be multiplied beyond necessity."* The interpretation of Occam's razor in terms of classifier models could be that, other things being equal, a simpler classifier should be preferred to a more complex one.

An assumption underpinning this heuristic is that if more data are available, the classifier will be more accurate.

Although theoretically disputed, the Occam's razor philosophy and the resulting heuristic approach toward avoiding over-fitting enjoy empirical success. Duda et al. [107] argue that we owe this success to the type of problems we are likely to face. Our data usually favor simpler methods over more complex ones.

Interestingly, classifier ensembles are based upon a certain degree of violation of Occam's razor. In order to obtain a collection of diverse classifier members, we allow and even encourage overtraining. In a situation where an individual classifier model and an ensemble give the same generalization error, the (more complex) ensemble may still be preferable because of its expected robustness.

2.7.2 The Triangle Diagram

Given that there is no universal winner of the all-classifier contest, we can try to find out for what type of data a given classifier model works well. Consider the diagram in Figure 2.25 that takes into account several characteristics of the data set. The x-axis is the prior probability of the largest class and y-axis is the prior probability of the smallest class. The feasible space is within a triangle, as shown in the figure. The right edge corresponds to the two-class problems because for this case the smallest and the largest prior probabilities sum up to 1. The number of classes increases from this edge toward the origin (0,0). The left edge of the triangle corresponds to equiprobable classes. The largest prior on this edge is equal to the smallest prior probability, which means that all classes have the same prior probabilities. This edge can be thought of

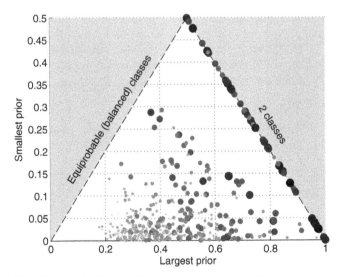

FIGURE 2.25   The triangle diagram. The size and the gray-level intensity of the dots indicate the classification accuracy. Larger and darker dots signify higher accuracy.

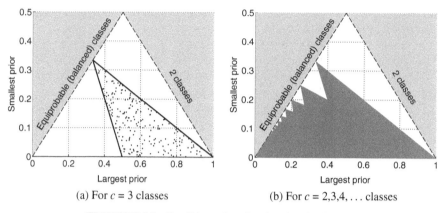

FIGURE 2.26   Feasible regions for the triangle diagram.

as the edge of balanced problems. The balance disappears toward the bottom right corner. The pinnacle of the triangle corresponds to two equiprobable classes.

To illustrate the application of the triangle diagram, consider a simulation where 500 problems were generated with 2–30 classes with Dirichlet random priors.[5] The classes were two-dimensional Gaussians, with random means drawn uniformly from the square $[0, 4] \times [0, 4]$ and identity covariance matrices. The number of points sampled from class i was $\lceil 100 \times p_i \rceil$, where $\lceil \cdot \rceil$ denotes "ceiling," the nearest larger integer. A linear classifier was trained on each data set and tested on an independently generated data set from the same distribution. Each data set is depicted as a point on the triangle diagram. The classification accuracy is indicated by the size and the colour of the dot—larger and darker dots are associated with higher accuracy. The diagram shows that higher accuracy of LDC is obtained for fewer classes.

Comparisons of classifiers can be illustrated by the triangle diagram. Suppose that we are interested in the type of data where classifier A is better than classifier B. We can plot the data dots where A wins and where B wins with different markers.

Not all regions of the triangle are accessible. Figure 2.26a shows the feasible region for $c = 3$ classes, and Figure 2.26b shows the feasible coverage for $c = 2, 3, 4, \dots$ classes.

For example, consider three classes such that the largest prior probability is $a \in [0, 1], a > 1/3$. Then there is a limit on the smallest probability. The *largest* smallest probability in this case is when the two minority classes share the remainder $1 - a$ in equal measure. Therefore, the upper limit of the smallest prior probability will be $\frac{1-a}{2}$. Note also that the largest prior cannot be smaller than $\frac{1}{3}$, which marks the end of the upper bound segment between points $(1, 0)$ and $\left(\frac{1}{3}, \frac{1}{3}\right)$. The largest prior a also restricts the smallest prior from below. For example, if $a = 0.35$, the smallest prior cannot be 0 because the remaining prior will be $0.65 > a$. The minimum cannot

---

[5]Each set of numbers $p_i, i = 1, \dots, c, p_i > 0, \sum_i p_i = 1$, has the same chance of being selected as the set of prior probabilities.

be smaller than $1 - 2a$. It can be shown that for c classes, the upper bound on the smallest prior probability is $\frac{1-a}{c-1}$ and the lower bound is $\max\{0, 1 - (c - 1)a\}$.

2.7.3   Choosing a Base Classifier for Ensembles

Classification accuracy is the most important criterion when choosing a classifier. In addition, we may look at the classifier's ability to deal with mixed types of features (quantitative and qualitative), handle noise, outliers, missing values, and irrelevant features and give interpretable solutions [179]. When choosing a classifier model for building ensembles, the desiderata changes. The classifier should be sensitive to changes in the data, initialization or training run. The classifiers chosen most often in this context are the decision tree classifier and the MLP. They are both sensitive to data resampling and feature subsampling which is a prerequisite for diversity in the ensemble. MLP depends also on the initialization of the weights at the beginning of the training run, while the decision tree classifier is sensitive to rotation of the space. Some rule-based classifiers have been used within classifier ensembles. A notable example is the RIPPER classifier proposed by Cohen [76] (JRip in WEKA). Ensembles of LDC, QDC, Naïve Bayes, and variants of SVM and k-nn have also been tried in spite of the lesser sensitivity of these models to the various diversifying heuristics.

APPENDIX

2.A.1   MATLAB CODE FOR THE FISH DATA

```
1  %-------------------------------------------------------------%
2  function [x,y,labels] = fish_data(grid_size,noise)
3  % --- generates the fish data set
4  % Input: -------------------------------------
5  %    grid_size:  number of points on one side
6  %    noise:      percent label noise
7  % Output: -------------------------------------
8  %    x,y:        point coordinates
9  %    labels:     labels = 1 (black) or 2 (gray)
10
11 % Create the 2d data as all the points on a square grid
12 [X,Y] = meshgrid(1:grid_size,1:grid_size);
13 x = X(:)/grid_size; y = Y(:)/grid_size; % scale in [0,1]
14
15 % Generate the class labels
16 lab1 = x.^3 - 2*x.*y + 1.6*y.^2 < 0.4; % component 1
17 lab2 = -x.^3 + 2*sin(x).*y + y < 0.7; % component 2
18 la = xor(lab1,lab2);
19
```

```
20  if noise > 0
21       N = grid_size^2;
22       rp = randperm(N); % shuffle
23       to_change = rp(1:round(noise/100 * N));
24       la(to_change) = 1 - la(to_change); % flip labels
25  end
26
27  labels = 2 - la;
28  %-----------------------------------------------------------%
```

2.A.2   MATLAB CODE FOR INDIVIDUAL CLASSIFIERS

2.A.2.1   Decision Tree

Function `tree_build` below is an example of a decision tree builder with the following choices:

- Impurity criterion: Gini.
- Stopping criterion (leaf creating): χ^2.
- CART algorithm: all features are considered continuous-values and a threshold is sought.

The function takes as its input the data, the labels, the number of classes, and the threshold for the pre-pruning stopping criterion. The output is a decision tree T of size $K \times 4$, where K is the number of nodes, including the leaves. The nodes in T are numbered as the respective row, from 1 to K. A row in T corresponding to a leaf node appears as $[a, 0, 0, 0]$, where a is the class label at this leaf. The root is the first row of T. The root and all intermediate nodes have the format $[b, c, d, e]$, where b is the split feature index, c is the split threshold, d is the node number for the left child ($x_b \leq c$) and e is the node number for the right child ($x_b > c$).

Using the tree builder above, the decision tree classifier for the data in Example 2.6 is

$$T = \begin{bmatrix} 1 & 0.3655 & 2 & 3 \\ 1 & 0 & 0 & 0 \\ 2 & 1.6298 & 4 & 5 \\ 2 & 0 & 0 & 0 \\ 1 & 0 & 0 & 0 \end{bmatrix}.$$

```
1  %-----------------------------------------------------------%
2  function T = tree_build(data, labels, classN, chi2_threshold)
3  % --- train tree classifier
```

```
4   if numel(unique(labels)) == 1 % all data are of the same class
5       T = [labels(1),0,0,0]; % make a leaf
6   else
7       [chosen_feature,threshold] = tree_select_feature(data,labels);
8       leftIndex = data(:,chosen_feature) <= threshold;
9       chi2 = tree_chi2(leftIndex,labels,classN);
10      if chi2 > chi2_threshold % accept the split
11          leftIndex = data(:,chosen_feature) <= threshold;
12          Tl = tree_build(data(leftIndex,:),labels(leftIndex),...
13              classN,chi2_threshold); % left subtree
14          Tr = tree_build(data(~ leftIndex,:),labels(~ leftIndex),...
15              classN,chi2_threshold); % right subtree
16          % merge the two trees
17          Tl(:,[3 4]) = Tl(:,[3 4]) + (Tl(:,[3 4]) > 0) * 1;
18          Tr(:,[3 4]) = Tr(:,[3 4]) + (Tr(:,[3 4]) > 0) * (size(Tl,1)+1);
19          T = [chosen_feature, threshold, 2, size(Tl,1)+2; Tl; Tr];
20      else % make a leaf
21          T = [mode(labels), 0, 0, 0];
22      end
23  end
24
25  %.......................................................................
26  function [top_feature, top_thre] = tree_select_feature(data,labels)
27  % --- select the best feature
28  [n, m] = size(data);
29  i_G = Gini(labels); % Gini index of impurity at the parent node
30  [D, s] = deal(zeros(1, m)); % preallocate for speed
31  for j = 1 : m % check each feature
32      if numel(unique(data(:,j))) == 1 % the feature has only 1 value
33          D(j) = 0; s(j) = -999; % missing
34      else
35          Dsrt = sort(data(:,j)); % sort j-th feature
36          dde_i = zeros(1, n); % preallocate for speed
37          for i = 1 : n-1 % check the n-1 split points
38              sp = (Dsrt(i) + Dsrt(i+1)) / 2;
39              left = data(:,j) <= sp;
40              % Make sure that there are points in both children nodes
41              if sum(left) > 0 && sum(left) < n
42                  i_GL = Gini(labels(left));i_GR = Gini(labels(~ left));
43                  dde_i(i) = i_G - mean(left)*i_GL - mean(~ left)*i_GR;
44              else % one child node is empty
45                  dde_i(i)=0;
46              end
47          end
48          [D(j), index_s] = max(dde_i); % best impurity reduction
49          s(j) = (Dsrt(index_s) + Dsrt(index_s+1)) / 2; % threshold
50      end
51  end
52  [~ , top_feature] = max(D); top_thre = s(top_feature);
```

```
53
54  %..........................................................
55  function chi2 = tree_chi2(left, labels, classN)
56  % --- calculate chi^2 statistic for the split of labels on "left"
57  n = numel(labels); chi2 = 0; n_L = sum(left);
58  for i = 1 : classN
59      n_i = sum(labels == i); n_iL = sum(labels(left) == i);
60      if n_i > 0 && n_L > 0 % add only for non-empty children nodes
61          chi2 = chi2 + (n * n_iL - n_i * n_L)^2 /...
62              (2 * n_i * (n_L) * (n - n_L));
63      end
64  end
65
66  %..........................................................
67  function i_G = Gini(labels)
68  % --- calculate Gini index
69  for i = 1 : max(labels)
70      P(i) = mean(labels == i);
71  end
72  i_G = 1 - P * P';
73  %--------------------------------------------------------------%
```

The function below is the classification part of the decision tree code. The output consists of the labels of the supplied data using tree *T* trained through `tree_build`.

```
1   %--------------------------------------------------------------%
2   function labels = tree_classify(T, test_data)
3   % classify test_data using the tree classifier T
4   for i = 1 : size(test_data,1)
5       index = 1; leaf = 0;
6       while leaf == 0,
7           if T(index,3) == 0, % leaf is found
8               labels(i) = T(index,1); leaf = 1;
9           else
10              if test_data(i,T(index,1)) <= T(index,2)
11                  index = T(index,3); %left
12              else
13                  index = T(index,4); %right
14              end
15          end
16      end
17  end
18  %--------------------------------------------------------------%
```

The example below shows the use of the decision tree training and classification functions. It builds a rather conservative decision tree for the fish data with 10% label

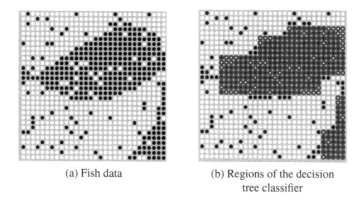

(a) Fish data   (b) Regions of the decision
                             tree classifier

FIGURE 2.A.1   An example of (a) the fish data with 10% label noise and (b) the MATLAB output from the TREE code. The points which the tree classifier labels as class "black dot" are circled.

noise. The output is a scatter plot of the data set with circles as the points labeled by the tree as "dots." The results will differ from one run to the next because the noise in the data set is randomly generated. An example of the fish data and the MATLAB output is shown in Figure 2.A.1.

```
1  %-----------------------------------------------------------%
2  % The TREE code
3  [x,y,labels] = fish_data(30,10);
4  T = tree_build([x y], labels, 2, 5);
5  la = tree_classify(T,[x y]);
6  ax = axes; hold on
7  scatter(ax,x,y,12,labels,'linewidth',3);
8  colormap gray, axis square off
9  plot(x(la==1),y(la==1),'bo','linewidth',3);
10 %-----------------------------------------------------------%
```

2.A.2.2   Naïve Bayes

Function `naive_bayes_train` trains a Naïve Bayes classifier. The function returns the classifier C, which can be used with `naive_bayes_classify` to label a data set. The TREE code above can be re-used to try the NB classifier. Lines 4 and 5 should be replaced with

```
1  %-----------------------------------------------------------%
2  C = naive_bayes_train([x y], labels);
3  la = naive_bayes_classify(C,[x y]);
4  %-----------------------------------------------------------%
```

```
1  %-----------------------------------------------------------%
2  function C = naive_bayes_train(data, labels)
3  % --- train parametric NB classifier
4  for i = 1:max(labels)
5      c_index = labels == i;
6      C(i).prior = mean(c_index);
7      C(i).mean = mean(data(c_index,:),1);
8      if sum(c_index) > 1 % class with 1 object
9          C(i).std  = std(data(c_index,:));
10     else
11         C(i).std = zeros(1,size(data,2));
12     end
13 end
14 %-----------------------------------------------------------%
```

```
1  %-----------------------------------------------------------%
2  function labels = naive_bayes_classify(C, data)
3  % --- classify with the trained NB classifier
4  g = zeros(numel(C),size(data,1));
5  for i = 1:numel(C)
6      Ms = repmat(C(i).mean,size(data,1),1);
7      Ss = repmat(C(i).std,size(data,1),1);
8      t = 1./Ss .* exp(-(data - Ms).^2 ./(2* Ss.^2));
9      g(i,:) = prod(t') * C(i).prior;
10 end
11 [~ ,labels] = max(g); labels = labels(:);
12 %-----------------------------------------------------------%
```

2.A.2.3 Multi-Layer Perceptron

Function mlp_train below uses the error backpropagation algorithm to train an MLP with one hidden layer. The number of nodes at the hidden layer is the input parameter M, and the maximal number of epochs is the input parameter MaxEpochs. The function returns the classifier C, which can be used with function mlp_classify to label a data set. The TREE code above can be re-used to try the MLP classifier. Lines 4 and 5 should be replaced with

```
1  %-----------------------------------------------------------%
2  C = mlp_train([x y], labels,10,1000);
3  la = mlp_classify(C,[x y]);
4  %-----------------------------------------------------------%
```

```
1  %-----------------------------------------------------------%
2  function C = mlp_train(data, labels, M, MaxEpochs)
```

```
3  % --- train an MLP with M hidden nodes
4  %
5  %* initialization
6  [n,m] = size(data); c = max(labels); % number of classes
7  % form c-component binary (target) vectors from the labels
8  bin_labZ = repmat(labels,1,c) == repmat(1:c,n,1);
9  % weights and biases input-hidden:
10 W1 = randn(M,m); B1 = randn(M,1);
11 % weights and biases hidden-output
12 W2 = randn(c,M); B2 = randn(c,1);
13 E = inf; % criterion value
14 t = 1; % epoch counter
15 eta = 0.1; % learning rate
16 epsilon = 0.0001; % termination criterion
17 %
18 %* normalization
19 mZ = mean(data); sZ = std(data);
20 data = (data - repmat(mZ,n,1))./repmat(sZ,n,1);
21 % store for the normalization of the testing data
22 C.means = mZ; C.std = sZ;
23 %
24 %* calculation
25 while E > epsilon && t <= MaxEpochs
26     % outputs of the hidden layer
27     oh = 1./(1 + exp(-[W1 B1] * [data ones(n,1)]'));
28     % outputs of the output layer
29     o = 1./(1+exp(-[W2 B2]*[oh; ones(1,n)]));
30     E = sum(sum((o'-bin_labZ).^2));
31     t = t + 1; % increment epoch counter
32     del_o = (o-bin_labZ').*o.*(1-o);
33     del_h = ((del_o'*W2).*oh'.*(1-oh'))';
34     for i = 1:c % update W2 and B2
35         for j = 1:M
36             W2(i,j) = W2(i,j)-eta*del_o(i,:)*oh(j,:)';
37         end
38         B2(i) = B2(i)-eta*del_o(i,:)*ones(n,1);
39     end
40     for i = 1:M % update W1 and B1
41         for j = 1:m
42             W1(i,j) = W1(i,j)-eta*del_h(i,:)*data(:,j);
43         end
44         B1(i) = B1(i)-eta*del_h(i,:)*ones(n,1);
45     end
46 end
47 %
```

```
48  %* store the output
49  C.W1 = W1; % weights input-hidden
50  C.W2 = W2; % weights hidden-output
51  C.B1 = B1; % bias input-hidden
52  C.B2 = B2; % bias hidden-output
53  C.oh = oh; % output of the hidden layer
54  %----------------------------------------------------------------%
```

```
1   %----------------------------------------------------------------%
2   function labels = mlp_classify(C,test_data)
3   % classify test_data using the MLP classifier C
4   AvailableLabels = 1:size(C.W2,1);
5   %* normalization
6   n = size(test_data,1);
7   test_data = (test_data - repmat(C.means,n,1))./repmat(C.std,n,1);
8   oh = 1./(1+exp(-[C.W1 C.B1]*[test_data ones(n,1)]'));
9   %* outputs of the hidden layer
10  o = 1./(1+exp(-[C.W2 C.B2]*[oh; ones(1,n)]));
11  %* outputs of the output layer
12  [~ ,Index] = max(o);
13  labels = AvailableLabels(Index);
14  labels = labels(:);
15  %----------------------------------------------------------------%
```

2.A.2.4   1-nn Classifier

Given below are three different MATLAB implementations of the 1-nn classifier. The first one is included for its readability, and the latter two for their efficiency. The results from the functions are identical.

```
1   %----------------------------------------------------------------%
2   function labels = one_nn_classify(ref_set,ref_lab,test_data)
3   % classify test_data using the given labeled reference set
4   labels = zeros(size(test_data,1),1);
5   for i = 1:size(test_data,1)
6       x = repmat(test_data(i,:),size(ref_set,1),1) - ref_set;
7       [~ ,ind] = min(sum(x.*x,2));
8       labels(i) = ref_lab(ind);
9   end
10  %----------------------------------------------------------------%
```

```
1   %----------------------------------------------------------------%
2   function labels = one_nn_classify_1(ref_set,ref_lab,test_data)
3   % Courtesy of Cameron Gray
```

```
4  % (does not require the Statistics toolbox)
5  [~ ,A] = cellfun(@(x) min(sum(bsxfun(@minus,x,ref_set).^2'))),...
6      num2cell(test_data,2));
7  labels = ref_lab(A);
8  %-----------------------------------------------------------%
```

```
1  %-----------------------------------------------------------%
2  function labels = one_nn_classify_2(ref_set,ref_lab,test_data)
3  % (requires the Statistics toolbox)
4  [~ ,A] = pdist2(ref_set,test_data,'euclidean','Smallest',1);
5  labels = ref_lab(A);
6  %-----------------------------------------------------------%
```

3

AN OVERVIEW OF THE FIELD

3.1 PHILOSOPHY

A classifier ensemble is sketched in Figure 3.1a. Several classifiers are employed to make a classification decision about the object submitted at the input, and the individual decisions are subsequently aggregated. The output of the ensemble is a class label for the object.

Classifier ensembles are justly receiving increasing attention and accolade and generating a wealth of research [53, 183, 311, 321, 335, 355, 357, 397, 425, 439]. Theoretical and empirical studies have demonstrated that an ensemble of classifiers is typically more accurate than a single classifier. Research on classifier ensembles permeate many strands of machine learning including streaming data [160, 326], biometrics [312], concept drift, and incremental learning [119].

Intuitive as this concept may be, there is no rigorous definition of a classifier ensemble. Figures 3.1b–d illustrate the uncertainty of the generic definition. Any classifier ensemble is, in fact, a classifier (Figure 3.1b). We can think of the constituent classifiers (called "base classifiers") as fancy feature extractors, while the combiner would be a simple classifier that aggregates the "fancy features." On the other hand, what is stopping us from proclaiming that a standard neural network classifier is a classifier ensemble (Figure 3.1c)? The neurons at the penultimate layer can be regarded as classifiers, whose decisions must be "deciphered" at the top layer. The top layer itself can be thought of as a combiner. Further still, why cannot we say that the features are actually some form of primitive classifiers while the classifier is a sophisticated combiner (Figure 3.1d)?

Combining Pattern Classifiers: Methods and Algorithms, Second Edition. Ludmila I. Kuncheva.
© 2014 John Wiley & Sons, Inc. Published 2014 by John Wiley & Sons, Inc.

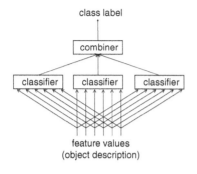

(a) A classifier ensemble

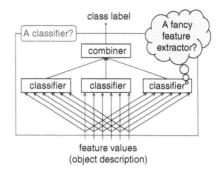

(b) Is the ensemble just a classifier?

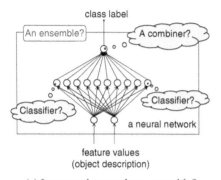

(c) Is a neural network an ensemble?

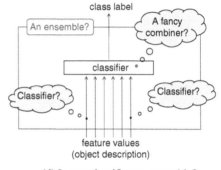

(d) Is any classifier an ensemble?

FIGURE 3.1   What is a classifier ensemble?

By combining classifiers we are aiming at an accurate classification decision which is achievable using simple trainable classifiers. But is this a valid pursuit?

In her critical review article "Multiple classifier combination: lessons and next steps" published in 2002 [183], Tin Ho writes:

> *Instead of looking for the best set of features and the best classifier, now we look for the best set of classifiers and then the best combination method. One can imagine that very soon we will be looking for the best set of combination methods and then the best way to use them all. If we do not take the chance to review the fundamental problems arising from this challenge, we are bound to be driven into such an infinite recurrence, dragging along more and more complicated combination schemes and theories and gradually losing sight of the original problem.*

The lesson is that we should make the best use of the tools and methodologies that we have at present, before setting off for new complicated designs. It is known that neural network classifiers are "universal approximators," which means that any classification boundary, however complicated, can be approximated to any desired precision with a finite neural network. This knowledge, however, does not give us a method for setting up or training such a network. It makes sense to compose a solution from manageable building blocks, which is the idea behind combining classifiers.

Maybe the most natural reasons for considering classifier ensembles are our own experiences in everyday decision making. Consulting several (hopefully independent) sources is a common approach to making decisions, be they as mundane as choosing a color for the living room walls, or as vital as picking a university to go to or accepting experimental medical treatment. Day [84] reviews the evolution of the concept of consensus in the context of electoral theory, traced back to the era of ancient Greek city states and the Roman Senate. The majority criterion was established in 1356 for the election of German kings. By 1450 it was adopted for the elections of the British House of Commons, and by 1500, as a rule to be followed in the House itself. Rokach [335] points at the wisdom of crowds as an argument in favor of classifier ensembles. An example of crowd wisdom is the popular television game "Who Wants to be a Millionaire", where asking the audience to help with choosing the correct answer is offered as a lifeline [311].

A decade ago, most papers on combining classifiers started with a justification argument. We can safely assume that the research community's view has moved forward since then. The common understanding is that classifier ensembles work for the following general reasons [95, 311] (Figure 3.2):

1. *Statistical reasons.* The empirical estimate of the classification performance is a random value depending on the given data and the training algorithm. As clever and stringent as the experimental protocol might be, there is always uncertainty associated with the performance estimate.

   Suppose we have a labeled data set Z and a number of different classifiers with a good performance on Z. We can pick a single classifier as the solution, running onto the risk of making a bad choice for the problem. For example, suppose that we run the 1-nn classifier or a decision tree classifier for L different subsets of features thereby obtaining L classifiers with zero resubstitution error. Although these classifiers are indistinguishable with respect to their (resubstitution) training error, they may have different generalization performances.

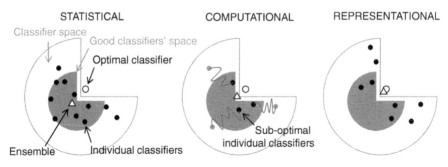

FIGURE 3.2 Three reasons for using classifier ensembles.

Instead of picking just one classifier, a safer option would be to use them all and average their outputs.

2. *Computational reasons.*

   (a) *Imperfect training algorithm.* Suppose that the quality of the estimate of the classification performance depends entirely upon the training algorithm. Some types of algorithms, such as the error-backpropagation algorithm for training neural networks, are only guaranteed to converge to a locally optimal solution (the curves in the middle subplot in Figure 3.2 show hypothetical trajectories for the classifiers during training). A combination of the outputs of several diverse suboptimal classifiers may lead to a better overall classifier.

   (b) *Too much data.* Instead of trying to accommodate all data in a single training algorithm, we may choose to train many classifiers on small "bites" of data and aggregate their outputs.

   (c) *Small sample size.* Re-sampling seems to be a good approach when the sample size is small. Different data sets generated through re-sampling can be used as training data for the base classifiers in an ensemble.

   (d) *Divide and conquer.* Regardless of the amount of data, a classification problem may benefit from splitting into smaller and easier-to-handle problems. A classifier is trained for each small problem and subsequently authorized to make a decision when the data point falls into its remit.

   (e) *Data fusion.* Sometimes data come from different sources, and the features may be of different nature (distinct pattern representation). Instead of pooling all features and trying to build a single classifier on the whole set, it may be better to build separate classifiers on the different groups of features and combine the classifier outputs. In emotion recognition, for example, data may come from different modalities: sensor readings from the peripheral nervous system, behavioral cues extracted from videos or voice recordings, image features describing facial expressions, and so on. Training a single classifier using all features may be impractical, and the results are likely to be worse than if different classifiers are trained on the different modalities.

3. *Representational reasons.* A complex classification boundary (of any shape) can be approximated with a desired precision by simple boundaries. Classifier ensembles enable such an approach. For example, an ensemble of linear classifiers can approximate a highly nonlinear classification boundary. The right subplot in Figure 3.2 illustrates the possibility of combining classifiers that are not even in the "good" space, and still reaching a solution close to the optimal one.

An improvement over the single best classifier or even on the average of the individual accuracies is not generally guaranteed. However, the experimental work published so far demonstrates the success of the ensemble approach to classification in a variety of application domains [297].

3.2   TWO EXAMPLES

3.2.1   The Wisdom of the "Classifier Crowd"

Consider the fish data set used in the previous chapter. Nine training data sets were prepared using all data points on the grid and flipping 10% randomly chosen labels. An MLP classifier was trained on each training data set. The classification boundaries of the individual MLPs are shown in Figure 3.3. The classification errors with respect to the true (undistorted) labels are shown underneath each plot. Each MLP had 10 hidden neurons whose weights were initialized with small random numbers. The training was run for 500 epochs.

Taking the majority vote, the testing error is 5.84%. For comparison, the mean of the individual errors is 8.56% and the minimum error is 6.44% (plot (i)). The ensemble boundary is plotted in Figure 3.3j. In this example, the ensemble decision is more accurate than the decision of any of the individual classifiers.

3.2.2   The Power of Divide-and-Conquer

The second example shows how the divide-and-conquer ensemble approach can boost the accuracy of a classifier as simple as the *largest prior* classifier. This classifier always assigns the label of the majority class. For the fish data set, the largest class (grey dots) contains 1520 out of the 2500 grid points. If all points were labeled as grey, the error rate would be 39.20%.

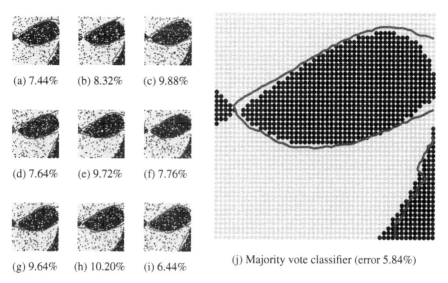

(a) 7.44%   (b) 8.32%   (c) 9.88%

(d) 7.64%   (e) 9.72%   (f) 7.76%

(g) 9.64%   (h) 10.20%   (i) 6.44%

(j) Majority vote classifier (error 5.84%)

FIGURE 3.3   (a)–(i) Classification boundaries of nine MLP classifiers trained on the fish data set with 10% label noise. Displayed underneath each plot is the testing error. (j) Boundary of the majority vote ensemble classifier.

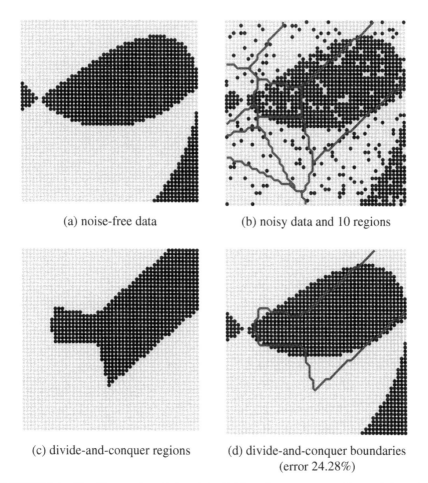

(a) noise-free data

(b) noisy data and 10 regions

(c) divide-and-conquer regions

(d) divide-and-conquer boundaries
(error 24.28%)

FIGURE 3.4   Classification boundaries of the divide-and-conquer ensemble consisting of 10 largest-prior classifiers.

The original (noise-free) fish data is shown in Figure 3.4a. Suppose that we divide the space randomly into 10 regions, as shown in Figure 3.4b. The regions are depicted over the training data set which includes 10% label noise. We subsequently apply the largest prior classifier in each region. The resultant class pattern is displayed in subplot Figure 3.4c and overlaid on the original set in subplot Figure 3.4d. The ensemble labels a data point by first identifying the region where this point belongs and then retrieving the label assigned to this region by the largest prior classifier. The error rate of the divide-and-conquer ensemble in this example is 24.28%, which is better than using the largest prior classifier on the whole data space.

This intuitive result can be easily proved. The beauty of the divide-and-conquer ensemble strategy (called further "a classifier selection approach") is that the result does not depend critically on the way the data space is split into regions.

3.3   STRUCTURE OF THE AREA

3.3.1   Terminology

Historically, pattern recognition and machine learning evolved with only little inter-action, separated by each field's strive for identity, the immaturity of electronic communications, and an ocean. Nonetheless, they are arguably facets of the same field and have been growing closer in the past decades [41, 252, 280]. Classifier combination, like other topics of common interest for the two areas, suffers from the curse of the tower of Babel: "Come, let us go down and confound their speech." Different terms for similar or identical concepts are still in operation:

$$
\begin{array}{r|l}
\text{feature} & = \text{attribute} \\
\text{classifier} & = \text{hypothesis} = \text{learner} = \text{inducer} = \text{generalizer} = \text{expert} \\
\text{data point} = \text{object} & = \text{example} = \text{instance} = \text{case} = \text{observation} \\
\text{ensemble} & = \text{team} = \text{pool} = \text{committee} = \text{meta learner}
\end{array}
$$

The series of annual International Workshops on Multiple Classifier Systems (MCS), held since 2000, has played a pivotal role in organizing, systematizing, and developing further the knowledge in the area of combining classifiers, as well as unifying the terminology [338]. To keep terminology consistent throughout this book, we will translate the material brought from the literature using the notions shown on the left in the list above.

3.3.2   A Taxonomy of Classifier Ensemble Methods

Figure 3.5 shows four levels of questions which need to be answered in order to construct a classifier ensemble.

Rokach [334] offers a comprehensive review of the classifier ensemble literature and proposes a taxonomy with five dimensions, accommodating a wide spectrum of existing classifier ensemble methods. Below we reproduce this taxonomy with some simplifications, minor alterations, and including comments and interpretations relevant to this book.

1. The combiner. (a) Some ensemble methods do not specify a combiner. (b) For those methods that do, the combiner can be
    i. Nontrainable. An example of this group is majority voting.
    ii. Trainable. This group includes the weighted majority voting and the Naïve Bayes combiner, as well as the "classifier selection" approach where one classifier in the ensemble is authorized to make the decision for a given object.
    iii. Meta classifier. The outputs of the individual classifiers are treated as inputs into a new trainable classifier, which itself constitutes the combiner. This approach is called "stacked generalization." Constructing a training set for the meta classifier is one of the main concerns of this combiner.

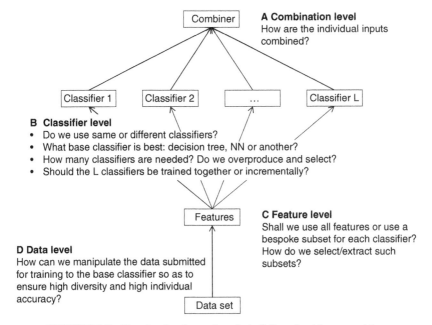

FIGURE 3.5   Four levels of questions in building classifier ensembles.

2. Building the ensemble. Can the base classifiers be trained independently or do they need to be trained in a sequence? An example of the latter is AdaBoost, where the training set of each added classifier depends upon the ensemble created thus far.

3. Diversity. How is diversity introduced into the ensemble? The following routes are proposed:

   (a) Use different approaches/parameters in the training of the individual classifiers. For example, use different random initialization of the multi-layer perceptron classifiers.

   (b) Manipulate the training sample for each ensemble member; for example, take a bootstrap sample from the training data or induce label noise, as in the first example above.

   (c) Choose different label targets. The idea is that each individual classifier solves a different classification task. Example of a classifier ensemble approach in this category is the Error Correcting Code (ECOC) ensembles, where each classifier is solving a dichotomy, separating two groups of classes.

   (d) Partitioning the training set. Horizontal partitioning means that different subsets (chunks, bites) of the training data are used as the training data for each base classifier. Vertical partitioning means that the training sets for the individual classifiers use different subsets of features.

   (e) Different classifier models or hybrid ensembles.

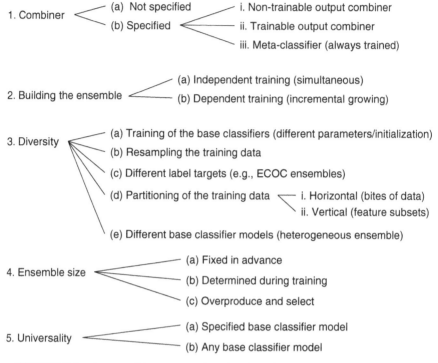

FIGURE 3.6   An ensemble taxonomy based on five dimensions (after Rokach [334]).

4. Ensemble size. How do we determine the number of classifiers in the ensemble? Is the ensemble built by simultaneous training of a desired number of classifiers, or iteratively, by adding/removing classifiers?

5. Universality (with respect to the base classifier). Some ensemble approaches can be used with any classifier model while others are tied to a specific classifier type. An example of a classifier-specific ensemble is the random forest, whose hallmark is its base classifier—the random tree [50].

The taxonomy is shown in Figure 3.6. Any classifier ensemble methodology can be described in terms of the five dimensions.

Ho [183] distinguishes between decision optimization and coverage optimization. Decision optimization refers to methods for choosing and optimizing the combiner for a fixed ensemble of base classifiers (answering the questions at level A in Figure 3.5). The alternative—coverage optimization—refers to methods for creating diverse base classifiers, assuming a fixed combiner (answering the questions at levels B, C, and D). This de-coupling of the ensemble construction is deemed reasonable because it reduces the complexity of the task [211]. This ensemble grouping is reflected in the "combiner" dimension of the five-point taxonomy.

▣ Example 3.1   Bagging as a taxonomy entry

Bagging ensembles [47] (explained in detail later) are created by drawing bootstrap samples from the training data sets, training a classifier on each sample, and combining the label outputs through majority voting. If the individual outputs are continuous valued, the combination is done by averaging. Figure 3.7 shows the answers to the four sets of questions for the bagging ensemble methods.

Underneath these answers is the signature of bagging with respect to the five-dimensional taxonomy in Figure 3.6, 1bi2a3b4a5b. The signature should be interpreted as follows

1bi	Combiner. Specified. Nontrainable output combiner.
2a	Building the ensemble. Independent training.
3b	Diversity. Resampling the training data.
4a	Ensemble size. Fixed in advance.
5b	Universality. Any base classifier model.

BAGGING

A Combination level

How are the individual outputs combined? VOTING/AVERAGE

B Classifier level

- Do we use same or different classifiers? SAME CLASSIFIERS
- What base classifier is best? typically DECISION TREES but can be any
- How many classifiers are needed? typically 100+, chosen in advance
- Should the L classifiers be trained together or incrementally? TOGETHER

C Feature level

Shall we use all features or use a bespoke subset for each classifier?
ALL FEATURES

D Data level

How can we manipulate the data submitted for training to the base classifier so as to ensure high diversity and high individual accuracy?
INDEPENDENT BOOTSTRAP SAMPLES

Signature in the 5-dimensional taxonomy: 1bi2a3b4a5b

FIGURE 3.7   Answers to the four sets of questions and the five-dimensional taxonomy signature for bagging.

The five dimensions are not independent, nor are the four sets of questions in Figure 3.5. For example, the ensemble size is related to the choice of a combiner. Small ensembles of strong classifiers may need a sophisticated trained combiner in order to improve on the single best ensemble member. On the other hand, large ensembles perform equally well for a variety of simple combiners. It is interesting to identify ensemble-creating approaches that share more than one property from the ensemble taxonomy.

3.3.3   Classifier Fusion and Classifier Selection

The above taxonomy draws upon the technological choices and heuristics in designing a classifier ensemble. Alternatively, we can group classifier ensemble methods with regard to the overall strategy that governs a specific design. Classifier fusion and classifier selection are two such strategies.

In classifier fusion, each ensemble member is supposed to have knowledge of the whole feature space. In classifier selection, each ensemble member is supposed to know well a part of the feature space and be responsible for objects in this part only. There are combination schemes lying in between the two pure strategies. Such a scheme, for example, is taking the average of the outputs with coefficients which depend on the input. The local competence of the classifiers determines their weights in the combination formula. The fusion-selection concept can be likened to competitive-cooperative ensembles, ensemble approach versus modular approach [360] and multiple topology versus hybrid topology [247].

Figure 3.8 gives a view of classifier combination approaches in the space of ensemble size and strength of the individual classifiers. The *perfect* classifier is marked by the dot in the upper left corner. Imperfect but strong classifiers can be combined with an ingenious combination rule to draw upon their diversity, however small this might be. Indeed, nearly perfect classifiers will make a small number of labeling mistakes, and an ensemble may improve on the individual accuracy if these errors are not made synchronously.

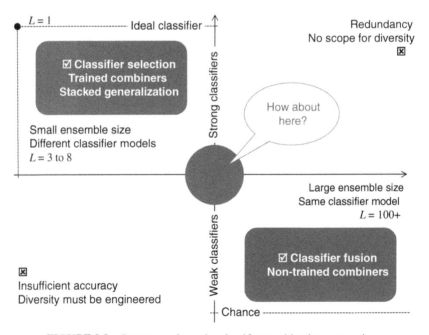

FIGURE 3.8   Summary of popular classifier combination approaches.

The approach where 3–8 different classifier models are combined is often termed *stacked generalization* [112, 420]. The term refers to the idea of stacking layers of classifiers. The combiner is a classifier itself, built upon the outputs of the individual classifiers. The individual outputs, be they class labels or degrees of support for the classes, are treated as *meta features* or intermediate features [223].

Classifier selection is one of the preferred approaches for heterogeneous ensembles with strong classifiers. Even if the competence regions of the classifiers are not very precisely determined, the ensemble accuracy will not drop dramatically.

The bottom right corner of the diagram reflects the bulk of the classifier ensemble research: large ensembles of diverse weak classifiers. Popular ensemble methods in this group are bagging, AdaBoost, and random forest. Almost invariably, the approaches in this quadrant fall in the classifier fusion group.

The top right quadrant corresponds to large ensembles with strong classifiers. This is a waste of resource because there is no scope for diversity, and many of the ensemble members will give identical output, regardless of how different the models and their parameters may be.

The bottom left quadrant is more interesting. Is it possible to build a good ensemble out of a small number of weak classifiers? The answer is yes! However, diversity must be engineered in such a way that no accuracy is wasted, and the classifiers complement one another. Such an ensemble must be able to rectify individual labeling mistakes. While possible in theory (shown later under the name "pattern of success"), this approach is not easy to bring to life. Instead of looking of ways to ensure complementarity, it is much easier to generate a large number of classifiers and move over to the bottom right quadrant.

How about the "middle ground?" Is there a gain in building classifier ensembles of moderate sizes (20–50 classifiers)? Does it pay off to have a heterogeneous ensemble with, say, two or three different classifier models? For example, consider an ensemble with 30 classifiers, 10 of which are decision trees, another 10 are neural networks, and the remaining 10 are Naïve Bayes classifiers. In a way, this part of the diagram calls for a joint consideration of the four levels of questions A–D in Figure 3.5. The top left quadrant looks for answers mostly at levels A and B, while the bottom right quadrant draws upon answers at level D and less so, C.

3.4   QUO VADIS?

3.4.1   Reinventing the Wheel?

We might pride ourselves for working in a relatively modern area of pattern recognition and machine learning, but, in fact, combining classifiers was proposed over 50 years ago. Take for example the idea of viewing the classifier output as a new feature vector. This could be traced back to Sebestyen [353] in his book *Decision Making Processes in Pattern Recognition* published in 1962. Sebestyen proposes cascade machines where the output of a classifier is fed as the input of the next classifier in the sequence, and so on. In 1979, Dasarathy and Sheela [82] propose a compound classifier where the decision is switched between two different classifier models depending on where the input is located. The book by Rastrigin and Erenstein [322] published in 1981 contains what is now known as dynamic classifier selection [421]. Unfortunately, Rastrigin and Erenstein's book only reached the Russian-speaking reader, and so did the book by Barabash, published in 1983 [26], containing interesting theoretical results about the majority vote for classifier combination. Zuev's study of 1987 [443], proposing a probabilistic model of a committee of classifiers, suffered the same fate.

3.4.2 The Illusion of Progress?

How far have we gone? About a decade ago, a curious parallel between the views of two experts in the field revealed how complex and prone to subjective bias the answer to this question is. In his invited lecture at the 3rd International Workshop on Multiple Classifier Systems, 2002, Ghosh proposed that [151]

> "... *our current understanding of ensemble-type multiclassifier systems is now quite mature ...*".

And yet, in an invited book chapter, also published in 2002, Ho states that [183]

> "*Many of the above questions are there because we do not yet have a scientific understanding of the classifier combination mechanisms.*"

Ho proceeds to nominate the stochastic discrimination theory by Kleinberg [210] as the only consistent and theoretically sound explanation of the success of classifier ensembles, criticizing other theories as being incomplete and assumption bound. However, as the practice invariably shows, ingenious heuristic developments are the heart, the soul, and the engine of many branches of science and research.

In a provocative paper published in 2006, Hand scrutinizes the claims of progress in classification methodologies [173]. He argues that much of the purported advance offered by modern (complex) classification methodologies, including ensemble approaches, may well be illusory.

An interesting visual representation of the typical evolution of any technology is the Gartner, Inc.'s *hype cycle*.[1] According to this model, shown in Figure 3.9, a peak of hype is followed by a valley of disillusionment and cynicism, eventually reaching what is called the "plateau of productivity" (asymptote of reality).

Is the evolution of the classifier ensemble field following this pattern? If so, where are we and what lies ahead? Could Hand's paper be heralding the downward slope leading to the trough of disillusionment? As we shall see next, there is no evidence of dramatical decrease of the visibility of the classifier ensemble field. On the other hand, the application research is taking over, which suggests that we may be approaching the plateau of productivity without experiencing a pronounced adverse dip.

---

[1] http://en.wikipedia.org/wiki/Hype_cycle

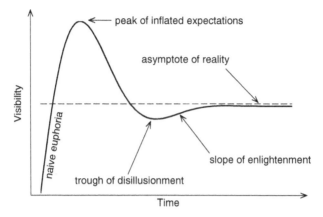

FIGURE 3.9　Gartner's hype cycle: a typical evolution pattern of a new technology.

3.4.3　A Bibliometric Snapshot

Thomson Reuters (formerly ISI) Web of Knowledge (WoK) has been described as the premier research platform for information and the largest accessible citation database.[2] It gives structured reference information gathered from journals and conference proceedings. WoK's searches are more restricted than those of Google Scholar, hence the number of returned citations is significantly smaller. Acknowledging that no system is perfect, WoK was chosen for this example because of its high reputation in the academic world and its reliability.

This example was carried out on January 4, 2013, so the publication and citation records are correct as of this date. The search filter was designed to accommodate the rich and evolving terminology of the classifier ensemble field:

```
"combining classifiers" OR
"classifier combination" OR
"classifier ensembles" OR
"ensemble of classifiers" OR
"combining multiple classifiers" OR
"committee of classifiers"  OR
"classifier committee" OR
"committees of neural networks"  OR
"consensus aggregation"  OR
"mixture of experts"  OR
"bagging predictors"  OR
adaboost  OR
(( "random subspace" OR "random forest"
   OR "rotation forest" OR boosting)
   AND "machine learning")
```

[2] http://wok.mimas.ac.uk/

Curiously, the oldest paper returned by the search is Zuev's 1986 study on a probabilistic model of a committee of classifiers [443] with no citations. The next oldest entry is Xu et al.'s seminal paper [425] with 864 citations. Notably, only this paper was returned for 1992, while the total number of returns was 4693. Hence the study was restricted to the period 1991–2012. The total number of papers covered by WoK was also retrieved for each year. Figure 3.10 gives the proportion (per mil) publications returned by the search ((returned/total) × 1000) over the 22-year span.

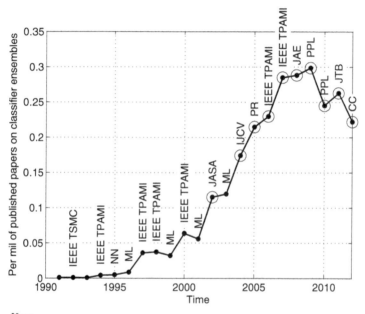

Key:

IEEE TSMC	IEEE Transactions on Systems, Man and Cybernetics
IEEE TPAMI	IEEE Transactions on Pattern Analysis and Machine Intelligence
NN	Neural Networks
ML	Machine Learning
JASA	Journal of the American Statistical Association
IJCV	International Journal of Computer Vision
PR	Pattern Recognition
JAE	Journal of Animal Ecology
PPL	Protein and Peptide Letters
JTB	Journal of Theoretical Biology
CC	Cerebral Cortex

FIGURE 3.10   Proportion (per mil) of published studies returned by the WoK search. The circled points correspond to application-oriented papers.

The journal signature of the most cited paper for the year is shown next to the year point on the graph. The circled points correspond to application-oriented papers while the noncircled ones correspond to more generic papers.

For example, the returned number of publications for 2004 was 1,519,596. Of these, 265 responded to the chosen combination of keywords. The per mil value on the graph for 2004 is therefore

$$\frac{265 \times 1000}{1,519,596} \approx 0.1744.$$

The most cited paper among the 265 papers for 2004 was the Viola and Jones face detection algorithm [405] published in the *International Journal of Computer Vision*. The journal signature (IJCV) is shown next to the data point on the graph. The point is circled because the paper is devoted to an application of an ensemble technology.

The upward trend up to 2009 shows the growing interest in the subject of classifier ensembles. Although the record for 2012 may be incomplete, the figure reveals a small decline after the peak in 2009. Whether or not this is symptomatic of saturating the area and starting a downward trend toward the "asymptote of reality" in Figure 3.9 remains to be seen.

Figure 3.11 plots the citation number returned by WoK of the most cited paper for the respective year. (Note that Google Scholar may return a much higher number.) Naturally, the citation counts of the more recent papers are low.

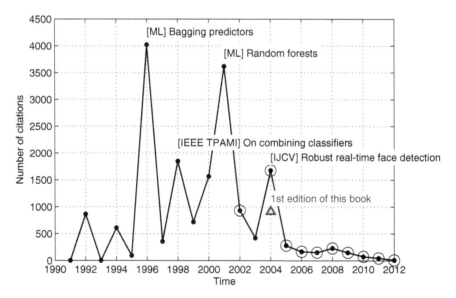

FIGURE 3.11 Number of citations of the most cited paper returned by the WoK search in the respective year. The circled points correspond to application-oriented papers. The search was carried out on January 4, 2013.

While the pattern of the citations graph may change tomorrow, the publication curve in Figure 3.10 is stable. The bibliometric snapshot suggests that the area of classifier ensembles has established itself and branched out to answer practical challenges in image analysis, proteomics, biometrics, medicine, ecology, etc. The fact that application-oriented papers hold the top citation places in recent years does not mean that there are no new theoretical or generic results that will peak in the future.

4

COMBINING LABEL OUTPUTS

How do we combine the outputs of the individual classifiers in the ensemble? Numerous theoretical analyses [145,207,235,241,249,260,394], experimental comparisons [109, 152, 222, 382, 434, 436], and reviews [392, 439] look for the answer to this question.

4.1 TYPES OF CLASSIFIER OUTPUTS

Consider a classifier ensemble consisting of L classifiers in the set $D = \{D_1, \dots, D_L\}$ and a set of classes $\Omega = \{\omega_1, \dots, \omega_c\}$. Xu et al. [425] distinguish between three types of classifier outputs:

- *Class labels.* (The abstract level.) Each classifier D_i produces a class label $s_i \in \Omega, i = 1, \dots, L$. Thus, for any object $\mathbf{x} \in \mathbb{R}^n$ to be classified, the L classifier outputs define a vector $\mathbf{s} = [s_1, \dots, s_L]^T \in \Omega^L$. At the abstract level, there is no information about the certainty of the guessed labels, nor are any alternative labels suggested. By definition, any classifier is capable of producing a label for $\mathbf{x}$, so the abstract level is universal.
- *Ranked class labels.* The output of each D_i is a subset of the class labels Ω, ranked in order of plausibility [184,391]. This type is especially suitable for problems with a large number of classes, such as character, face, and speaker recognition.

- *Numerical support for the classes.* (The measurement level.) Each classifier D_i produces a c-dimensional vector $[d_{i,1}, \ldots, d_{i,c}]^T$. The value $d_{i,j}$ represents the support for the hypothesis that the vector $\mathbf{x}$ submitted for classification comes from class ω_j. The outputs $d_{i,j}$ are functions of the input $\mathbf{x}$, but to simplify the notation we will use just $d_{i,j}$ instead of $d_{i,j}(\mathbf{x})$. Without loss of generality, we can assume that the outputs contain values between 0 and 1, spanning the space $[0, 1]^c$.

We add to this list one more output type:

- *Oracle.* The output of classifier D_i for a given $\mathbf{x}$ is only known to be either *correct* or *wrong*. We deliberately disregard the information as to which class label has been assigned. The oracle output is artificial because we can only apply it to a labeled data set. For a given data set $\mathbf{Z}$, classifier D_i produces an output vector $\mathbf{y}_i$ such that

$$
y_{ij} = \begin{cases} 1, & \text{if } D_i \text{ classifies object } \mathbf{z}_j \text{ correctly,} \\ 0, & \text{otherwise.} \end{cases} \tag{4.1}
$$

4.2   A PROBABILISTIC FRAMEWORK FOR COMBINING LABEL OUTPUTS

Consider class label outputs $\mathbf{s} = [s_1, \ldots, s_L]^T \in \Omega^L$. We are interested in the probability

$$
P(\omega_k \text{ is the true class } |\mathbf{s}), \quad k = 1, \ldots, c,
$$

denoted for short $P(\omega_k|\mathbf{s})$. Assume that the classifiers give their decisions independently, conditioned upon the class label. Conditional independence means that[1]

$$
P(s_1, s_2, \ldots, s_L|\omega_k) = P(s_1|\omega_k)P(s_2|\omega_k) \ldots P(s_L|\omega_k).
$$

Therefore we can write

$$
P(\omega_k|\mathbf{s}) = \frac{P(\omega_k)}{P(\mathbf{s})} \prod_{i=1}^{L} P(s_i|\omega_k). \tag{4.2}
$$

Split the product into two parts depending on which classifiers suggested ω_k. Denote by I_+^k the set of indices of classifiers which suggested ω_k, and by I_-^k the set of

---

[1]However, this assumption precludes unconditional independence, that is,

$$
P(s_1, s_2, \ldots, s_L) \neq P(s_1)P(s_2) \cdots P(s_L).
$$

indices of classifiers which suggested another class label. The probability of interest becomes

$$P(\omega_k|\mathbf{s}) = \frac{P(\omega_k)}{P(\mathbf{s})} \times \prod_{i \in I_+^k} P(s_i|\omega_k) \times \prod_{i \in I_-^k} P(s_i|\omega_k). \qquad (4.3)$$

This decomposition allows us to define the optimality conditions for several combination rules [244]. Optimality is understood in a sense that the combiner guarantees the minimum possible classification error.

4.3   MAJORITY VOTE

Never underestimate the power of stupid people in large groups.
—George Carlin

4.3.1   "Democracy" in Classifier Combination

Dictatorship and majority vote are perhaps the two oldest strategies for decision making [84, 159]. Three consensus patterns: unanimity, simple majority, and plurality, are illustrated below. Assume that shapes correspond to class labels, and the decision makers are the individual classifiers in the ensemble. The final label in all three consensus patterns is ■

Unanimity	■ ■ ■ ■ ■ ■ ■ ■ ■ ■
Simple majority	■ ■ ■ ■ ■ ■ △ △ △ △
Plurality	■ ■ ■ ■ △ △ △ × × ×

Assume that the *label* outputs of the classifiers are given as c-dimensional binary vectors $[d_{i,1}, \ldots, d_{i,c}]^T \in \{0, 1\}^c$, $i = 1, \ldots, L$, where $d_{i,j} = 1$ if D_i labels $\mathbf{x}$ in ω_j, and 0, otherwise. The *plurality vote* will return class ω_k if

$$\sum_{i=1}^{L} d_{i,k} = \max_{j=1}^{c} \sum_{i=1}^{L} d_{i,j}. \qquad (4.4)$$

Ties are resolved arbitrarily. This rule is often called in the literature *the majority vote*. It will indeed coincide with the simple majority (50% of the votes +1) in the case of two classes ($c = 2$). Xu et al. [425] suggest a thresholded plurality vote. They augment the set of class labels Ω with one more class, ω_{c+1}, for all objects for which

---

MAJORITY (PLURALITY) VOTE COMBINER (MV)

Training: None

Operation: For each new object

1. Find the class labels $s_1, \ldots, s_L$, assigned to this object by the L base classifiers.
2. Calculate the number of votes for each class ω_k, $k = 1, \ldots, c$.

$$P(k) = \sum_{i=1}^{L} I(s_i, \omega_k),$$

where $I(a, b) = 1$ if $a = b$ and 0 otherwise.
3. Assign label k^* to the object, where

$$k^* = \arg \max_{k=1}^{c} P(k).$$

Return the ensemble label of the new object.

FIGURE 4.1   Training and operation algorithm for the majority vote combiner.

the ensemble either fails to determine a class label with a sufficient confidence or produces a tie. Thus the decision is

$$\begin{cases} \omega_k, & \text{if } \sum_{i=1}^{L} d_{i,k} \geq \alpha.L, \\ \omega_{c+1}, & \text{otherwise,} \end{cases} \tag{4.5}$$

where $0 < \alpha \leq 1$. For the simple majority, we can pick α to be $\frac{1}{2} + \epsilon$, where ϵ is arbitrarily small and $0 < \epsilon < \frac{1}{L}$. When $\alpha = 1$, Equation 4.5 becomes the *unanimity voting* rule: a decision is made for a class label only if all decision makers agree on that label; otherwise the ensemble refuses to decide and assigns label ω_{c+1} to **x**. The algorithm of the majority vote combiner is shown in Figure 4.1.

The plurality vote (Equation 4.4), called in a wide sense "the majority vote," is the most often used rule from the majority vote group [26, 28, 248, 250, 260, 343].

4.3.2   Accuracy of the Majority Vote

Assume that

- The number of classifiers, L, is odd.
- The probability for each classifier to give the correct class label is p for any $\mathbf{x} \in \mathbb{R}^n$.

TABLE 4.1   Tabulated Values of the Majority Vote Accuracy of L Independent Classifiers with Individual Accuracy p

	$L = 3$	$L = 5$	$L = 7$	$L = 9$
$p = 0.6$	0.6480	0.6826	0.7102	0.7334
$p = 0.7$	0.7840	0.8369	0.8740	0.9012
$p = 0.8$	0.8960	0.9421	0.9667	0.9804
$p = 0.9$	0.9720	0.9914	0.9973	0.9991

- The classifier outputs are independent, that is, for any set of classifiers $A \subseteq D$, $A = \{D_{i_1}, \ldots, D_{i_K}\}$,

$$P(D_{i_1} = s_{i_1}, \ldots, D_{i_K} = s_{i_K})$$
$$= P(D_{i_1} = s_{i_1}) \times \cdots \times P(D_{i_K} = s_{i_K}), \tag{4.6}$$

where s_{i_j} is the label output of classifier D_{i_j}.

According to Equation 4.4, the majority vote will give an accurate class label if at least $\lfloor L/2 \rfloor + 1$ classifiers give correct answers ($\lfloor a \rfloor$ denotes the "floor," which is the nearest integer smaller than a).[2] Then the accuracy of the ensemble is

$$P_{\text{maj}} = \sum_{m=\lfloor L/2 \rfloor+1}^{L} \binom{L}{m} p^m (1 - p)^{L-m}. \tag{4.7}$$

The probabilities of correct classification of the ensemble for $p = 0.6, 0.7, 0.8$, and 0.9 and $L = 3, 5, 7$, and 9, are displayed in Table 4.1.

The following result is also known as the Condorcet Jury Theorem (1785) [359]:

1. If $p > 0.5$, then P_{maj} in Equation 4.7 is monotonically increasing and

$$P_{\text{maj}} \rightarrow 1 \quad \text{as} \quad L \rightarrow \infty. \tag{4.8}$$

2. If $p < 0.5$, then P_{maj} in Equation 4.7 is monotonically decreasing and

$$P_{\text{maj}} \rightarrow 0 \quad \text{as} \quad L \rightarrow \infty. \tag{4.9}$$

3. If $p = 0.5$, then $P_{\text{maj}} = 0.5$ for any L.

This result supports the intuition that we can expect improvement over the individual accuracy p only when p is higher than 0.5. Lam and Suen [250] proceed to

---

[2]Notice that the majority (50% + 1) is necessary and sufficient in the case of two classes and is sufficient but not necessary for $c > 2$. Thus the accuracy of an ensemble using plurality when $c > 2$ could be greater than the majority vote accuracy.

analyze the case of even L and the effect on the ensemble accuracy of adding or removing classifiers.

Shapley and Grofman [359] note that the result is valid even for unequal p, provided the distribution of the individual accuracies p_i is symmetrical about the mean.

▣ Example 4.1   Majority and unanimity in medical diagnostics

An accepted practice in medicine is to confirm the diagnosis by several (supposedly independent) tests. Lachenbruch [245] studies the unanimity and majority rules on a sequence of three tests for HIV diagnosis.

Sensitivity and specificity are the two most important characteristics of a medical test. Sensitivity (denoted by U) is the probability that the test procedure declares an affected individual affected (probability of a *true positive*). Specificity (denoted by V) is the probability that the test procedure declares an unaffected individual unaffected (probability of a *true negative*).[3]

Denote by T the event "positive test result," and by A, the event "the person is affected by the disease." Then $U = P(T|A)$ and $V = P(\overline{T}|\overline{A})$, where the over-bar denotes negation. We regard the test as an individual classifier with accuracy $p = U \times P(A) + V \times (1 - P(A))$, where $P(A)$ is the probability for the occurrence of the disease among the examined individuals, or the *prevalence* of the disease. In testing for HIV, a unanimous positive result from three tests is required to declare the individual affected [245]. Since the tests are applied one at a time, encountering the first negative result will cease the procedure. Another possible combination is the majority vote which will stop if the first two readings agree or otherwise take the third reading to resolve the tie. Table 4.2 shows the outcomes of the tests and the overall decision for the unanimity and majority rules.

TABLE 4.2   Unanimity and Majority Schemes for Three Independent Consecutive Tests

Unanimity sequences for decision (+):	$\langle +++ \rangle$		
Unanimity sequences for decision (−):	$\langle - \rangle$	$\langle +- \rangle$	$\langle ++- \rangle$
Majority sequences for decision (+):	$\langle ++ \rangle$	$\langle -++ \rangle$	$\langle +-+ \rangle$
Majority sequences for decision (−):	$\langle -- \rangle$	$\langle -+- \rangle$	$\langle +-- \rangle$

Assume that the three tests are applied independently and all have the same sensitivity u and specificity v. Then the sensitivity and the specificity of the procedure with the unanimity vote become

$$U_{\text{una}} = u^3,$$

$$V_{\text{una}} = 1 - (1 - v)^3. \qquad (4.10)$$

---

[3]In social sciences, for example, sensitivity translates to "convicting the guilty" and specificity to "freeing the innocent" [359].

For the majority vote,

$$U_{\text{maj}} = u^2 + 2u^2(1 - u) = u^2(3 - 2u),$$

$$V_{\text{maj}} = v^2(3 - 2v). \tag{4.11}$$

For $0 < u < 1$ and $0 < v < 1$, by simple algebra we obtain

$$U_{\text{una}} < u \quad \text{and} \quad V_{\text{una}} > v, \tag{4.12}$$

and

$$U_{\text{maj}} > u \quad \text{and} \quad V_{\text{maj}} > v. \tag{4.13}$$

Thus, there is a certain gain on both sensitivity and specificity if majority vote is applied. Therefore, the combined accuracy $P_{\text{maj}} = U \times P(A) + V \times (1 - P(A))$ is also higher than the accuracy of a single test $p = u \times P(A) + v \times (1 - P(A))$. For the unanimity rule, there is a substantial increase of specificity at the expense of decreased sensitivity. To illustrate this point, consider the ELISA test used for diagnosing HIV. According to Lachenbruch [245], this test has been reported to have sensitivity $u = 0.95$ and specificity $v = 0.99$. Then

$$U_{\text{una}} \approx 0.8574 \quad V_{\text{una}} \approx 1.0000$$
$$U_{\text{maj}} \approx 0.9928 \quad V_{\text{maj}} \approx 0.9997.$$

The sensitivity of the unanimity scheme is dangerously low. This means that the chance of an affected individual being misdiagnosed as unaffected is above 14%. There are different ways to remedy this. One possibility is to apply a more expensive and more accurate second test in case ELISA gave a positive result, for example, the Western Blot test, for which $u = v = 0.99$ [245].

4.3.3 Limits on the Majority Vote Accuracy: An Example

Let $D = \{D_1, D_2, D_3\}$ be an ensemble of three classifiers with the same individual probability of correct classification $p = 0.6$. Suppose that there are 10 objects in a hypothetical data set, and that each classifier correctly labels exactly six of them. Each classifier output is recorded as correct (1) or wrong (0). Given these requirements, *all* possible combinations of distributing 10 elements into the 8 combinations of outputs of the three classifiers are shown in Table 4.3. The penultimate column of Table 4.3 shows the majority vote accuracy of each of the 28 possible combinations. It is obtained as the proportion (out of 10 elements) of the sum of the entries in columns "111," "101," "011," and "110" (two or more correct votes). The rows of the table are ordered by the majority vote accuracy. To clarify the entries in Table 4.3, consider as an example the first row. The number 3 in the column under the heading "101," means that exactly three objects are correctly recognized by D_1 and D_3 (the first and the third 1's of the heading) and misclassified by D_2 (the zero in the middle).

TABLE 4.3   All Possible Combinations of Correct/Incorrect Classification of 10 Objects by Three Classifiers so that Each Classifier Recognizes Exactly Six Objects

No	111	101	011	001	110	100	010	000	P_{maj}	$P_{maj} - p$
	a	b	c	d	e	f	g	h		
				Pattern of success						
1	0	3	3	0	3	0	0	1	0.9	0.3
2	2	2	2	0	2	0	0	2	0.8	0.2
3	1	2	2	1	3	0	0	1	0.8	0.2
4	0	2	3	1	3	1	0	0	0.8	0.2
5	0	2	2	2	4	0	0	0	0.8	0.2
6	4	1	1	0	1	0	0	3	0.7	0.1
7	3	1	1	1	2	0	0	2	0.7	0.1
8	2	1	2	1	2	1	0	1	0.7	0.1
9	2	1	1	2	3	0	0	1	0.7	0.1
10	1	2	2	1	2	1	1	0	0.7	0.1
11	1	1	2	2	3	1	0	0	0.7	0.1
12	1	1	1	3	4	0	0	0	0.7	0.1
				Identical classifiers						
13	6	0	0	0	0	0	0	4	0.6	0.0
14	5	0	0	1	1	0	0	3	0.6	0.0
15	4	0	1	1	1	1	0	2	0.6	0.0
16	4	0	0	2	2	0	0	2	0.6	0.0
17	3	1	1	1	1	1	1	1	0.6	0.0
18	3	0	1	2	2	1	0	1	0.6	0.0
19	3	0	0	3	3	0	0	1	0.6	0.0
20	2	1	1	2	2	1	1	0	0.6	0.0
21	2	0	2	2	2	2	0	0	0.6	0.0
22	2	0	1	3	3	1	0	0	0.6	0.0
23	2	0	0	4	4	0	0	0	0.6	0.0
24	5	0	0	1	0	1	1	2	0.5	−0.1
25	4	0	0	2	1	1	1	1	0.5	−0.1
26	3	0	1	2	1	2	1	0	0.5	−0.1
27	3	0	0	3	2	1	1	0	0.5	−0.1
				Pattern of failure						
28	4	0	0	2	0	2	2	0	0.4	−0.2

The entries in the table are the number of occurrences of the specific binary output of the three classifiers in the particular combination. The majority vote accuracy P_{maj} and the improvement over the single classifier, $P_{maj} - p$ are also shown. Three characteristic classifier ensembles are marked.

The table offers a few interesting facts:

- There is a case where the majority vote produces 90% correct classification. Although purely hypothetical, this vote distribution is *possible* and offers a dramatic increase over the individual rate $p = 0.6$.

TABLE 4.4   The 2 × 2 Relationship Table with Probabilities

	D_k correct (1)	D_k wrong (0)
D_i correct (1)	a	b
D_i wrong (0)	c	d

Total, $a + b + c + d = 1$

- On the other hand, the majority vote is not guaranteed to do better than a single member of the ensemble. The combination in the bottom row has a majority vote accuracy of 0.4.

The best and the worst possible cases illustrated above are named "the pattern of success" and the "pattern of failure" [241] and detailed next.

4.3.4   Patterns of Success and Failure

Consider two classifiers D_i and D_k, and a 2 × 2 table of probabilities that summarizes their combined outputs as in Table 4.4.

The three-classifier problem from the previous section can be visualized using two pairwise tables as in Table 4.5. For this case,

$$a + b + c + d + e + f + g + h = 1. \tag{4.14}$$

The probability of correct classification of the majority vote of the three classifiers is (two or more correct)

$$P_{\text{maj}} = a + b + c + e. \tag{4.15}$$

All three classifiers have the same individual accuracy p, which brings in the following three equations:

$$
\begin{aligned}
a + b + e + f &= p, \quad D_1 \text{ correct;} \\
a + c + e + g &= p, \quad D_2 \text{ correct;} \\
a + b + c + d &= p, \quad D_3 \text{ correct.}
\end{aligned}
\tag{4.16}
$$

Maximizing P_{maj} in Equation 4.15 subject to conditions 4.14, 4.16 and $a, b, c, d, e, f, g, h \geq 0$, for $p = 0.6$, we obtain $P_{\text{maj}} = 0.9$ with the pattern highlighted in

TABLE 4.5   The Probabilities in Two Two-way Tables Illustrating a Three-Classifier Voting Ensemble

D_3 correct (1) $D_2 \rightarrow$ $D_1 \downarrow$	1	0	D_3 wrong (0) $D_2 \rightarrow$ $D_1 \downarrow$	1	0
1	a	b	1	e	f
0	c	d	0	g	h

TABLE 4.6   The Pattern of Success

D_3 correct (1)			D_3 wrong (0)		
$D_2 \rightarrow$			$D_2 \rightarrow$		
$D_1 \downarrow$	1	0	$D_1 \downarrow$	1	0
1	0	α	1	α	0
0	α	0	0	0	$\gamma = 1 - 3\alpha$

Table 4.3: $a = d = f = g = 0$, $b = c = e = 0.3$, $h = 0.1$. This example, optimal for three classifiers, indicates the possible characteristics of the best combination of L classifiers. The "pattern of success" and "pattern of failure" defined later follow the same intuition although we do not include a formal proof for their optimality.

Consider the pool D of L (odd) classifiers, each with accuracy p. For the majority vote to give a correct answer we need $\lfloor L/2 \rfloor + 1$ or more of the classifiers to be correct. Intuitively, the best improvement over the individual accuracy will be achieved when exactly $\lfloor L/2 \rfloor + 1$ votes are correct. Any extra correct vote for the same **x** will be wasted because it is not needed to give the correct class label. Correct votes which participate in combinations not leading to a correct overall vote are also wasted. To use the above idea, we make the following definition: *The pattern of success* is a distribution of the L classifier outputs such that:

1. The probability of any combination of $\lfloor L/2 \rfloor + 1$ correct and $\lfloor L/2 \rfloor$ incorrect votes is α.
2. The probability of all L votes being incorrect is γ.
3. The probability of any other combination is zero.

For $L = 3$, the two-table expression of the pattern of success is shown in Table 4.6. Here no votes are wasted; the only combinations that occur are where all classifiers are incorrect or exactly $\lfloor L/2 \rfloor + 1$ are correct. To simplify notation, let $l = \lfloor L/2 \rfloor$. The probability of a correct majority vote (P_{maj}) for the pattern of success is the sum of the probabilities of each correct majority vote combination. Each such combination has probability α. There are $\binom{L}{l+1}$ ways of having $l + 1$ correct out of L classifiers. Therefore

$$P_{\text{maj}} = \binom{L}{l+1} \alpha. \tag{4.17}$$

The pattern of success is only possible when $P_{\text{maj}} \leq 1$

$$\alpha \leq \frac{1}{\binom{L}{l+1}}. \tag{4.18}$$

To relate the individual accuracies p to α and P_{maj}, consider the following argument. In the pattern of success, if D_i gives a correct vote, then the remaining $L - 1$ classifiers must give l correct votes. There are $\binom{L-1}{l}$ ways in which the remaining $L - 1$ classifiers can give l correct votes, each with probability α. So the overall accuracy p of D_i is

$$p = \binom{L-1}{l} \alpha. \tag{4.19}$$

Expressing α from Equation 4.19 and substituting in Equation 4.17 gives

$$P_{\text{maj}} = \frac{pL}{l+1} = \frac{2pL}{L+1}. \tag{4.20}$$

Feasible patterns of success have $P_{\text{maj}} \leq 1$, so Equation 4.20 requires

$$p \leq \frac{L+1}{2L}. \tag{4.21}$$

If $p > \frac{(L+1)}{2L}$, then $P_{\text{maj}} = 1$ can be achieved, but there is an excess of correct votes. The improvement over the individual p will not be as large as for the pattern of success but the majority vote accuracy will be 1 anyway. The final formula for P_{maj} is

$$P_{\text{maj}} = \min\left\{ 1, \frac{2pL}{L+1} \right\}. \tag{4.22}$$

The worst possible behavior of an ensemble of L classifiers each with accuracy p is described by the pattern of failure.

The pattern of failure is a distribution of the L classifier outputs such that:

1. The probability of any combination of $\lfloor L/2 \rfloor$ correct and $\lfloor L/2 \rfloor + 1$ incorrect votes is β.
2. The probability of all L votes being correct is δ.
3. The probability of any other combination is zero.

For $L = 3$, the two-table expression of the pattern of failure is shown in Table 4.7.

The worst scenario is when the correct votes are wasted, that is, grouped in combinations of exactly l out of L correct (one short for the majority to be correct). The excess of correct votes needed to make up the individual p are also wasted by all the votes being correct together, while half of them plus one would suffice.

TABLE 4.7   The Pattern of Failure

| D_3 correct (1) | | | D_3 wrong (0) | | |
| $D_2 \rightarrow$ | | | $D_2 \rightarrow$ | | |
$D_1 \downarrow$	1	0	$D_1 \downarrow$	1	0
1	$\delta = 1 - 3\beta$	0	1	0	β
0	0	β	0	β	0

The probability of a correct majority vote (P_{maj}) is δ. As there are $\binom{L}{l}$ ways of having l correct out of L classifiers, each with probability β, then

$$P_{maj} = \delta = 1 - \binom{L}{l}\beta. \tag{4.23}$$

If D_i gives a correct vote, then either all the remaining classifiers are correct (probability δ) or exactly $l - 1$ are correct out of the $L - 1$ remaining classifiers. For the second case there are $\binom{L-1}{l-1}$ ways of getting this, each with probability β. To get the overall accuracy p for classifier D_i we sum the probabilities of the two cases:

$$p = \delta + \binom{L-1}{l-1}\beta. \tag{4.24}$$

Combining Equations 4.23 and 4.24 gives

$$P_{maj} = \frac{pL - l}{l+1} = \frac{(2p-1)L+1}{L+1}. \tag{4.25}$$

For values of individual accuracy $p > 0.5$, the pattern of failure is always possible. Matan [277] gives tight upper and lower bounds of the majority vote accuracy in the case of unequal individual accuracies (see Appendix 4.A.1). Suppose that classifier D_i has accuracy p_i, and $\{D_1, \ldots, D_L\}$ are arranged so that $p_1 \leq p_2 \cdots \leq p_L$. Let $k = l + 1 = (L + 1)/2$. Matan proves that

1. The upper bound of the majority vote accuracy of the ensemble is

$$\max P_{maj} = \min\{1, \Sigma(k), \Sigma(k-1), \ldots, \Sigma(1)\}, \tag{4.26}$$

where

$$\Sigma(m) = \frac{1}{m} \sum_{i=1}^{L-k+m} p_i, \quad m = 1, \ldots, k. \tag{4.27}$$

2. The lower bound of the majority vote accuracy of the ensemble is

$$\min P_{\text{maj}} = \max\{0, \xi(k), \xi(k-1), \dots, \xi(1)\}, \tag{4.28}$$

where

$$\xi(m) = \frac{1}{m} \sum_{i=k-m+1}^{L} p_i - \frac{L-k}{m}, \quad m = 1, \dots, k. \tag{4.29}$$

▣ Example 4.2   Matan's limits on the majority vote accuracy
Let $D = \{D_1, \dots, D_5\}$ be a set of classifiers with accuracies (0.56, 0.58, 0.60, 0.60, 0.62), respectively. To find the upper bound of the majority vote accuracy of this ensemble, form the sums $\Sigma(m)$ for $m = 1, 2, 3$:

$$\Sigma(1) = 0.56 + 0.58 + 0.60 = 1.74;$$

$$\Sigma(2) = \frac{1}{2}(0.56 + 0.58 + 0.60 + 0.60) = 1.17;$$

$$\Sigma(3) = \frac{1}{3}(0.56 + 0.58 + 0.60 + 0.60 + 0.62) = 0.99. \tag{4.30}$$

Then

$$\max P_{\text{maj}} = \min\{1, 1.74, 1.17, 0.99\} = 0.99. \tag{4.31}$$

For the lower bound,

$$\xi(1) = 0.60 + 0.60 + 0.62 - (5 - 3) = -0.18;$$

$$\xi(2) = \frac{1}{2}(0.58 + 0.60 + 0.60 + 0.62) - \frac{5-3}{2} = 0.20;$$

$$\xi(3) = \frac{1}{3}(0.56 + 0.58 + 0.60 + 0.60 + 0.62) - \frac{5-3}{3} = 0.32. \tag{4.32}$$

Then

$$\min P_{\text{maj}} = \max\{0, -0.18, 0.20, 0.32\} = 0.32. \tag{4.33}$$

The range of possible results from the majority vote across D is wide, so without more knowledge about how the classifiers are related to each other we can only guess within this range. If we assume that the classifier outputs are independent, then $P_{\text{maj}} = 0.67$, which indicates that there is much more to be achieved from the majority vote with dependent outputs.

Matan's result leads to the pattern of success and the pattern of failure as the upper and the lower bounds, respectively, for $p_1 = \dots = p_L = p$. Demirekler and Altincay [87] and Ruta and Gabrys [343] give further insights into the behavior of the two limit patterns.

Hierarchical majority voting ensembles have been found to be very promising [277,343,359]. There is a potential gain in accuracy but this has only been shown by construction examples.

4.3.5 Optimality of the Majority Vote Combiner

Majority vote (plurality vote for more than two classes) is the optimal combiner when the individual classifier accuracies are equal, the "leftover probability" is uniformly distributed across the remaining classes, and the prior probabilities for the classes are the same. The following theorem states this result more formally.

Theorem 4.1   Let D be an ensemble of L classifiers. Suppose that

1. The classifiers give their decisions independently, conditioned upon the class label.
2. The individual classification accuracy is $P(s_i = \omega_k | \omega_k) = p$ for any classifier i and class ω_k, and also for any data point in the feature space.
3. The probability for incorrect classification is equally distributed among the remaining classes, that is $P(s_i = \omega_j | \omega_k) = \frac{1-p}{c-1}$, for any $i = 1, \ldots, L$, $k, j = 1, \ldots, c$ $j \neq k$.

Then the majority vote is the optimal combination rule.

Proof.   Substituting in the probabilistic framework defined in Equation 4.3,

$$P(\omega_k | \mathbf{s}) = \frac{P(\omega_k)}{P(\mathbf{s})} \times \prod_{i \in I_+^k} p \times \prod_{i \in I_-^k} \frac{1-p}{c-1} \qquad (4.34)$$

$$= \frac{P(\omega_k)}{P(\mathbf{s})} \times \prod_{i \in I_+^k} p \times \prod_{i \in I_-^k} \frac{1-p}{c-1} \times \frac{\prod_{i \in I_+^k} \frac{1-p}{c-1}}{\prod_{i \in I_+^k} \frac{1-p}{c-1}} \qquad (4.35)$$

$$= \frac{P(\omega_k)}{P(\mathbf{s})} \times \prod_{i \in I_+^k} \frac{p(c-1)}{1-p} \times \prod_{i=1}^{L} \frac{1-p}{c-1} . \qquad (4.36)$$

Notice that $P(\mathbf{s})$ and the last product term in Equation 4.36 do not depend on the class label. The prior probability, $P(\omega_k)$, does depend on the class label but not on the votes, so it can be designated as the *class constant*. Rearranging and taking the logarithm,

$$\log(P(\omega_k | \mathbf{s})) = \log\left(\frac{(1-p)^L}{P(\mathbf{s})(c-1)^L}\right) + \log\left(P(\omega_k)\right)$$

$$+ \log\left(\frac{p(c-1)}{1-p}\right) \times |I_+^k|, \qquad (4.37)$$

where $|.|$ denotes cardinality. Dividing by $\log\left(\frac{p(c-1)}{1-p}\right)$ and dropping all terms that do not depend on the class label or the vote counts, we can create the following class-support functions for the object $\mathbf{x}$:

$$\mu_k(\mathbf{x}) = \underbrace{\frac{\log\left(P(\omega_k)\right)}{\log\left(\frac{p(c-1)}{1-p}\right)}}_{\text{class constant } \zeta(\omega_k)} + \quad |I_+^k|. \qquad (4.38)$$

Note that $|I_+^k|$ is the number of votes for ω_k. Choosing the class label corresponding to the largest support function is equivalent to choosing the class most voted for, subject to a constant term. ∎

Interestingly, the standard majority vote rule does not include a class constant and is still one of the most robust and accurate combiners for classifier ensembles. The class constant may sway the vote, especially for highly unbalanced classes and uncertain ensemble decisions where the number of votes for different classes are close. However, including the class constant will make majority vote a trainable combiner, which defeats one of its main assets. To comply with the common interpretation, here we adopt the standard majority vote formulation, whereby the class label is obtained by

$$\omega = \arg\max_k |I_+^k|. \qquad (4.39)$$

4.4 WEIGHTED MAJORITY VOTE

The weighted majority vote is among the most intuitive and widely used combiners [204,261]. It is the designated combination method derived from minimizing a bound on the training error in AdaBoost [118, 133].

If the classifiers in the ensemble are not of identical accuracy, then it is reasonable to attempt to give the more competent classifiers more power in making the final decision. The label outputs can be represented as degrees of support for the classes in the following way:

$$d_{i,j} = \begin{cases} 1, & \text{if } D_i \text{ labels } \mathbf{x} \text{ in } \omega_j, \\ 0, & \text{otherwise.} \end{cases} \qquad (4.40)$$

The class-support function for class ω_j obtained through weighted voting is

$$\mu_j(\mathbf{x}) = \sum_{i=1}^{L} b_i d_{i,j}, \qquad (4.41)$$

where b_i is a coefficient for classifier D_i. Thus, the value of the class-support function 4.41 will be the sum of the weights for those members of the ensemble whose output for $\mathbf{x}$ is ω_j.

4.4.1 Two Examples

■ Example 4.3   Assigning weights to the classifiers
Consider an ensemble of three classifiers D_1, D_2, and D_3 with accuracies 0.6, 0.6, and 0.7, respectively, and with independent oracle outputs. An accurate ensemble vote will be obtained if any two classifiers are correct. The ensemble accuracy will be

$$P_{\text{maj}} = 0.6^2 \times 0.3 + 2 \times 0.4 \times 0.6 \times 0.7 + 0.6^2 \times 0.7 = 0.6960. \qquad (4.42)$$

Clearly, it will be better if we remove D_1 and D_2 and reduce the ensemble to the single and more accurate classifier D_3. We introduce weights or coefficients of importance b_i, $i = 1, 2, 3$, and rewrite Equation 4.4 as: choose class label ω_k if

$$\sum_{i=1}^{L} b_i d_{i,k} = \max_{j=1}^{c} \sum_{i=1}^{L} b_i d_{i,j}. \qquad (4.43)$$

For convenience we normalize the weights so that

$$\sum_{i=1}^{c} b_j = 1. \qquad (4.44)$$

Assigning $b_1 = b_2 = 0$ and $b_3 = 1$, we get rid of D_1 and D_2, leading to $P_{\text{maj}} = p_3 = 0.7$. In fact, any set of weights which makes D_3 the dominant classifier will yield the same P_{maj}, for example, $b_3 > 0.5$ and any b_1 and b_2 satisfying Equation 4.44.

In the above example the weighted voting did not improve on the single best classifier in the ensemble even for independent classifiers. The following example shows that, in theory, the weighting might lead to a result better than both the single best member of the ensemble and the simple majority.

■ Example 4.4   Improving the accuracy by weighting
Consider an ensemble of five classifiers $D_1, \ldots, D_5$ with accuracies (0.9, 0.9, 0.6, 0.6, 0.6).[4] If the classifiers are independent, the majority vote accuracy (at least three out of five correct votes) is

$$P_{\text{maj}} = 3 \times 0.9^2 \times 0.4 \times 0.6 + 0.6^3 + 6 \times 0.9 \times 0.1 \times 0.6^2 \times 0.4$$
$$\approx 0.877. \qquad (4.45)$$

Assume now that the weights given to the voters are (1/3, 1/3, 1/9, 1/9, 1/9). Then the two more competent classifiers agreeing will be enough to make the decision because the score for the class label they agree upon will become 2/3. If they disagree, that is, one is correct and one is wrong, the vote of the ensemble will be decided by

---

[4] After [359].

the majority of the remaining three classifiers. Then the accuracy for the weighted voting will be

$$P^w_{\text{maj}} = 0.9^2 + 2 \times 3 \times 0.9 \times 0.1 \times 0.6^2 \times 0.4 + 2 \times 0.9 \times 0.1 \times 0.6^3$$

$$\approx 0.927. \tag{4.46}$$

Again, any set of weights that satisfy Equation 4.44 and make the first two classifiers prevail when they agree will lead to the same outcome.

4.4.2 Optimality of the Weighted Majority Vote Combiner

Here we use the probabilistic framework 4.2 to derive the optimality conditions for the weighted majority vote. This type of combiner follows from relaxing the assumption about equal individual accuracies. Hence the majority vote combiner is a special case of the weighted majority combiner for equal individual accuracies.

Theorem 4.2   Let D be an ensemble of L classifiers. Suppose that

1. The classifiers give their decisions independently, conditioned upon the class label.
2. The individual classification accuracy is $P(s_i = \omega_k | \omega_k) = p_i$ for any class ω_k, and also for any data point in the feature space.
3. The probability for incorrect classification is equally distributed among the remaining classes, that is $P(s_i = \omega_j | \omega_k) = \frac{1-p_i}{c-1}$, for any $i = 1, \dots, L$, $k, j = 1, \dots, c, j \neq k$.

Then the weighted majority vote is the optimal combination rule with weights

$$w_i = \log \left(\frac{p_i}{1 - p_i} \right), \quad 0 < p_i < 1. \tag{4.47}$$

Proof.   Following the same derivation path as with the majority vote optimality, Equation 4.3 becomes

$$P(\omega_k | \mathbf{s}) = \frac{P(\omega_k)}{P(\mathbf{s})} \times \prod_{i \in I_+^k} p_i \times \prod_{i \in I_-^k} \frac{1 - p_i}{c - 1} \tag{4.48}$$

$$= \frac{P(\omega_k)}{P(\mathbf{s})} \times \prod_{i \in I_+^k} \frac{p_i(c - 1)}{1 - p_i} \times \prod_{i=1}^{L} \frac{1 - p_i}{c - 1} \tag{4.49}$$

$$= \frac{1}{P(\mathbf{s})} \times \prod_{i=1}^{L} \frac{1 - p_i}{c - 1} \times P(\omega_k) \times \prod_{i \in I_+^k} \frac{p_i(c - 1)}{1 - p_i}. \tag{4.50}$$

Then

$$\log(P(\omega_k|\mathbf{s})) = \log\left(\frac{\prod_{i=1}^{L}(1-p_i)}{P(\mathbf{s})(c-1)^L}\right) + \log\left(P(\omega_k)\right)$$

$$+ \sum_{i\in|I_+^k|} \log\left(\frac{p_i}{1-p_i}\right) + |I_+^k| \times \log(c-1). \qquad (4.51)$$

Dropping the first term, which will not influence the class decision, and expressing the classifier weights as

$$w_i = \log\left(\frac{p_i}{1-p_i}\right), \quad 0 < p_i < 1,$$

will transform Equation 4.51 to

$$\mu_k(\mathbf{x}) = \underbrace{\log(P(\omega_k))}_{\text{class constant } \zeta(\omega_k)} + \sum_{i\in|I_+^k|} w_i + |I_+^k| \times \log(c-1). \qquad (4.52)$$

■

If $p_i = p$ for all $i = 1, \ldots, L$, Equation 4.52 reduces to the majority vote Equation 4.37.

Similar proofs have been derived independently by several researchers in different fields of science such as democracy studies, pattern recognition, and automata theory. The earliest reference according to Ref. [26, 359] was Pierce, 1961 [309] .

The algorithm of the weighted majority vote combiner is shown in Figure 4.2. Note that the figure shows the conventional version of the algorithm which does not include the class constant or the last term in Equation 4.52.

4.5   NAÏVE-BAYES COMBINER

Exploiting the independence assumption further leads to a combiner called the "independence model" [385], "Naïve Bayes (NB)," "simple Bayes," [103] and even "idiot's Bayes" [107, 330]. Sometimes the first adjective is skipped and the combination method is called just "*Bayes* combination."

4.5.1   Optimality of the Naïve Bayes Combiner

We can derive this combiner by finally dropping the assumption of equal individual accuracies in the probabilistic framework 4.2.

WEIGHTED MAJORITY VOTE COMBINER (WMV)

Training

1. Obtain an array $E_{(N \times L)}$ with individual outputs of L classifiers for N objects. Entry $e(i, j)$ is the class label assigned by classifier D_j to object i. An array $T_{(N \times 1)}$ with the true labels is also provided.
2. Estimate the accuracy of each base classifier D_i, $i = 1, \ldots, L$, as the proportion of matches between column i of E and the the true labels T. Denote the estimates by $\hat{p}_i$.
3. Calculate the weights for the classifiers

$$v_i = \log \left(\frac{\hat{p}_i}{1 - \hat{p}_i} \right), \quad 0 < \hat{p}_i < 1, \quad i = 1 \ldots, L.$$

Operation: For each new object

1. Find the class labels $s_1, \ldots, s_L$ assigned to this object by the L base classifiers.
2. Calculate the score for all classes

$$P(k) = \sum_{s_i = \omega_k} v_i, \quad k = 1, \ldots, c.$$

3. Assign label k^* to the object, where

$$k^* = \arg \max_{k=1}^{c} P(k).$$

Return the ensemble label of the new object.

FIGURE 4.2   Training and operation algorithm for the weighted majority vote combiner.

Theorem 4.3   Let D be an ensemble of L classifiers. Suppose that the classifiers give their decisions independently, conditioned upon the class label. Then the Naïve Bayes combiner

$$\max \left\{ P(\omega_k) \prod_{i=1}^{L} P(s_i | \omega_k) \right\}$$

is the optimal combination rule.

Proof.   Think of $P(s_i = \omega_j | \omega_k)$ as the (j, k)th entry in a probabilistic confusion matrix for classifier i. In this case, Equation (4.2) can be used directly:

$$P(\omega_k | \mathbf{s}) = \frac{P(\omega_k)}{P(\mathbf{s})} \prod_{i=1}^{L} P(s_i | \omega_k) . \tag{4.53}$$

Dropping $P(\mathbf{s})$, which does not depend on the class label, the support for class ω_k is

$$\mu_k(\mathbf{x}) = P(\omega_k) \prod_{i=1}^{L} P(s_i | \omega_k) . \tag{4.54}$$

∎

4.5.2   Implementation of the NB Combiner

The implementation of the NB method on a data set $\mathbf{Z}$ with cardinality N is explained below. For each classifier D_i, a $c \times c$ confusion matrix CM^i is calculated by applying D_i to the training data set. The (k, s)th entry of this matrix, $\mathrm{cm}_{k,s}^i$, is the number of elements of the data set whose true class label was ω_k and were assigned by D_i to class ω_s. Denote by N_k the number of elements of $\mathbf{Z}$ from class ω_k, $k = 1, \ldots, c$. Taking $\mathrm{cm}_{k,s_i}^i / N_k$ to be an estimate of the probability $P(s_i | \omega_k)$ and N_k / N to be an estimate of the prior probability for class ω_s, the support for class ω_k in Equation 4.54 can be expressed as

$$\mu_k(\mathbf{x}) = \frac{1}{N_k^{L-1}} \prod_{i=1}^{L} \mathrm{cm}^i(k, s_i). \tag{4.55}$$

▉ Example 4.5   NB combination

Consider a problem with $L = 2$ classifiers, D_1 and D_2, and $c = 3$ classes. Let the number of training data points be $N = 20$. From these, let eight be from ω_1, nine from ω_2, and three from ω_3. Suppose that the following confusion matrices have been obtained for the two classifiers:

$$\mathrm{CM}^1 = \begin{bmatrix} 6 & 2 & 0 \\ 1 & 8 & 0 \\ 1 & 0 & 2 \end{bmatrix} \quad \text{and} \quad \mathrm{CM}^2 = \begin{bmatrix} 4 & 3 & 1 \\ 3 & 5 & 1 \\ 0 & 0 & 3 \end{bmatrix} . \tag{4.56}$$

Assume $D_1(\mathbf{x}) = s_1 = \omega_2$ and $D_2(\mathbf{x}) = s_2 = \omega_1$ for the input $\mathbf{x} \in \mathbb{R}^n$. Using Equation 4.55,

$$\mu_1(\mathbf{x}) \propto = \frac{1}{8} \times 2 \times 4 = 1;$$

$$\mu_2(\mathbf{x}) \propto \frac{1}{9} \times 8 \times 3 = \frac{8}{3} \approx 2.67;$$

$$\mu_3(\mathbf{x}) \propto \frac{1}{3} \times 0 \times 0 = 0. \tag{4.57}$$

As $\mu_2(\mathbf{x})$ is the highest of the three values, the maximum membership rule will label $\mathbf{x}$ in ω_2.

Notice that a zero as an estimate of $P(s_i | \omega_k)$ automatically nullifies $\mu_k(\mathbf{x})$ regardless of the rest of the estimates. Titterington et al. [385] study the NB classifier for independent categorical features. They discuss several modifications of the estimates to account for the possible zeros. For the NB combination, we can plug in Equation 4.54 in the following estimate:

$$\hat{P}(s_i|\omega_k) = \frac{cm^i_{k,s_i} + \frac{1}{c}}{N_k + 1},\tag{4.58}$$

where N_k is the number of elements in the training set $\mathbf{Z}$ from class ω_k, $k = 1, \ldots, c$. The algorithm for the training and the operation of the NB combiner is shown in Figure 4.3. MATLAB function `nb_combiner` is given in Appendix 4.A.2.

NAÏVE BAYES COMBINER (NB)

Training

1. Obtain an array $E_{(N \times L)}$ with individual outputs of L classifiers for N objects. Entry $e(i, j)$ is the class label assigned by classifier D_j to object i. An array $T_{(N \times 1)}$ with the true labels is also provided.
2. Find the number of objects in each class within T. Denote these numbers by $N_1, N_2, \ldots, N_c$.
3. For each classifier D_i, $i = 1, \ldots, L$, calculate a bespoke $c \times c$ confusion matrix C_i. The (j_1, j_2)th entry is

$$C_i(j_1, j_2) = \frac{K(j_1, j_2) + \frac{1}{c}}{N_{j_1} + 1},$$

where $K(j_1, j_2)$ is the number of objects with true class label j_1, labeled by classifier D_i in class j_2.

Operation: For each new object

1. Find the class labels $s_1, \ldots, s_L$ assigned to this object by the L base classifiers.
2. For each class ω_k, $k = 1, \ldots, c$
   (a) Set $P(k) = \frac{N_k}{N}$.
   (b) For $i = 1 \ldots L$, calculate $P(k) \leftarrow P(k) \times C_i(k, s_i)$.
3. Assign label k^* to the object, where

$$k^* = \arg\max_{k=1}^{c} P(k).$$

Return the ensemble label of the new object.

FIGURE 4.3   Training and operation algorithm for the NB combiner.

◧ Example 4.6   NB combination with a correction for zeros

Take the 20-point data set and the confusion matrices CM^1 and CM^2 from the previous example. The estimates of the class-conditional pmfs for the values $s_1 = \omega_2$ and $s_2 = \omega_1$ are

$$\mu_1(\mathbf{x}) \propto \frac{N_1}{N} \times \left(\frac{c_{1,2}^1 + \frac{1}{3}}{N_1 + 1} \right) \left(\frac{c_{1,1}^2 + \frac{1}{3}}{N_1 + 1} \right)$$

$$= \frac{8}{20} \times \left(\frac{2 + \frac{1}{3}}{8 + 1} \right) \left(\frac{4 + \frac{1}{3}}{8 + 1} \right) \approx 0.050$$

$$\mu_2(\mathbf{x}) \propto \frac{N_2}{N} \times \left(\frac{c_{2,2}^1 + \frac{1}{3}}{N_2 + 1} \right) \left(\frac{c_{2,1}^2 + \frac{1}{3}}{N_2 + 1} \right)$$

$$= \frac{9}{20} \times \left(\frac{8 + \frac{1}{3}}{9 + 1} \right) \left(\frac{3 + \frac{1}{3}}{9 + 1} \right) \approx 0.125$$

$$\mu_3(\mathbf{x}) \propto \frac{N_3}{N} \times \left(\frac{c_{3,2}^1 + \frac{1}{3}}{N_3 + 1} \right) \left(\frac{c_{3,1}^2 + \frac{1}{3}}{N_3 + 1} \right)$$

$$= \frac{3}{20} \times \left(\frac{0 + \frac{1}{3}}{3 + 1} \right) \left(\frac{0 + \frac{1}{3}}{3 + 1} \right) \approx 0.001. \tag{4.59}$$

Again, label ω_2 will be assigned to $\mathbf{x}$. Notice that class ω_3 now has a small nonzero support.

Despite the condescending names it has received, the NB combiner has been acclaimed for its rigorous statistical underpinning and robustness. It has been found to be surprisingly accurate and efficient in many experimental studies. The surprise comes from the fact that the combined entities are seldom independent. Thus, the independence assumption is nearly always violated, sometimes severely. However, it turns out that the classifier performance is quite robust, even in the case of dependence. Furthermore, attempts to amend the NB by including estimates of some dependencies do not always pay off [103].

4.6   MULTINOMIAL METHODS

In this group of methods we estimate the posterior probabilities $P(\omega_k|\mathbf{s})$ for all $k = 1, \ldots, c$ and every combination of votes $\mathbf{s} \in \Omega^L$. The highest posterior probability determines the class label for $\mathbf{s}$. Then, given an $\mathbf{x} \in \mathbb{R}^n$, first the labels $s_1, \ldots, s_L$ are assigned by the classifiers in the ensemble D, and then the final label is retrieved for $\mathbf{s} = [s_1, \ldots, s_L]^T$.

"Behavior Knowledge Space" (BKS) is a fancy name for the multinomial combination. The label vector $\mathbf{s}$ is regarded as an index to a cell in a look-up table (the BKS table) [190]. The table is designed using a labeled data set $\mathbf{Z}$. Each $\mathbf{z}_j \in \mathbf{Z}$ is placed in the cell indexed by the $\mathbf{s}$ for that object. The number of elements in each cell are tallied, and the most representative class label is selected for this cell. The highest score corresponds to the highest estimated posterior probability $P(\omega_k|\mathbf{s})$. Ties are resolved arbitrarily. The empty cells are labeled in some appropriate way. For example, we can choose a label at random or use the result from a majority vote between the elements of $\mathbf{s}$.

To have a reliable multinomial combiner, the data set should be large. The BKS combination method is often overtrained: it works very well on the training data but poorly on the testing data. Raudys [324,325] carried out a comprehensive analysis of the problems and solutions related to the training of BKS (among other combiners) for large and small sample sizes.

The BKS combiner is the optimal combiner for any dependencies between the classifier outputs. The caveat here is that it is hardly possible to have reliable estimates of the posterior probabilities for all possible c^L output combinations $\mathbf{s}$, even for the most frequently occurring combinations.

▣ Example 4.7   BKS combination method
Consider a problem with three classifiers and two classes. Assume that $D_1, D_2,$ and D_3 produce output $(s_1, s_2, s_3) = (\omega_2, \omega_1, \omega_2)$. Suppose that there are 100 objects in $\mathbf{Z}$ for which this combination of labels occurred: 60 having label ω_1 and 40 having label ω_2. Hence the table cell indexed by $(\omega_2, \omega_1, \omega_2)$ will contain label ω_1 no matter that the majority of the classifiers suggest otherwise.

From an implementation point of view, the BKS combiner can be regarded as the nearest neighbor classifier in the space of the ensemble outputs over the training data set. The concept of distance is replaced by exact match. If there are more than one nearest neighbors (exact matches) in the training set, the labels of the matches are tallied, and the label of the largest class representation is assigned. The algorithm is shown in Figure 4.4.

▣ Example 4.8   BKS combiner for the fish data
A MATLAB function `bks_combiner` implementing the algorithm in Figure 4.4 is given in Appendix 4.A.2. A MATLAB script which uses the function to label the fish data is also provided. Note that the script needs function `fish_data`, given in Appendix 2.A.1.

Fifty random linear classifiers were generated in the data space. The grid space was scaled to the unit square. To generate a linear classifier, a random point $P(p_1, p_2)$ was selected within the square (not necessarily a node on the grid). Two random numbers were drawn from a standard normal distribution, to be used as coefficients

---

BEHAVIOR KNOWLEDGE SPACE (BKS) COMBINER

Training

1. Obtain a reference array $E_{(N \times L)}$ with individual outputs of L classifiers for N objects. Entry $e(i, j)$ is the class label assigned by classifier D_j to object i. An array $T_{(N \times 1)}$ with the true labels is also provided.
2. Find the prevalent class label within T, say ω_p.

Operation: For each new object

1. Find the class labels assigned to this object by the L base classifiers and place them in a vector row **r**.
2. Compare **r** with each row of the reference array. Record in a set S the labels of the objects whose rows match **r**.
3. If there is no match ($S = \emptyset$), assign label ω_p to the new object. Otherwise, assign the prevalent class label within S. If there is a label tie, choose at random among the tied classes.

Return the ensemble label of the new object.

FIGURE 4.4   Operation algorithm for the BKS combiner for a given reference ensemble with labeled data and a set of new objects to be labeled by the ensemble.

a and b in the line equation $ax + by + c = 0$. Then the constant c was calculated so that P lies on the line: $c = -ap_1 - bp_2$.

BKS has been applied to combine the outputs of the 50 classifiers. Figure 4.5 shows the classifier boundaries and the regions labeled as the fish (class black dots) by the ensemble with three levels of noise: 0%, 20%, and 35%. The accuracy displayed as the plot title is calculated with respect to the original (noise-free) class labels.

The accuracy of the BKS combiner is very high, even for large amount of label noise. The real problem with this combiner comes when the testing data evokes

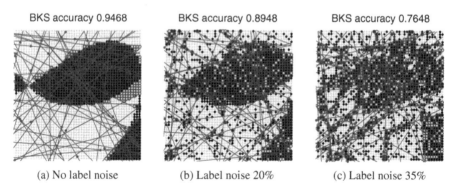

| BKS accuracy 0.9468 | BKS accuracy 0.8948 | BKS accuracy 0.7648 |

| (a) No label noise | (b) Label noise 20% | (c) Label noise 35% |

FIGURE 4.5   BKS classifier combiner for ensembles with $L = 50$ random linear classifiers for the fish data, with different amount of label noise.

TABLE 4.8 Scopes of Optimality (Denoted by a Black Square) and the Number of Tunable Parameters of the Four Combiners for a Problem with c Classes and an Ensemble of L Classifiers

Combiner	1	2	3	4	Number of parameters
Majority vote (not trained)	■	–	–	–	none (requires equal priors for the classes)
Weighted majority vote	■	■	–	–	$L + c$
Naïve Bayes	■	■	■	–	$L * c^2 + c$
BKS	■	■	■	■	c^L

Column headings:
1. Equal p
2. Classifier-specific p_i
3. Full confusion matrix
4. Independence is not required

ensemble outputs which do not appear in the reference ensemble. If the reference data is sufficiently representative, unmatched outputs will be relatively rare. Our implementation of the BKS combiner does not look beyond the exact match. It is possible to combat the brittleness of the method by considering distances between the (nominal) label vectors.

4.7 COMPARISON OF COMBINATION METHODS FOR LABEL OUTPUTS

Table 4.8 shows the optimality scopes and the number of tunable parameters for each combiner.

In practice, the success of a particular combiner will depend partly on the validity of the assumptions and partly on the availability of sufficient data to make reliable estimates of the parameters.

The optimality of the combiners is asymptotic, and holds for sample size approaching infinity. For finite sample sizes, the accuracy of the estimates of the parameters may be the primary concern. A combiner with fewer tunable parameters may be preferable even though its optimality assumption does not hold.

■ **Example 4.9 Label output combiners for the fish data set**
Consider the following experiment. Fifty linear classifiers were randomly generated in the grid space of the fish data set. An example of 50 linear boundaries is shown in Figure 4.6a.

The labels of the two regions for each linear classifier were assigned randomly. The accuracy of the classifier was evaluated. Note that the accuracy estimate is exact because we have all possible data points (nodes on the grid). If the accuracy was less than 50%, the regions were swapped over. Knowing the exact value of the classification accuracy eliminates the estimation error. Thus the only source of error in the ensemble error estimate came from the assumptions being incorrect.

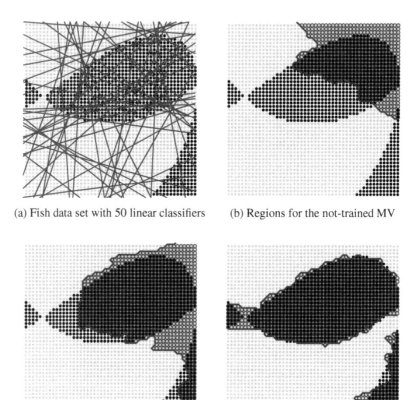

(a) Fish data set with 50 linear classifiers (b) Regions for the not-trained MV

(c) Regions for the NB combiner (d) Regions for the BKS combiner

FIGURE 4.6 A random linear ensemble for the fish data set.

The shaded regions in plots (b), (c), and (d) in Figure 4.6b,c,d show the ensemble classification regions for class "black dots" for three combination rules: the majority vote (Equation 4.39), the NB combiner (Equation 4.54) with the correction for zeros, and the BKS combiner. The individual and ensemble accuracies are detailed in Table 4.9.

Two hundred runs were carried out with different random "bunch of straws" (50 random classifiers) thrown in the unit square. Table 4.9 shows the average accuracies together with the standard deviations. The accuracies are ranked as expected for the 200-run experiment. Progressively alleviating the assumption of equal individual accuracies pays off. Weighted majority vote is better than the majority vote, and NB is better than both. BKS is always the best combiner because there is no parameter estimation error. However, this ranking is not guaranteed. Violation of the assumptions may affect the ensemble accuracy to various degrees, and disturb the ranking.

TABLE 4.9   Classification Accuracies in Percentage of the Individual and Ensemble Classifiers for Different Label Combiners

Classifier/ensemble	Example	Average of 200 runs $\pm \sigma$
Largest prior classifier	64.48	64.48 ± 0.00
Average individual classifier	59.76	59.91 ± 0.81
Majority vote (not trained)	^{a}70.60	68.53 ± 2.69
Weighted majority	68.84	69.63 ± 1.91
Naïve Bayes	^{a}82.92	75.21 ± 2.85
BKS	^{a}95.72	94.42 ± 1.00

^{a}Plot appears in Figure 4.6.

APPENDIX

4.A.1   MATAN'S PROOF FOR THE LIMITS ON THE MAJORITY VOTE ACCURACY

Here we give a sketch of the proof as offered in Ref. [277].

Theorem 4.A.1   Given is a classifier ensemble $D = \{D_1, \dots, D_L\}$. Suppose that classifiers D_i have accuracies p_i, $i = 1, \dots, L$, and are arranged so that $p_1 \leq p_2 \cdots \leq p_L$. Let $k = l + 1 = (L + 1)/2$. Then

1. *The upper bound of the majority vote accuracy of the ensemble is*

$$\max P_{\text{maj}} = \min\{1, \Sigma(k), \Sigma(k-1), \dots, \Sigma(1)\}, \qquad (4.A.1)$$

   where

$$\Sigma(m) = \frac{1}{m} \sum_{i=1}^{L-k+m} p_i, \quad m = 1, \dots, k. \qquad (4.A.2)$$

2. *The lower bound of the majority vote accuracy of the ensemble is*

$$\min P_{\text{maj}} = \max\{0, \xi(k), \xi(k-1), \dots, \xi(1)\}, \qquad (4.A.3)$$

   where

$$\xi(m) = \frac{1}{m} \sum_{i=k-m+1}^{L} p_i - \frac{L-k}{m}, \quad m = 1, \dots, k. \qquad (4.A.4)$$

Proof (sketch).   ■

Upper Bound. The accuracy p_i is the average accuracy of D_i across the whole feature space $\mathbb{R}^n$ and can be written as

$$p_i = \int_{\mathbb{R}^n} \mathcal{I}_i(\mathbf{x})p(\mathbf{x})d\mathbf{x}, \tag{4.A.5}$$

where $\mathcal{I}_i$ is an indicator function for classifier D_i, defined as

$$\mathcal{I}_i(\mathbf{x}) = \begin{cases} 1, & \text{if } D_i \text{ recognizes correctly } \mathbf{x}, \\ 0, & \text{otherwise}, \end{cases} \tag{4.A.6}$$

and $p(\mathbf{x})$ is the probability density function of $\mathbf{x}$. The majority vote accuracy is the probability of having k or more correct votes, averaged over the feature space $\mathbb{R}^n$.

$$P_{\text{maj}} = \int_{\sum \mathcal{I}_i(\mathbf{x}) \geq k} p(\mathbf{x})d\mathbf{x}. \tag{4.A.7}$$

First we note that $P_{\text{maj}} \leq 1$ and then derive a series of inequalities for P_{maj}. For any $\mathbf{x}$ where the majority vote is correct, at least k of the classifiers are correct. Thus,

$$\sum_{i=1}^{L} p_i = \sum_{i=1}^{L} \int_{\mathbb{R}^n} \mathcal{I}_i(\mathbf{x})p(\mathbf{x}) \, d\mathbf{x} = \int_{\mathbb{R}^n} \sum_{i=1}^{L} \mathcal{I}_i(\mathbf{x})p(\mathbf{x}) \, d\mathbf{x}$$

$$\geq \int_{\sum \mathcal{I}_i(\mathbf{x}) \geq k} kp(\mathbf{x})d\mathbf{x} = kP_{\text{maj}}. \tag{4.A.8}$$

Then

$$P_{\text{maj}} \leq \frac{1}{k} \sum_{i=1}^{L} p_i. \tag{4.A.9}$$

Let us now remove the most accurate member of the ensemble, D_L, and consider the remaining $L - 1$ classifiers:

$$\sum_{i=1}^{L-1} p_i = \sum_{i=1}^{L-1} \int_{\mathbb{R}^n} \mathcal{I}_i(\mathbf{x})p(\mathbf{x}) \, d\mathbf{x}. \tag{4.A.10}$$

For each point $\mathbf{x} \in \mathbb{R}^n$ where the majority vote (using the whole ensemble) has been correct, that is, $\sum \mathcal{I}_i(\mathbf{x}) \geq k$, there are now at least $k - 1$ correct individual votes. Thus,

$$\sum_{i=1}^{L-1} p_i = \int_{\mathbb{R}^n} \sum_{i=1}^{L-1} \mathcal{I}_i(\mathbf{x})p(\mathbf{x}) \, d\mathbf{x},$$

$$\geq \int_{\sum_{i=1}^{L} \mathcal{I}_i(\mathbf{x}) \geq k} (k-1)p(\mathbf{x}) \, d\mathbf{x} = (k-1)P_{\text{maj}}. \tag{4.A.11}$$

Then

$$P_{\text{maj}} \le \frac{1}{k-1} \sum_{i=1}^{L-1} p_i. \qquad (4.A.12)$$

Similarly, by dropping from the remaining set the most accurate classifier at a time, we can derive the series of inequalities (Equation 4.A.1). Note that we can remove *any* classifier from the ensemble at a time, not just the most accurate one, and arrive at a similar inequality. Take for example the step where we remove D_L. The choice of the most accurate classifier is dictated by the fact that the remaining ensemble of $L-1$ classifiers will have the smallest sum of the individual accuracies. So as P_{maj} is less than $\frac{1}{2} \sum_{i=1}^{L-1} p_i$, it will be less than any other sum involving $L-1$ classifiers which includes p_L and excludes a smaller p_i from the summation.

The next step is to show that the upper bound is achievable. Matan suggests to use induction on both L and k for that [277].

Lower Bound. To calculate the lower bound, Matan proposes to invert the concept, and look again for the upper bound but of $(1 - P_{\text{maj}})$.

4.A.2   SELECTED MATLAB CODE

```
1  %----------------------------------------------------------------%
2  function oul = nb_combiner(otl,ree,rel)
3  % --- Naive Bayes (NB) combiner for label outputs
4  % Input: ------------------------------------
5  %      otl:   outputs to label
6  %             = array N(objects)-by-L(classifiers)
7  %             entry (i,j) is the label of object i
8  %             by classifier j (integer labels)
9  %      ree:   reference ensemble
10 %             = array M(objects)-by-L(classifiers)
11 %             entry (i,j) is the label of object i
12 %             by classifier j (integer labels)
13 %      rel:   reference labels
14 %             = array M(objects)-by-1
15 %             true labels (integers)
16 % Output:    ------------------------------------
17 %      oul:   output labels
18 %             = array N(objects)-by-1
19 %             assigned labels (integers)
20
21 % Training ------------------------------------
22 c = max(rel); % number of classes, assuming that the
```

```
23  % class labels are integers 1,2,3,...,c
24  L = size(ree,2); % number of classifiers
25
26  for i = 1:c
27      cN(i) = sum(rel == i); % class counts
28  end
29
30  for i = 1:L
31      % cross-tabulate the classes to find the
32      % confusion matrices
33      for j1 = 1:c
34          for j2 = 1:c
35              CM(i).cm(j1,j2) = (sum(rel == j1 & ree(:,i) ...
36                  == j2) + 1/c) / (cN(j1) + 1);
37              % correction for zeros included
38          end
39      end
40  end
41
42  % Operation -------------------------------------
43  N = size(otl,1);
44  oul = zeros(N,1); % pre-allocate for speed
45  for i = 1:N
46      P = cN/numel(rel);
47      for j = 1:c % calculate the score for each class
48          for k = 1:L
49              P(j) = P(j) * CM(k).cm(j,otl(i,k));
50          end
51      end
52      [~,oul(i)] = max(P);
53  end
54  %------------------------------------------------------------------%
```

```
1   %------------------------------------------------------------------%
2   function oul = bks_combiner(otl,ree,rel)
3   % --- BKS combiner for label outputs
4   % Input: --------------------------------------
5   %      otl:  outputs to label
6   %            = array N(objects)-by-L(classifiers)
7   %            entry (i,j) is the label of object i
8   %            by classifier j (integer labels)
9   %      ree:  reference ensemble
10  %            = array M(objects)-by-L(classifiers)
11  %            entry (i,j) is the label of object i
12  %            by classifier j (integer labels)
```

```
13  %       rel:   reference labels
14  %              = array M(objects)-by-1
15  %              true labels (integers)
16  % Output:   ------------------------------------
17  %     oul:   output labels
18  %              = array N(objects)-by-1
19  %              assigned labels (integers)
20
21  N = size(otl,1);
22  M = size(ree,1);
23  largest_class = mode(rel);
24  oul = zeros(N,1); % pre-allocate for speed
25  for i = 1:N
26      matches = sum(ree ~= repmat(otl(i,:),M,1),2) == 0;
27      if sum(matches)
28          oul(i) = mode(rel(matches));
29      else % there is no match in the reference
30          % ensemble output; use the largest prior
31          oul(i) = largest_class;
32      end
33  end
34  %--------------------------------------------------------------------%
```

An example of using the BKS combiner function is shown below. Note that, with a minor edit of lines 31–35, the BKS function can be replaced by the NB combiner function or any other combiner function.

```
1   %--------------------------------------------------------------------%
2   clear all, close all
3   clc
4
5   % Generate and plot the data
6   [~,~,labtrue] = fish_data(50,0);
7   [x,y,lb] = fish_data(50,20); figure, hold on
8   plot(x(lb == 1),y(lb == 1),'k.','markers',14)
9   plot(x(lb == 2),y(lb == 2),'k.','markers',14,...
10      'color',[0.87, 0.87, 0.87])
11  axis([0 1 0 1]) % cut the figure to the unit square
12  axis square off % equalize and remove the axes
13
14  % Generate and plot the ensemble of linear classifiers
15  L = 50; % ensemble size
16  N = numel(x); % number of data points
17  ensemble = zeros(N,L); % pre-allocate for speed
18  for i = 1:L
```

```
19      p = rand(1,2); % random point in the unit square
20      w = randn(1,2); % random normal vector to the line
21      w0 = p * w'; % the free term (neg)
22      plot([0 1],[w0, (w0-w(1))]/w(2),'r-',...
23          'linewidth',1.4) % plot the linear boundary
24      plot(p(1),p(2),'r.','markersize',15)
25      pause(0.08)
26      t = 2  - ([x y] * w' - w0 > 0);
27      if mean(t == lb) < 0.5, t = 3-t; end % revert labels
28      ensemble(:,i) = t; % store output of classifier i
29 end
30 % Find and plot the BKS combiner output
31 output_bks = bks_combiner(ensemble,ensemble,lb);
32 accuracy_bks = mean(output_bks == labtrue);
33 plot(x(output_bks==1),y(output_bks==1),'bo','linewidth',1.5)
34 title(['BKS accuracy ',num2str(accuracy_bks)])
35 %------------------------------------------------------------------%
```

5

COMBINING CONTINUOUS-VALUED OUTPUTS

5.1 DECISION PROFILE

Consider the canonical model of a classifier illustrated in Figure 1.9. The degrees of support for a given input $\mathbf{x}$ can be interpreted in different ways, the two most common being *confidences* in the suggested labels and *estimates of the posterior probabilities* for the classes.

Let $\mathbf{x} \in \mathbb{R}^n$ be a feature vector and $\Omega = \{\omega_1, \omega_2, \dots, \omega_c\}$ be the set of class labels. Each classifier D_i in the ensemble $\mathcal{D} = \{D_1, \dots, D_L\}$ outputs c degrees of support. Without loss of generality we can assume that all c degrees are in the interval $[0, 1]$, that is, $D_i : \mathbb{R}^n \to [0, 1]^c$. Denote by $d_{i,j}(\mathbf{x})$ the support that classifier D_i gives to the hypothesis that $\mathbf{x}$ comes from class ω_j. The larger the support, the more likely the class label ω_j. The L classifier outputs for a particular input $\mathbf{x}$ can be organized in a **decision profile** ($DP(\mathbf{x})$) as the matrix shown in Figure 5.1.

The methods described in this chapter use $DP(\mathbf{x})$ to find the overall support for each class, and subsequently label the input $\mathbf{x}$ in the class with the largest support. There are two general approaches to this task. First, we can use the knowledge that the values in column j are the individual supports for class ω_j and derive an overall support value for that class. Simple algebraic expressions, such as average or product, can be used for this. Alternatively, we may ignore the context of $DP(\mathbf{x})$ and treat the values $d_{i,j}(\mathbf{x})$ as features in a new feature space, which we call the *intermediate feature space*. The final decision is made by another classifier that takes the intermediate feature space as input, and produces a class label (stacked generalization). The important question is

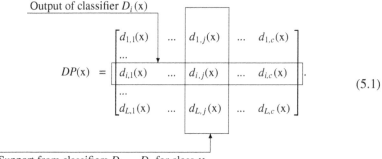

Support from classifiers $D_1 \ldots D_L$ for class ω_j

FIGURE 5.1   Decision profile for an input **x**.

how we train such architectures to make sure that the increased complexity is justified by a corresponding gain in accuracy.

5.2   HOW DO WE GET PROBABILITY OUTPUTS?

Calibrating the classifiers' outputs is important, especially for heterogeneous ensembles [31]. Some of the base classifiers described in Chapter 2 produce soft labels right away. An example of such outputs is the discriminant scores of the linear discriminant classifier (LDC). It is more convenient though if these degrees were in the interval [0, 1], with 0 meaning no support and 1 meaning full support. We can simply normalize the values so that $\mathbb{R}$ is mapped to [0, 1]. In addition, to comply with the probability context, we can scale the degrees of support so that their sum is one. The standard solution to this problem is the *softmax* method [107]. Let $g_1(\mathbf{x}), \ldots, g_c(\mathbf{x})$ be the output of classifier D. Then the new support scores $g_1'(\mathbf{x}), \ldots, g_c'(\mathbf{x}), g_j'(\mathbf{x}) \in [0, 1]$, $\sum_{j=1}^{c} g_j'(\mathbf{x}) = 1$, are obtained as

$$g_j'(\mathbf{x}) = \frac{\exp\left\{g_j(\mathbf{x})\right\}}{\sum_{k=1}^{c} \exp\left\{g_k(\mathbf{x})\right\}}. \tag{5.2}$$

It is desirable that $g_j'(\mathbf{x})$ are credible estimates of the probabilities for the classes given the input **x**. Some ways to obtain continuous outputs as estimates of the posterior probabilities $P(\omega_j|\mathbf{x}), j = 1, \ldots, c$, are detailed below.

5.2.1   Probabilities Based on Discriminant Scores

Consider the LDC and the quadratic discriminant classifiers. These classifiers are optimal for normal class-conditional densities (a seldom valid but very useful assumption). The discriminant function for class ω_i, denoted $g_i(\mathbf{x})$, is derived from

$\log P(\omega_i)p(\mathbf{x}|\omega_i)$ by dropping all the additive terms that do not depend on the class label as shown by Equations 2.8, 2.9, 2.10, and 2.11.

For the LDC, we arrived at Equation 2.11:

$$g_i(\mathbf{x}) = \log(P(\omega_i)) - \frac{1}{2}\boldsymbol{\mu}_i^T \Sigma^{-1} \boldsymbol{\mu}_i + \boldsymbol{\mu}_i^T \Sigma^{-1} \mathbf{x} = w_{i0} + \mathbf{w}_i^T \mathbf{x}. \quad (5.3)$$

Returning the dropped terms into the starting equation, we have

$$\log(P(\omega_i)p(\boldsymbol{\mu}|\omega_i)) = \underbrace{\log(P(\omega_i)) - \frac{1}{2}(\mathbf{x} - \boldsymbol{\mu}_i)^T \Sigma^{-1}(\mathbf{x} - \boldsymbol{\mu}_i)}_{g_i(\mathbf{x})}$$

$$-\frac{n}{2}\log(2\pi) - \frac{1}{2}\log(|\Sigma|). \quad (5.4)$$

Denote by C the constant (possibly depending on $\mathbf{x}$ but not on i) absorbing all dropped additive terms

$$C = -\frac{n}{2}\log(2\pi) - \frac{1}{2}\log(|\Sigma|). \quad (5.5)$$

Then

$$P(\omega_j)p(\mathbf{x}|\omega_j) = \exp(C)\ \exp\{g_j(\mathbf{x})\}. \quad (5.6)$$

The posterior probability for class ω_j for the given $\mathbf{x}$ is

$$P(\omega_j|\mathbf{x}) = \frac{P(\omega_j)p(\mathbf{x}|\omega_j)}{p(\mathbf{x})} \quad (5.7)$$

$$= \frac{\exp(C) \times \exp\{g_j(\mathbf{x})\}}{\sum_{k=1}^{c} \exp(C) \times \exp\{g_k(\mathbf{x})\}} = \frac{\exp\{g_j(\mathbf{x})\}}{\sum_{k=1}^{c} \exp\{g_k(\mathbf{x})\}}, \quad (5.8)$$

which is the softmax transform (Equation 5.2).

For a two-class problem, instead of comparing $g_1(\mathbf{x})$ with $g_2(\mathbf{x})$, we can form a single discriminant function $g(\mathbf{x}) = g_1(\mathbf{x}) - g_2(\mathbf{x})$, which we compare with the threshold 0. In this case,

$$P(\omega_1|\mathbf{x}) = \frac{1}{1 + \exp\{-g(\mathbf{x})\}} \quad (5.9)$$

and

$$P(\omega_2|\mathbf{x}) = 1 - P(\omega_1|\mathbf{x}) = \frac{1}{1 + \exp\{g(\mathbf{x})\}}. \quad (5.10)$$

This is also called the *logistic link function* [310].

Alternatively, we can think of $g(\mathbf{x})$ as a new feature, and estimate the two class-conditional probability density functions (pdf), $p(g(\mathbf{x})|\omega_1)$ and $p(g(\mathbf{x})|\omega_2)$ in this new one-dimensional space. The posterior probabilities are calculated from the Bayes formula. The same approach can be extended to a c-class problem by estimating a class-conditional pdf $p(g_j(\mathbf{x})|\omega_j)$ on $g_j(\mathbf{x})$, $j = 1, \dots, c$, and calculating subsequently $P(\omega_j|\mathbf{x})$.

Next, consider a neural network (NN) with c outputs, each corresponding to a class. Denote the NN output by

$$(y_1, \dots, y_c) \in \mathbb{R}^c$$

and the target by

$$(t_1, \dots, t_c) \in \{0, 1\}^c.$$

The target for an object $\mathbf{z}_j$ from the data set $\mathbf{Z}$ is typically a binary vector with 1 at the position of the class label of $\mathbf{z}_j$ and zeros elsewhere. It is known that, if trained to optimize the squared error between the NN output and the target, in the asymptotic case, the NN output y_j will be an approximation of the posterior probability $P(\omega_j|\mathbf{x})$, $j = 1, \dots, c$ [40, 330]. Wei et al. [414] argue that the theories about the approximation are based on several assumptions that might be violated in practice: (i) that the network is sufficiently complex to model the posterior distribution accurately, (ii) that there are sufficient training data, and (iii) that the optimization routine is capable of finding the global minimum of the error function. The typical transformation which forms a probability distribution from $(y_1, \dots, y_L)$ is the softmax transformation (Equation 5.2) [107]. Wei et al. [414] suggest a histogram-based remapping function. The parameters of this function are tuned separately from the NN training. In the operation phase, the NN output is fed to the remapping function and calibrated to give more adequate posterior probabilities.

■ Example 5.1 SVM output calibration

Figure 5.2a shows a two-dimensional (2D) data set with two classes plotted with different markers. Each class contains 3000 data points. A training set of 120 points was randomly sampled. The training set is marked on the plot with thicker and brighter markers. The SVM classifier was trained using this training data. The classification boundary is shown on the scatterplot.

To examine how accurately the calibrated SVM output matches the posterior probabilities, the whole data set of 6000 objects was fed to the trained SVM classifier and the outputs were calibrated into posterior probabilities using Equation 5.10. Next, a histogram with 100 bins was created and the 6000 probability estimates for class 1 were distributed in the respective bins. The true class labels of the objects were recovered and used to calculate the probability for class 1 in each bin. Figure 5.2b shows the nearly perfect correspondence between the SVM probabilities and the probabilities calculated from the data labels.

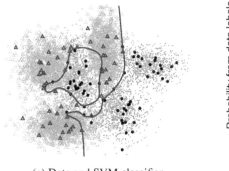
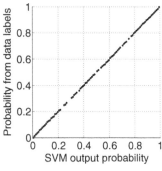

(a) Data and SVM classifier   (b) SVM output probability

FIGURE 5.2 Calibrated output for the SVM classifier on a two-dimensional two-class data set.

5.2.2 Probabilities Based on Counts: Laplace Estimator

Consider finding probability estimates from decision trees. Each leaf of the tree defines a set of posterior probabilities. These are assigned to any point that reaches the leaf. Provost and Domingos [314] analyze the reasons for the insufficient capability of the standard decision tree classifiers to provide adequate estimates of the probabilities and conclude that the very heuristics that help us build small and accurate trees are responsible for that. Special amendments were proposed which led to the so-called *probability estimating trees* (PETs). These trees still have high classification accuracy but their main purpose is to give more accurate estimates of the posterior probabilities.

We calculate estimates of $P(\omega_j|\mathbf{x})$, $j = 1, \ldots, c$, as the class proportions of the training data points that reached the leaf (the *maximum likelihood* (ML) estimates). Let $k_1, \ldots, k_c$ be the number of training points from classes $\omega_1, \ldots, \omega_c$, respectively, at some leaf node t, and let $K = k_1 + \cdots + k_c$. The ML estimates are

$$\hat{P}(\omega_j|\mathbf{x}) = \frac{k_j}{K}, \quad j = 1, \ldots, c. \tag{5.11}$$

The problem is that when the total number of points, K, is small, the estimates of these probabilities are unreliable. Besides, the tree-growing strategies try to make the leaves as pure as possible. Thus, most probability estimates will be pushed toward 1 and 0 [430].

To remedy this, the *Laplace estimate* or *Laplace correction* can be applied [314, 430, 384]. The idea is to adjust the estimates so that they are less extreme. For c classes, the Laplace estimate used in [314] is

$$\hat{P}(\omega_j|\mathbf{x}) = \frac{k_j + 1}{K + c}. \tag{5.12}$$

Zadrozny and Elkan [430] apply a different version of the Laplace estimate using a parameter m which controls the degree of regularization of the estimate (called *m-estimation*). The idea is to smooth the posterior probability toward the (estimate of the) prior probability for the class

$$\hat{P}(\omega_j|\mathbf{x}) = \frac{k_j + m \times \hat{P}(\omega_j)}{K + m}.$$

(5.13)

If m is large, then the estimate is close to the prior probability. If $m = 0$, then we have the ML estimates and no regularization. Zadrozny and Elkan suggest that m should be chosen so that $m \times P(\omega_j) \approx 10$ and also point out that practice has shown that the estimate 5.13 is quite robust with respect to the choice of m.

Suppose that ω^* is the majority class at node t. Ting and Witten [384] propose the following version of the Laplace estimator:

$$\hat{P}(\omega_j|\mathbf{x}) = \begin{cases} 1 - \frac{\sum_{l \neq j} k_l + 1}{K+2}, & \text{if } \omega_j = \omega^* , \\ (1 - \hat{P}(\omega^*|\mathbf{x})) \times \frac{k_j}{\sum_{l \neq j} k_l}, & \text{otherwise.} \end{cases}$$

(5.14)

The general consensus in the PET studies is that for good estimates of the posterior probabilities, the tree should be grown without pruning, and a form of the Laplace correction should be used for calculating the probabilities.

The same argument can be applied for smoothing the estimates of the k nearest neighbor classifier (k-nn) discussed in Chapter 2. There are many weighted versions of k-nn whereby the posterior probabilities are calculated using distances. While the distance-weighted versions have been found to be asymptotically equivalent to the nonweighted versions in terms of classification accuracy [24], there is no such argument when class ranking is considered. It is possible that the estimates of the soft k-nn versions are more useful for ranking than for labeling. A simple way to derive $\hat{P}(\omega_j|\mathbf{x})$ from k-nn is to average the similarities between $\mathbf{x}$ and its nearest neighbors from class ω_j. Let k be the number of neighbors, $\mathbf{x}^{(i)}$ be the ith nearest neighbor of $\mathbf{x}$, and $d(\mathbf{x}, \mathbf{x}^{(i)})$ be the distance between $\mathbf{x}$ and $\mathbf{x}^{(i)}$. Then

$$\hat{P}(\omega_j|\mathbf{x}) = \frac{\sum_{\mathbf{x}^{(j)} \in \omega_j} \frac{1}{d(\mathbf{x}, \mathbf{x}^{(j)})}}{\sum_{i=1}^{k} \frac{1}{d(\mathbf{x}, \mathbf{x}^{(i)})}}.$$

(5.15)

Albeit intuitive, these estimates are not guaranteed to be good approximations of the posterior probabilities.

▣ Example 5.2   Laplace corrections and soft k-nn

Figure 5.3 shows a point in a 2D feature space (the cross, $\mathbf{x}$) and its seven nearest neighbors from ω_1 (open circles), ω_2 (bullets), and ω_3 (triangle).

The Euclidean distances between $\mathbf{x}$ and its neighbors are as follows:

$\mathbf{x}$	1	2	3	4	5	6	7
Distance	1	1	$\sqrt{2}$	2	$2\sqrt{2}$	$2\sqrt{2}$	$\sqrt{13}$
Label	ω_2	ω_1	ω_1	ω_3	ω_1	ω_1	ω_2

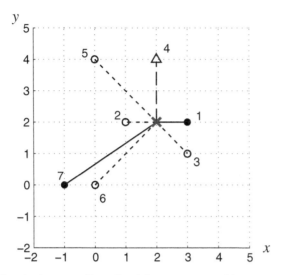

FIGURE 5.3 A point in a two-dimensional feature space and its seven nearest neighbors from ω_1 (open circles), ω_2 (bullets), and ω_3 (triangle).

TABLE 5.1 Probability Estimates for the Example in Figure 5.3 Using the Laplace Corrections and Distance-based k-nn

Method	$\mu_1(\mathbf{x})$	$\mu_2(\mathbf{x})$	$\mu_3(\mathbf{x})$
ML	$\frac{4}{7} = 0.571$	$\frac{2}{7} = 0.286$	$\frac{1}{7} = 0.143$
Standard Laplace [314]	$\frac{5}{10} = 0.500$	$\frac{3}{10} = 0.300$	$\frac{2}{10} = 0.200$
m-estimation [430]	$\frac{8}{19} = 0.421$	$\frac{6}{19} = 0.316$	$\frac{5}{19} = 0.263$
($m = 12$, equiprobable classes) Ting and Witten [384]	$\frac{12}{27} = 0.444$	$\frac{10}{27} = 0.370$	$\frac{5}{27} = 0.185$
Distance based	0.576	0.290	0.134

The probability estimates $\mu_j(\mathbf{x})$, $j = 1, 2, 3$, using the Laplace corrections and the distance-based formula are shown in Table 5.1. As seen in the table, all the corrective modifications of the estimates bring them closer to one another compared to the standard ML estimates, that is, the modifications smooth the estimates away from the 0/1 bounds.

Accurate estimates are a sufficient but not a necessary condition for a high classification accuracy. The final class label will be correctly assigned as long as the degree of support for the correct class label exceeds the degrees for the other classes. Investing effort into refining the probability estimates will be justified in problems with a large number of classes c, where the ranking of the classes by their likelihood is more important than identifying just one winning label. Examples of such tasks are person identification, text categorization, and fraud detection.

5.3   NONTRAINABLE (FIXED) COMBINATION RULES

5.3.1   A Generic Formulation

Simple fusion methods are the most obvious choice when constructing a multiple classifier system [207, 237, 382, 394, 399]. A degree of support for class ω_j is calculated from the L entries in the jth column of $DP(\mathbf{x})$

$$\mu_j(\mathbf{x}) = \mathcal{F}\left(d_{1,j}(\mathbf{x}), \dots d_{L,j}(\mathbf{x})\right), \tag{5.16}$$

where $\mathcal{F}$ is a combination function. The class label of $\mathbf{x}$ is found as the index of the maximum $\mu_j(\mathbf{x})$. $\mathcal{F}$ can be chosen in many different ways:

- Average (Sum)

$$\mu_j(\mathbf{x}) = \frac{1}{L} \sum_{i=1}^{L} d_{i,j}(\mathbf{x}). \tag{5.17}$$

- Minimum/maximum/median combiner, for example,

$$\mu_j(\mathbf{x}) = \max_i \{d_{i,j}(\mathbf{x})\}. \tag{5.18}$$

- Trimmed mean combiner (competition jury). For a $K\%$ trimmed mean the L degrees of support are sorted and $\frac{K}{2}\%$ of the values are dropped on each side. The overall support $\mu_j(\mathbf{x})$ is found as the mean of the remaining degrees of support.
- Product combiner

$$\mu_j(\mathbf{x}) = \prod_{i=1}^{L} d_{i,j}(\mathbf{x}). \tag{5.19}$$

AVERAGE COMBINER

Training: None

Operation: For each new object
1. Classify the new object $\mathbf{x}$ to find its decision profile $DP(\mathbf{x})$, as in Equation 5.1.
2. Calculate the support for each class by

$$P(k) = \frac{1}{L} \sum_{i=1}^{L} d_{j,i}, \quad k = 1, \dots, c.$$

3. Assign label k^* to the object, where

$$k^* = \arg\max_{k=1}^{c} P(k).$$

Return the ensemble label of the new object.

FIGURE 5.4   Training and operation algorithm for the average combiner.

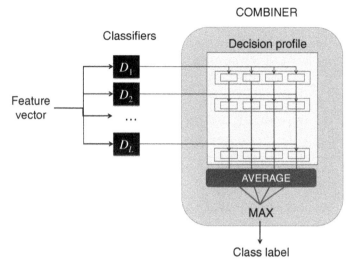

FIGURE 5.5   Operation of the average combiner.

Represented by the average combiner, the category of simple nontrainable combiners is described in Figure 5.4, and illustrated diagrammatically in Figure 5.5. These combiners are called nontrainable, because once the individual classifiers are trained, their outputs can be fused to produce an ensemble decision, without any further training.

■ Example 5.3   Simple nontrainable combiners
The following example helps to clarify simple combiners. Let $c = 3$ and $L = 5$. Assume that for a certain **x**

$$DP(\mathbf{x}) = \begin{bmatrix} 0.1 & 0.5 & 0.4 \\ 0.0 & 0.0 & 1.0 \\ 0.4 & 0.3 & 0.4 \\ 0.2 & 0.7 & 0.1 \\ 0.1 & 0.8 & 0.2 \end{bmatrix}. \tag{5.20}$$

Applying the simple combiners column wise, we obtain:

Combiner	$\mu_1(\mathbf{x})$	$\mu_2(\mathbf{x})$	$\mu_3(\mathbf{x})$
Average	0.16	0.46	0.42
Minimum	0.00	0.00	0.10
Maximum	0.40	0.80	1.00
Median	0.10	0.50	0.40
40% trimmed mean	0.13	0.50	0.33
Product	0.00	0.00	0.0032

Note that we do not require that $d_{i,j}(\mathbf{x})$ for classifier D_i sum up to one. We only assume that they are measured in the same units. If we take the class with the maximum support to be the class label of $\mathbf{x}$, the minimum, maximum, and product will label $\mathbf{x}$ in class ω_3, whereas the average, the median, and the trimmed mean will put $\mathbf{x}$ in class ω_2.

5.3.2   Equivalence of Simple Combination Rules

5.3.2.1   *Equivalence of MINIMUM and MAXIMUM Combiners for Two Classes*
Let $D = \{D_1, \ldots, D_L\}$ be the classifier ensemble and $\Omega = \{\omega_1, \omega_2\}$ be the set of class labels. The individual outputs are estimates of the posterior probabilities. The output $d_{i,j}$ of classifier D_i (supporting the hypothesis that $\mathbf{x}$ comes from class ω_j) is an estimate of $P(\omega_j|\mathbf{x})$, $j = 1, 2$. Here we prove that the minimum and the maximum combiners are equivalent for $c = 2$ classes and any number of classifiers L, provided the two outputs from each classifier satisfy

$$\hat{P}(\omega_1|\mathbf{x}) + \hat{P}(\omega_2|\mathbf{x}) = 1.$$

This equivalence means that the class label assigned by the minimum and the maximum combiners will be the same. In case of a tie for one of the rules, there will be a tie for the other rule as well, and any of the two class labels could be assigned in both cases.

Proposition 5.1   Let $a_1, \ldots, a_L$ be the L outputs for class ω_1, and $1 - a_1, \ldots, 1 - a_L$ be the L outputs for class ω_2, $a_i \in [0, 1]$. Then the class label assigned to $\mathbf{x}$ by the MINIMUM and MAXIMUM combination rules is the same.

Proof.   Without loss of generality assume that $a_1 = \min_i a_i$, and $a_L = \max_i a_i$. Then the minimum combination rule will pick a_1 and $1 - a_L$ as the support for ω_1 and ω_2, respectively, and the maximum rule will pick a_L and $1 - a_1$. Consider the three possible relationships between a_1 and $1 - a_L$.

(a)   If $a_1 > 1 - a_L$ then $a_L > 1 - a_1$, and the selected class is ω_1
   with both methods.

(b)   If $a_1 < 1 - a_L$ then $a_L < 1 - a_1$, and the selected class is ω_2
   with both methods.

(c)   If $a_1 = 1 - a_L$ then $a_L = 1 - a_1$, and we will pick a class at random
   with both methods.

Note: A discrepancy between the error rates of the two combination methods might occur in numerical experiments due to the random tie break in (c). If we agree to always assign class ω_1 when the support for the two classes is the same (a perfectly justifiable choice), the results for the two methods will coincide.   ∎

5.3.2.2   Equivalence of MAJORITY VOTE and MEDIAN Combiner for Two Classes and Odd L   Consider again the case of two classes, and L classifiers with outputs for a certain $\mathbf{x}$, $a_1, \ldots, a_L$, for class ω_1, and $1 - a_1, \ldots, 1 - a_L$, for class ω_2, where L is odd.

Proposition 5.2   The class label assigned to $\mathbf{x}$ by the MAJORITY VOTE rule and MEDIAN combination rule is the same.

Proof.   Assume again that $a_1 = \min_i a_i$, and $a_L = \max_i a_i$. Consider the median rule first. The median of the outputs for class ω_1 is $a_{\frac{L+1}{2}}$.

(a) If $a_{\frac{L+1}{2}} > 0.5$, then the median of the outputs for ω_2, $1 - a_{\frac{L+1}{2}} < 0.5$, and class ω_1 will be assigned. The fact that $a_{\frac{L+1}{2}} > 0.5$ means that all $a_{\frac{L+1}{2}+1}, \ldots, a_L$ are strictly greater than 0.5. This makes at least $\frac{L+1}{2}$ posterior probabilities for ω_1 greater than 0.5, which, when "hardened," will give label ω_1. Then the majority vote rule will assign to $\mathbf{x}$ class label ω_1.

(b) Alternatively, if $a_{\frac{L+1}{2}} < 0.5$, then $1 - a_{\frac{L+1}{2}} > 0.5$, and class ω_2 will be assigned by the median rule. In this case, at least $\frac{L+1}{2}$ posterior probabilities for ω_2 are greater than 0.5, and the majority vote rule will assign label ω_2 as well.

(c) For $a_{\frac{L+1}{2}} = 0.5$ a tie occurs, and any of the two labels can be assigned by the median rule. The same applies for the majority vote, as all the soft votes at 0.5 (same for both classes) can be "hardened" to any of the two class labels.

Again, a difference in the estimated errors of the two methods might occur in experiments due to the arbitrary "hardening" of label 0.5. For example, if we agree to always assign class ω_1 when the posterior probabilities are both 0.5, the results for the two methods will coincide. ∎

5.3.3   Generalized Mean Combiner

The *generalized mean* [105] is a useful aggregation formula governed by a parameter. Applied in the classifier combination context, the ensemble output for class ω_j is

$$\mu_j(\mathbf{x}, \alpha) = \left(\frac{1}{L} \sum_{i=1}^{L} d_{i,j}(\mathbf{x})^\alpha \right)^{\frac{1}{\alpha}}, \tag{5.21}$$

where α is the parameter. Some special cases of the generalized mean are shown in Table 5.2.

Observe that the geometric mean is equivalent to the product combiner. Raising to the power of $\frac{1}{L}$ is a monotonic transformation which does not depend on the class label j, and therefore will not change the order of $\mu_j(\mathbf{x})$s. Hence the winning label

TABLE 5.2 Special Cases of the Generalized Mean

$\alpha \to -\infty$	$\Rightarrow$	$\mu_j(\mathbf{x}, \alpha) = \min_i\{d_{i,j}(\mathbf{x})\}$	minimum
$\alpha = -1$	$\Rightarrow$	$\mu_j(\mathbf{x}, \alpha) = \left(\frac{1}{L}\sum_{i=1}^{L}\frac{1}{d_{i,j}(\mathbf{x})}\right)^{-1}$	harmonic mean
$\alpha \to 0$	$\Rightarrow$	$\mu_j(\mathbf{x}, \alpha) = \left(\prod_{i=1}^{L} d_{i,j}(\mathbf{x})\right)^{1/L}$	geometric mean
$\alpha = 1$	$\Rightarrow$	$\mu_j(\mathbf{x}, \alpha) = \frac{1}{L}\sum_{i=1}^{L} d_{i,j}(\mathbf{x})$	arithmetic mean
$\alpha \to \infty$	$\Rightarrow$	$\mu_j(\mathbf{x}, \alpha) = \max_i\{d_{i,j}(\mathbf{x})\}$	maximum

obtained from the product combiner will be the same as the winning label from the geometric mean combiner.

As we are considering nontrainable combiners here, we assume that the system designer chooses α beforehand. This parameter can be thought of as the "level of optimism" of the combiner. The *minimum* combiner ($\alpha \to -\infty$) is the most pessimistic choice. With this combiner, we know that ω_j is supported by *all* members of the ensemble at least as much as $\mu_j(\mathbf{x})$. At the other extreme, *maximum* is the most optimistic combiner. Here we accept an ensemble degree of support $\mu_j(\mathbf{x})$ on the ground that *at least one* member of the team supports ω_j with this degree. If we choose to tune α with respect to the ensemble performance, then we should regard the generalized mean combiner as a trainable combiner as discussed later. The generalized mean combiner is detailed in Figure 5.6.

GENERALIZED MEAN COMBINER

Training: Choose the level of optimism α (see Table 5.2).

Operation: For each new object:
1. Classify the new object $\mathbf{x}$ to find its decision profile $DP(\mathbf{x})$, as in Equation 5.1.
2. Calculate the support for each class by

$$P(k) = \left(\frac{1}{L}\sum_{i=1}^{L} d_{i,k}(\mathbf{x})^\alpha\right)^{\frac{1}{\alpha}}, \quad k = 1, \ldots, c.$$

3. Assign label k^* to the object, where

$$k^* = \arg\max_{k=1}^{c} P(k).$$

Return the ensemble label of the new object.

FIGURE 5.6 Training and operation algorithm for the generalized mean combiner.

▣ Example 5.4   Effect of the level of optimism α.

To illustrate the effect of the level of optimism α we used the 2D rotated checker board data set. Examples of a training and a testing data set are shown in Figure 5.7a and 5.7b, respectively.

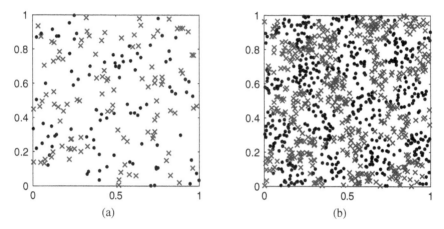

(a)   (b)

FIGURE 5.7   An example of a training (a) and a testing (b) set for the rotated checker board data. One hundred randomly sampled training/testing sets were used in the experiment.

One hundred training/testing sets were generated from a uniform distribution within the unit square. The labels of the points were assigned as in the rotated checker board example. In each experiment, the training set consisted of 200 examples and the testing set consisted of 1000 examples. Each ensemble was formed by taking 10 bootstrap samples of size 200 from the training data (uniform sampling with replacement) and training a classifier on each sample. We chose SVM as the base ensemble classifier.[1] A Gaussian kernel with spread $\sigma = 0.3$ was applied, with a penalizing constant $C = 50$. The generalized mean formula (5.21) was used, where the level of optimism α was varied from -50 to 50 with finer discretization from -1 to 1. The ensemble error, averaged across the 100 runs, is plotted against α in Figure 5.8a.

A zoom window of the ensemble error for $\alpha \in [-2, 5]$ is shown in Figure 5.8b. The average, product, and harmonic mean combiners are identified on the curve. For this example, the average combiner gave the best result.

The results from the illustration above should not be taken as evidence that the average combiner is always the best. The shape of the curve will depend heavily on the problem and on the base classifier used. The average and the product are the two most popular combiners. Yet, there is no guideline as to which one is better for a specific problem. The current understanding is that the average, in general, might be less accurate than the product for some problems, but is the more stable of the two [8, 109, 207, 281, 382, 383].

---

[1]The version we used is the SVM implementation within the Bioinformatics toolbox of MATLAB.

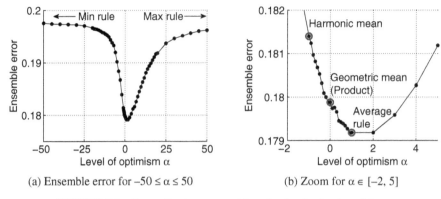

(a) Ensemble error for $-50 \leq \alpha \leq 50$ (b) Zoom for $\alpha \in [-2, 5]$

FIGURE 5.8   Generalized mean combiner for the checker board data.

Even though Figure 5.8 shows a clear trend, the actual difference between the ensemble classification errors for the different combiners is negligible. Much more accuracy can be gained (or lost) by changing the width of the Gaussian kernel σ, or the penalizing constant C.

5.3.4   A Theoretical Comparison of Simple Combiners

Can we single out a combiner that performs best in a simple scenario? Consider the following set-up [9, 235]:

- There are only two classes, $\Omega = \{\omega_1, \omega_2\}$.
- All classifiers produce soft class labels, $d_{j,i}(\mathbf{x}) \in [0, 1]$, $i = 1, 2$, $j = 1, \dots, L$, where $d_{j,i}(\mathbf{x})$ is an estimate of the posterior probability $P(\omega_i|\mathbf{x})$ by classifier D_j for an input $\mathbf{x} \in \mathbb{R}^n$. We consider the case where for any $\mathbf{x}$, $d_{j,1}(\mathbf{x}) + d_{j,2}(\mathbf{x}) = 1$, $j = 1, \dots, L$.
- Let $\mathbf{x} \in \mathbb{R}^n$ be a data point to classify. Without loss of generality, we assume that the true posterior probability is $P(\omega_1|\mathbf{x}) = p > 0.5$. Thus, the Bayes-optimal class label for $\mathbf{x}$ is ω_1, and a classification error occurs if label ω_2 is assigned.

Assumption.   The classifiers commit independent and identically distributed errors in estimating $P(\omega_1|\mathbf{x})$ such that

$$d_{j,1}(\mathbf{x}) = P(\omega_1|\mathbf{x}) + \eta(\mathbf{x}) = p + \eta(\mathbf{x}), \tag{5.22}$$

and respectively

$$d_{j,2}(\mathbf{x}) = 1 - p - \eta(\mathbf{x}), \tag{5.23}$$

where $\eta(\mathbf{x})$ has

   (i) a normal distribution with mean 0 and variance σ^2 (we take σ to vary between 0.1 and 1) and

   (ii) a uniform distribution spanning the interval $[-b, +b]$ (b varies from 0.1 to 1).

Thus $d_{j,1}(\mathbf{x})$ is a random variable with normal or uniform distribution and so is $d_{j,2}(\mathbf{x})$.

The combiners we compare are minimum (same as maximum), average, and median combiners [235] (Equation 5.16). As the median and the majority vote combiners are also identical for two classes, the comparison includes majority vote as well. Finally, we include the individual classification rate and an "oracle" combiner which predicts the correct class label if at least one classifier in the ensemble gives a correct prediction. The performance of the combination rules is expected to be better than that of the individual classifier but worse than that of the oracle.

Table 5.3 shows the analytical expressions of the probability of error for the combiners under the normal distribution assumption, and Table 5.4, for the uniform distribution. The derivations of the expressions are shown in Appendix 5.A.1. As explained in the appendix, there is no closed-form expression for the Min/Max combiner for the normal distribution of the estimation error, so these combiners were taken into the comparison only for the uniform distribution.

Figures 5.9 and 5.10 show the ensemble error rate for the normal and uniform distributions, respectively, as a function of two arguments: the true posterior probability $P(\omega_1 | \mathbf{x}) = p$ and the parameter of the distribution. For the normal distribution (Figure 5.9), σ took values from 0.1 to 1, and for the uniform

TABLE 5.3 **The Theoretical Error P_e for the Single Classifier and the Six Fusion Methods for the Normal Distribution**

Method	Ensemble error rate, P_e
Single classifier	$\Phi\left(\dfrac{0.5 - p}{\sigma}\right)$ (Individual error rate)
Min/Max	—
Average (Sum)	$\Phi\left(\dfrac{\sqrt{L}(0.5 - p)}{\sigma}\right)$
Median/Majority	$\displaystyle\sum_{j=\frac{L+1}{2}}^{L} \binom{L}{j} \times \Phi\left(\dfrac{0.5 - p}{\sigma}\right)^j \times \left[1 - \Phi\left(\dfrac{0.5 - p}{\sigma}\right)\right]^{L-j}$
Oracle	$\Phi\left(\dfrac{0.5 - p}{\sigma}\right)^L$

Notes:
L is the number of classifiers in the ensemble;
p is the true posterior probability $P(\omega_1 | \mathbf{x})$ for class ω_1 for the given object $\mathbf{x}$;
$\Phi(.)$ is the cumulative distribution function for the standard normal distribution $N(0, 1)$.

TABLE 5.4   The Theoretical Error P_e for the Single Classifier and the Six Fusion Methods for the Uniform Distribution $(p - b < 0.5)$

Method	Ensemble error rate, P_e
Single classifier	$\dfrac{0.5 - p + b}{2b}$ (Individual error rate.)
Min/Max	$\dfrac{1}{2}\left(\dfrac{1 - 2p}{2b} + 1\right)^L$
Average	$\Phi\left(\dfrac{\sqrt{3L}(0.5 - p)}{b}\right)$
Median/Majority	$\displaystyle\sum_{j=\frac{L+1}{2}}^{L}\binom{L}{j} \times \left(\dfrac{0.5 - p + b}{2b}\right)^j \times \left[1 - \dfrac{0.5 - p + b}{2b}\right]^{L-j}$
Oracle	$\left(\dfrac{0.5 - p + b}{2b}\right)^L$

Notes:

L is the number of classifiers in the ensemble;

p is the true posterior probability $P(\omega_1|\mathbf{x})$ for class ω_1 for the given object $\mathbf{x}$;

$\Phi(.)$ is the cumulative distribution function for the standard normal distribution $N(0, 1)$.

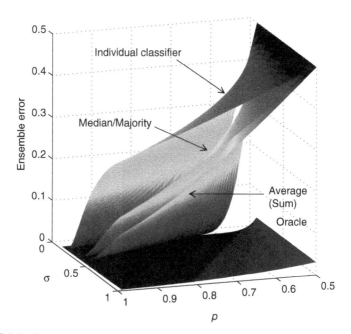

FIGURE 5.9   Ensemble error rate of the individual classifier and the simple combiners for normal distribution of the estimation error η.

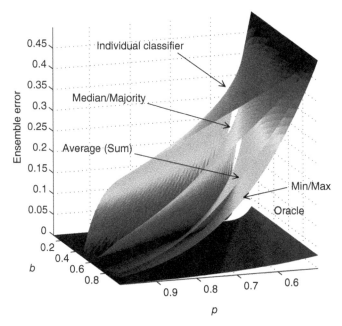

FIGURE 5.10 Ensemble error rate of the individual classifier and the simple combiners for uniform distribution of the estimation error η.

distribution, b took values from 0.1 to 1, ensuring that $p - b < 0.5$. The posterior probability p was varied from 0.5 to 1 for both figures. The ensemble size for this example was $L = 5$ classifiers.

The surfaces in both figures are clearly layered beneath one another. Expectedly, the top surface (largest error) is the individual classifier while the bottom layer (smallest error) is the oracle combiner. Further on, when p is close to 0.5, the Bayes error is high, and so is the ensemble error. The ensemble error goes down to 0 for a higher p and a lower variability of the estimate (low σ and low b), and does so quicker for the better combiners.

The average and the median/vote methods have a closer performance for normally distributed than for the uniformly distributed η, the average being the better of the two. Finally, for the uniform distribution, the average combiner is outperformed by the minimum/maximum combiner.

Figure 5.11 shows the behavior of the combiners as a function of the ensemble size L. We chose a fairly difficult problem where the true posterior probability is 0.55 (high uncertainty), and the spread parameter of the distributions is large ($\sigma = 0.9$ for the normal distribution, and $b = 0.9$ for the uniform distribution). The figure confirms that the above observations hold for any number of classifiers. It also indicates that larger ensembles secure a smaller classification error, and amplify the performance differences of the combiners. Even though this analysis is based on assumptions and theory, it suggests that checking several combiners for a set of trained classifiers may pay off.

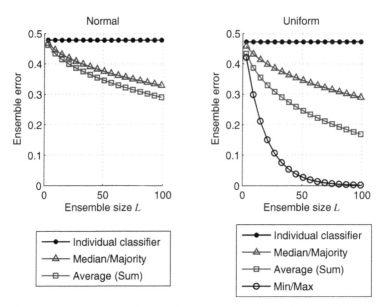

FIGURE 5.11 Ensemble error for the simple combiners as a function of the ensemble size L.

Similar analyses can be carried out for distributions other than normal or uniform. Kittler and Alkoot [206], Chen and Cheng [71], and Cabrera [62] studied the behavior of various combiners for nonnormal distributions and large ensemble sizes. The conclusion is that for nonsymmetrical, bimodal, or heavy-tailed distributions, the combiners may have very different performances. Even though the assumptions may not hold in real-life problems, these analyses suggest that choosing a suitable combiner is important.

We should be aware of the following caveat. Although the combiners in this section are considered nontrainable, any comparison for the purpose of picking one among them is, in fact, a form of training. Choosing a combiner is the same as tuning the level of optimism α of the generalized mean combiner. We look at the question "to train or not to train" later in this chapter.

5.3.5 Where Do They Come From?

5.3.5.1 Intuition and Common Sense Many simple combiners come from intuition and common sense. Figure 5.12 shows an example. Suppose that we want to bet on a horse, and have three choices. One of the horses will win the race, and the classification task is to predict which horse. The only information we have access to is the opinions of four friends. Each friend offers a guess of the probability for each horse to win the race. The friends are the classifiers in the ensemble, and the probabilities they predicted are arranged in a decision profile, as shown in the figure. The decision maker has to decide which combination rule to apply. If there

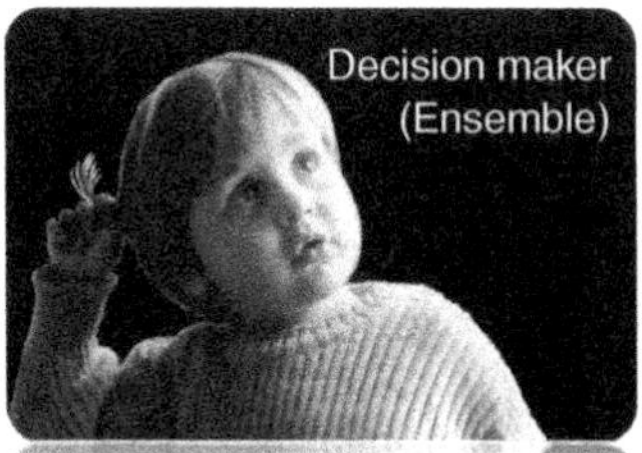

Which horse shall I put my money on?

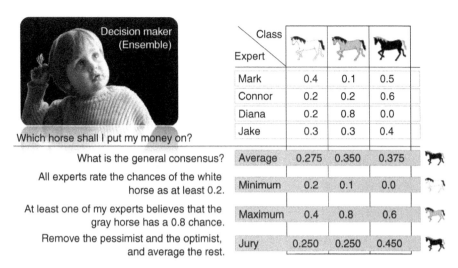

	Class / Expert			
	Mark	0.4	0.1	0.5
	Connor	0.2	0.2	0.6
	Diana	0.2	0.8	0.0
	Jake	0.3	0.3	0.4
What is the general consensus?	Average	0.275	0.350	0.375
All experts rate the chances of the white horse as at least 0.2.	Minimum	0.2	0.1	0.0
At least one of my experts believes that the gray horse has a 0.8 chance.	Maximum	0.4	0.8	0.6
Remove the pessimist and the optimist, and average the rest.	Jury	0.250	0.250	0.450

FIGURE 5.12 Simple combiners come from intuition and common sense.

is no further information about how accurate the predictions might be, the decision maker can choose their level of optimism, and pick the respective combiner. Take the over-conservative minimum combiner, for example. Given a horse, the support for this "class," denoted μ, can be interpreted in the following way. All experts agree to at least a degree μ that the horse will win. It makes sense therefore, to choose the horse with the largest μ. The opposite strategy is to choose the horse that achieved the highest degree of support among all horses and all experts. In this case, there is at least one expert that believes in this horse with a degree this high. Both minimum and maximum combiners disregard the consensus opinion. Conversely, the average (sum), median, and the jury combiners measure a central tendency of the support for the classes. The decision maker might reach a different conclusion depending on the combination rule they apply. In the absence of a ground truth, we cannot judge whether the decision was right. This example merely demonstrates the flexibility of the simple combiners.

Interestingly, many simple combiners can be derived as the *optimal* combiner under various scenarios and assumptions.

5.3.5.2 Conditional Independence of Features We can regard $d_{i,j}(\mathbf{x})$ as an estimate of the posterior probability $P(\omega_j|\mathbf{x})$ produced by classifier D_i. Finding an optimal (in the Bayesian sense) combination of these estimates is not straightforward. Here we give a brief account of some of the theories underpinning the most common simple combiners.

Consider L different conditionally independent feature subsets. Each subset generates a part of the feature vector, $\mathbf{x}^{(i)}$, so that $\mathbf{x} = [\mathbf{x}^{(1)}, \ldots, \mathbf{x}^{(L)}]^T$, $\mathbf{x} \in \mathbb{R}^n$. For example, suppose that there are $n = 10$ features altogether and these are grouped in the following $L = 3$ conditionally independent subsets $(1,4,8)$, $(2,3,5,6,10)$, and $(7,9)$. Then

$$\mathbf{x}^{(1)} = [x_1, x_4, x_8]^T, \quad \mathbf{x}^{(2)} = [x_2, x_3, x_4, x_6, x_{10}]^T, \quad \mathbf{x}^{(3)} = [x_7, x_9]^T.$$

From the assumed independence, the class-conditional pdf for class ω_j is a product of the class-conditional pdf on each feature subset (representation):

$$p(\mathbf{x}|\omega_j) = \prod_{i=1}^{L} p(\mathbf{x}^{(i)}|\omega_j). \tag{5.24}$$

Deriving the product rule weighted by the prior probabilities as the optimal combiner for this case is straightforward [44, 231, 256]. The jth output of classifier D_i is an estimate of the probability

$$P(\omega_j|\mathbf{x}^{(i)}, D_i) = \frac{P(\omega_j)p(\mathbf{x}^{(i)}|\omega_j)}{p(\mathbf{x}^{(i)})}, \tag{5.25}$$

hence

$$p(\mathbf{x}^{(i)}|\omega_j) = \frac{P(\omega_j|\mathbf{x}^{(i)})p(\mathbf{x}^{(i)})}{P(\omega_j)}. \tag{5.26}$$

The posterior probability using the whole of $\mathbf{x}$ is

$$P(\omega_j|\mathbf{x}) = \frac{P(\omega_j)p(\mathbf{x}|\omega_j)}{p(\mathbf{x})} = \frac{P(\omega_j)}{p(\mathbf{x})} \prod_{i=1}^{L} p(\mathbf{x}^{(i)}|\omega_j). \tag{5.27}$$

Substituting Equation 5.26 into Equation 5.27,

$$P(\omega_j|\mathbf{x}) = P(\omega_j)^{(1-L)} \prod_{i=1}^{L} P(\omega_j|\mathbf{x}^{(i)}) \times \frac{\prod_{i=1}^{L} p(\mathbf{x}^{(i)})}{p(\mathbf{x})}. \tag{5.28}$$

The fraction at the end does not depend on the class label k therefore we can ignore it when calculating the support $\mu_j(\mathbf{x})$ for class ω_j. Taking the classifier output $d_{i,k}(\mathbf{x}^{(i)})$ as the estimate of $P(\omega_j|\mathbf{x}^{(i)})$ and estimating the prior probabilities for the classes from the data, the support for ω_j is calculated as the product combination rule

$$P(\omega_j|\mathbf{x}) \propto P(\omega_j)^{(1-L)} \prod_{i=1}^{L} P(\omega_j|\mathbf{x}^{(i)}) \tag{5.29}$$

$$= \hat{P}(\omega_j)^{(1-L)} \prod_{i=1}^{L} d_{i,k}(\mathbf{x}^{(i)}) = \mu_j(\mathbf{x}). \tag{5.30}$$

Kittler et al. [207] take this formula further to derive the average combiner. They investigate the error sensitivity of the two combiners and show that the average combiner is much more resilient to estimation errors of the posterior probabilities than the product combiner. The product combiner is over-sensitive to estimates close to zero. The presence of such estimates has the effect of veto on that particular class regardless of how large some of the estimates of other classifiers might be.

5.3.5.3 Kullback–Leibler Divergence

Miller and Yan [281] offer a theoretical framework for the average and product combiners based on the *Kullback–Leibler divergence* (KL). KL divergence measures the distance between two probability distributions, q (prior distribution) and p (posterior distribution). KL divergence is also called "relative entropy" or "cross-entropy," denoted by $KL(p \parallel q)$.[2] It can be interpreted as the amount of information necessary to change the prior probability distribution q into posterior probability distribution p. For a discrete x,

$$KL(p \parallel q) = \sum_x p(x) \log_2 \left(\frac{p(x)}{q(x)} \right). \tag{5.31}$$

For identical distributions, the KL divergence is zero. We regard each row of $DP(\mathbf{x})$ as a prior probability distribution on the set of class labels Ω and use $d_{i,j}(\mathbf{x})$ to denote the estimate of the probability $P(\omega_j|\mathbf{x}, D_i)$. Denote by $P_{(i)}$ the probability distribution on Ω provided by classifier D_i, that is, $P_{(i)} = (d_{i,1}(\mathbf{x}), \dots, d_{i,c}(\mathbf{x}))$. For example, let $DP(\mathbf{x})$ be

$$DP(\mathbf{x}) = \begin{bmatrix} 0.3 & 0.7 \\ 0.6 & 0.4 \\ 0.5 & 0.5 \end{bmatrix}.$$

Then $P_{(1)} = (0.3, 0.7)$ is the pmf on $\Omega = \{\omega_1, \omega_2\}$ due to classifier D_1.

Given the L sets of probability estimates, one for each classifier, our first hypothesis is that the true values of $P(\omega_i|\mathbf{x})$ (posterior probabilities) are the ones most agreed upon by the ensemble $D = \{D_1, \dots, D_L\}$. Denote these agreed values by $P_{ens} = (\mu_1(\mathbf{x}), \dots, \mu_c(\mathbf{x}))$. Then the averaged KL divergence across the L ensemble members is

$$KL_{av} = \frac{1}{L} \sum_{i=1}^{L} KL(P_{ens} \parallel P_{(i)}). \tag{5.32}$$

We seek P_{ens} which minimizes Equation 5.32. To simplify the notation, we shall drop the $(\mathbf{x})$ from $\mu_j(\mathbf{x})$ and $d_{i,j}(\mathbf{x})$ keeping in mind that we are operating on a specific point $\mathbf{x}$ in the feature space $\mathbb{R}^n$. Take $\partial KL_{av}/\partial \mu_j$, include the term with the Lagrange

---

[2]It is assumed that for any x, if $q(x) = 0$ then $p(x) = 0$, and also $0 \times \log 0 = 0$.

multiplier to ensure that P_{ens} is a pmf, and set to zero

$$\frac{\partial}{\partial \mu_j}\left[KL_{\text{av}} + \lambda\left(1 - \sum_{k=1}^{c} \mu_k\right)\right]$$

$$= \frac{1}{L}\sum_{i=1}^{L}\frac{\partial}{\partial \mu_j}\left[\sum_{k=1}^{c}\mu_k \log_2\left(\frac{\mu_k}{d_{i,k}}\right)\right] - \lambda \qquad (5.33)$$

$$= \frac{1}{L}\sum_{i=1}^{L}\left(\log_2\left(\frac{\mu_j}{d_{i,j}}\right) + C\right) - \lambda = 0, \qquad (5.34)$$

where $C = \frac{1}{\ln(2)}$. Solving for μ_j, we obtain

$$\mu_j = 2^{(\lambda - C)}\prod_{i=1}^{L}(d_{i,j})^{\frac{1}{L}}. \qquad (5.35)$$

Substituting Equation 5.35 in $\sum_{k=1}^{c}\mu_j = 1$ and solving for λ we arrive at

$$\lambda = C - \log_2\left(\sum_{k=1}^{c}\prod_{i=1}^{L}(d_{i,k})^{\frac{1}{L}}\right). \qquad (5.36)$$

Substituting λ back in Equation 5.35 yields the final expression for the ensemble probability for class ω_j given the input **x** as the normalized geometric mean

$$\mu_j = \frac{\prod_{i=1}^{L}(d_{i,j})^{\frac{1}{L}}}{\sum_{k=1}^{c}\prod_{i=1}^{L}(d_{i,k})^{\frac{1}{L}}}. \qquad (5.37)$$

Note that the denominator of μ_j does not depend on j. Also, the power $\frac{1}{L}$ in the numerator is only a monotone transformation of the product and will not change the ordering of the discriminant functions obtained through product. Therefore, the ensemble degree of support for class ω_j, $\mu_j(\mathbf{x})$ reduces to the *product combination rule*

$$\mu_j = \prod_{i=1}^{L} d_{i,j}. \qquad (5.38)$$

If we swap the places of the prior and posterior probabilities in Equation 5.32 and again look for a minimum with respect to μ_j, we obtain

$$\frac{\partial}{\partial \mu_j} \left[KL_{av} + \lambda \left(1 - \sum_{k=1}^{c} \mu_k \right) \right]$$

$$= \frac{1}{L} \sum_{i=1}^{L} \frac{\partial}{\partial \mu_j} \left[\sum_{k=1}^{c} d_{i,k} \log_2 \left(\frac{d_{i,k}}{\mu_k} \right) \right] - \lambda \qquad (5.39)$$

$$= -\frac{1}{C L \mu_j} \sum_{i=1}^{L} d_{i,j} - \lambda = 0, \qquad (5.40)$$

where C is again $\frac{1}{\ln(2)}$. Solving for μ_j gives

$$\mu_j = -\frac{1}{\lambda C L} \sum_{i=1}^{L} d_{i,j}. \qquad (5.41)$$

Substituting Equation 5.41 in $\sum_{k=1}^{c} \mu_k = 1$ and solving for λ leads to

$$\lambda = -\frac{1}{C L} \sum_{k=1}^{c} \sum_{i=1}^{L} d_{i,k} = -\frac{L}{C L} = -\frac{1}{C}. \qquad (5.42)$$

The final expression for the ensemble probability for class ω_j, given the input $\mathbf{x}$, is the normalized arithmetic mean:

$$\mu_j = \frac{1}{L} \sum_{i=1}^{L} d_{i,j}, \qquad (5.43)$$

which is *average combination rule* (the same as average or sum combiner).

The average combiner was derived in the same way as the product combiner under a slightly different initial assumption. We assumed that P_{ens} is some unknown prior pmf which needs to be transformed into the L posterior pmfs suggested by the L ensemble members. Thus, to derive the average rule, we minimized the average information necessary to transform P_{ens} to the individual pmfs.

Miller and Yan go further and propose weights which depend on the "critic" for each classifier and each $\mathbf{x}$ [281]. The "critic" estimates the probability that the classifier is correct in labeling $\mathbf{x}$. Miller and Yen derive the product rule with the critic probability as the power of $d_{i,j}$ and the sum rule with the critic probabilities as weights. Their analysis and experimental results demonstrate the advantages of the weighted rules. The authors admit that there is no reason why one set-up should be preferred to another.

5.4   THE WEIGHTED AVERAGE (LINEAR COMBINER)

Given an object **x**, this combiner aggregates the class supports from the decision profile to arrive at a single support value for each class. Three groups of average combiners can be distinguished based on the respective number of weights:

- *L weights.* In this model there is one weight per classifier. The support for class ω_j is calculated as

$$\mu_j(\mathbf{x}) = \sum_{i=1}^{L} w_i \, d_{i,j}(\mathbf{x}). \tag{5.44}$$

- *c × L weights.* The weights are class-specific and classifier-specific. The support for class ω_j is calculated as

$$\mu_j(\mathbf{x}) = \sum_{i=1}^{L} w_{ij} \, d_{i,j}(\mathbf{x}). \tag{5.45}$$

Again, only the *j*th column of the decision profile is used in the calculation, that is, the support for class ω_j is obtained from the individual supports for ω_j.
- *c × c × L weights.* The support for each class is obtained by a linear combination of the entire decision profile $DP(\mathbf{x})$,

$$\mu_j(\mathbf{x}) = \sum_{i=1}^{L} \sum_{k=1}^{c} w_{ikj} \, d_{i,k}(\mathbf{x}), \tag{5.46}$$

where w_{ikj} is the (i, k)th weight for class ω_j. The whole of the decision profile is used as the intermediate feature space.

The following subsections present different ways to calculate the weights.

5.4.1   Consensus Theory

The weights may be set so as to express the quality of the classifiers. Accurate and robust classifiers should receive larger weights. Such weight assignments may come from subjective estimates or theoretical set-ups.

Berenstein et al. [35] bring to the attention of the Artificial Intelligence community the so-called *consensus theory* which has enjoyed a considerable interest in social and management sciences but remained not well known elsewhere. The theory looks into combining expert opinions and in particular combining *L* probability distributions on Ω (in our case, the rows of the decision profile $DP(\mathbf{x})$) into a single distribution $(\mu_1(\mathbf{x}), \dots, \mu_c(\mathbf{x}))$. A *consensus rule* defines the way this combination is carried out. Consensus rules are derived to satisfy a set of desirable theoretical properties [34, 44, 289].

Based on an experimental study, Ng and Abramson [289] advocate using simple consensus rules such as the weighted average, called the *linear opinion pool* (Equation 5.44), and the weighted product called the *logarithmic opinion pool*. The approach taken to assigning weights in consensus theory is based on the decision maker's knowledge of the importance of the experts (classifiers). The weights are assigned on the basis of some subjective or objective measure of importance of the experts [35].

5.4.2 Added Error for the Weighted Mean Combination

Extending the theoretical study of Tumer and Ghosh [393], Fumera and Roli derive the added error for the weighted average combination rule [143, 145].

The ensemble estimate of $P(\omega_j|x)$ is

$$\hat{P}(\omega_j|x) = \sum_{i=1}^{L} w_i\, d_{i,j}, \quad i = 1, \ldots, c, \tag{5.47}$$

where $d_{i,j}$ is the respective entry in the decision profile and w_i are classifier-specific weights such that

$$\sum_{i=1}^{L} w_i = 1, \quad w_i \geq 0. \tag{5.48}$$

Under a fairly large list of assumptions, a set of optimal weights for independent classifiers can be calculated from the *added errors* of the individual classifiers, E_{add}^i, $m = 1, \ldots, L$. The added error is the excess above the Bayes error for the problem for the specific classifier. The weights are

$$w_i = \frac{\dfrac{1}{E_{\text{add}}^i}}{\sum_{k=1}^{L} \dfrac{1}{E_{\text{add}}^k}}, \quad i = 1, \ldots, L. \tag{5.49}$$

Since we do not have a way to estimate the added error, we can use as a proxy the estimates of the classification errors of the base classifiers.

Despite the appealing theoretical context, this way of calculating the weights was not found to be very successful [144]. This can be due to the unrealistic and restrictive assumptions which define the optimality conditions giving rise to these weights. Fumera and Roli's experiments suggested that for large ensembles, the advantage of weighted averaging over simple averaging disappears. Besides, in weighted averaging we have to estimate the L weights, which is a potential source of error and may cancel the already small advantage.

REGRESSION COMBINER

Training: Given is a set of L trained classifiers and a labeled data set.

1. Find the outputs (decision profiles) of the classifiers for each point in the data set.
2. For each class $j, j = 1, \dots, c$, train a regression of the type 5.46. We can choose to fit the regression with or without an intercept term.
3. Return the coefficients of the c regressions. The coefficients of the regression for class j are denoted by w_{ikj} as in Equation 5.46. If there was an intercept term, the number of returned coefficients for each regression is $L \times c + 1$.

Operation: For each new object

1. Classify the new object $\mathbf{x}$ and find its decision profile $DP(\mathbf{x})$ as in Equation 5.1.
2. Calculate the support for each class $P(j) = \mu_j(\mathbf{x})$ as in Equation 5.46.
3. Assign label i^* to the object, where

$$i^* = \arg \max_{j=1}^{c} P(j).$$

Return the ensemble label of the new object.

FIGURE 5.13   Training and operation algorithm for the linear regression combiner.

5.4.3 Linear Regression

One way to set the weights is to fit a linear regression to the posterior probabilities. Take $d_{i,j}(\mathbf{x})$, $i = 1, \dots, L$, to be estimates of the posterior probability $P(\omega_j | \mathbf{x})$. For classification problems, the target output is given only in the form of a class label. So the target values for $P(\omega_j | \mathbf{x})$ are either 1 (in ω_j) or 0 (not in ω_j). Figure 5.13 shows the training and the operation of the regression combiner.

Classifier combination through linear regression has received significant attention. The following questions have been discussed:

- Should the weights be nonnegative? If they are, the value of the weight may be interpreted as the importance of a classifier.
- Should the weights be constrained to sum up to one?
- Should there be an intercept term?

It is believed that these choices have only a marginal impact on the final outcome [193, 384]. The important question is what criterion should be optimized. Minimum squared error (MSE) is the traditional criterion for regression [175–177, 388]. Different criteria have been examined in the context of classifier combination through linear regression, for both small [120] and large ensembles [327], an example of which is the hinge function, which is responsible for the classification margins [120, 395].

Consider the largest regression model, where the whole decision profile is involved in approximating each posterior probability as in Equation 5.46. Given a data set

$Z = \{z_1, \ldots, z_N\}$ with labels $\{y_1, \ldots, y_N\}$, $y_j \in \Omega$, Ergodan and Sen [120] formulate the optimization problem as looking for a weight vector $\mathbf{w}$ which minimizes

$$\Psi(\mathbf{w}) = \underbrace{\frac{1}{N}\sum_{j=1}^{N}}_{\text{objects}} \underbrace{\sum_{i=1}^{c}}_{\text{classes}} \mathcal{L}(\mu_i(z_j), y_j, \omega_i, \mathbf{w}) + R(\mathbf{w}), \qquad (5.50)$$

objects classes

where $\mathcal{L}(\mu_i(z_j), y_j, \omega_i, \mathbf{w})$ is the loss incurred when labeling object $z_j \in Z$, with true label y_j, as belonging to class ω_i. $R(\mathbf{w})$ is a regularization term which serves to penalize very large weights.[3] Why is the penalty term needed? Say there are five classifiers and four classes. For this small problem, the regression 5.46 will need $L \times c \times c = 5 \times 4 \times 4 = 80$ weights. The chance of over-training cannot be ignored, hence the need for a regularization term.

To use this optimization set-up, two choices must be made: the type of loss function $\mathcal{L}$ and the regularization function R.

Let us simplify the notation to $\mathcal{L}(a, b)$ where $a \in \{-1, 1\}$ is the true label, and b is the predicted quantity. The classification loss is $\mathcal{L}(a, b) = 0$ if the signs of a and b match and $\mathcal{L}(a, b) = 1$, otherwise. Minimizing this loss is ideal but mathematically awkward, hence Rosasco et al. [339] analyze several alternatives:

- The square loss:

$$\mathcal{L}(a, b) = (a - b)^2 = (1 - ab)^2. \qquad (5.51)$$

- The hinge loss:

$$\mathcal{L}(a, b) = \max\{1 - ab, 0\}. \qquad (5.52)$$

This is the criterion function that is minimized for training the SVM classifier.
- The logistic loss:

$$\mathcal{L}(a, b) = \frac{1}{\ln 2} \ln(1 + \exp\{-ab\}). \qquad (5.53)$$

Based on its theoretical properties, Ergodan and Sen [120] recommend the hinge loss function.

---

[3] An intercept term b can be added to the regression in Equation 5.46, and included in the weight vector $\mathbf{w}$.

Reid and Grudic [327] study the effect of different regularization functions.

- L_2 regularization, which, used with the square loss function 5.51, is called *ridge regression*:

$$R(\mathbf{w}) = \lambda \sum_k w_k^2 = \lambda ||\mathbf{w}||_2^2. \tag{5.54}$$

- L_1 (LASSO)[4] regularization:

$$R(\mathbf{w}) = \lambda \sum_k |w_k| = \lambda ||\mathbf{w}||_1. \tag{5.55}$$

- The elastic net regularization, which combines the above two. The regularization term is

$$R(\mathbf{w}) = \lambda ||\mathbf{w}||_2^2 + (1 - \lambda)||\mathbf{w}||_1. \tag{5.56}$$

Ridge regression arrives at dense models (using all classifiers in the ensemble) whereas LASSO produces sparse ensembles. Applying the three penalty terms with the square loss for large ensembles, Reid and Grudic [327] draw the following conclusions. Ridge regression outperforms nonregularized regression, and improves on the performance of the single best classifier in the ensemble. LASSO was not as successful as the ridge regression, leading the authors to conclude that dense models were better than the sparse models.

Calculating the solution of the optimization problem with the hinge loss function is not straightforward. However, MATLAB Statistics Toolbox offers a ridge regression code, which we will use for the illustration here.

▣ Example 5.5   Ridge regression for posterior probabilities
We used again the letter data set from the UCI Machine Learning Repository [22]. The set consists of $N = 20,000$ data points described by $n = 16$ features and labeled into the $c = 26$ classes of the letters of the Latin alphabet. Since the data set is reasonably large, we used the hold-out method for this example. The data set was randomly split into training, validation, and testing parts. The training part was used to train $L = 51$ linear classifiers, the validation part, for training the ridge regression with a pre-specified value of the parameter λ, and the testing part was used to estimate the testing error of the ensemble. Each classifier was trained on a bootstrap sample from the training set. The data set was chosen on purpose. The number of classes is large, $c = 26$, which means that the dimensionality of the intermediate space is $L \times c = 51 \times 26 = 1326$. This makes classification in the intermediate space challenging and sets the scene for demonstrating the advantages of ridge regression. Twenty-six sets of coefficients were fitted on the 1326 features, one regression for each class, and the ensemble outputs were calculated as explained in Figure 5.13.

---

[4]LASSO stands for least absolute shrinkage and selection operator.

TABLE 5.5   Ensemble Error for a Ridge Regression with Parameter λ

λ	Training/validation/testing split in %			
	4/16/80	12/48/40	16/64/20	8/72/20
0.01	0.1728	0.1034	0.1012	0.0985
0.02	0.1714	0.1031	0.1007	0.0985
0.50	0.1559	0.1029	0.1012	0.0975
0.80	0.1536	0.1029	0.1014	0.0985
LDC on training + validation	0.3056	0.2913	0.2944	0.3108
Decision tree on ensemble	0.3925	0.2993	0.2801	0.2834

Table 5.5 shows the ensemble error for a ridge regression on the whole decision profile (Equation 5.46), minimizing MSE with L_2 penalty term (Equation 5.54).

Along with the ridge regression results, we show the classification error for:

1. The LDC trained on the training plus validation data, and tested on the testing data.[5]
2. Decision tree classifier built on the validation set, using as inputs the classifier outputs. The classifiers were trained on the training data. The decision tree was tested on the classifier outputs for the testing data. Thus, the decision tree is the combiner, trained on unseen data, and tested on another unseen data set.[6]

    What does the example show?

    (i) *The regression combiner was invariably better than the decision tree combiner.* In all four splits of the data into training/validation/testing, the ensemble errors for the ridge regression were smaller that those for the decision tree combiner.

    (ii) The regression combiner was invariably better than the individual LDC. Interestingly, the decision tree combiner failed miserably in comparison with the regression combiner for this problem, and barely managed to improve on the classification error of the individual LDC for the two larger validation sets.

    (iii) *Larger validation sets led to smaller ensemble errors.* The training set was kept small on purpose. By doing so we aimed at creating an ensemble of fairly weak but diverse linear classifiers. For such an ensemble, the combiner would be important, and clear differences between the combiners could be expected.

    (iv) *The penalty constant λ had a marked effect for the smallest validation set and a little effect for larger sets.* This was also to be expected, as λ is supposed to correct for the instability of the regression trained on a small sample.

---

[5]Function `classify` from the Statistics Toolbox of MATLAB was used for the LDC.
[6]Function `classregtree` from the Statistics Toolbox of MATLAB was used for the decision tree classifier.

This example shows that the regression combiner may work well, especially for problems with a large number of classes, and large ensemble sizes, resulting in a high-dimensional intermediate feature space. Its success will likely depend on the data set, the ensemble size, the way the individual classifiers are trained, and so on.

Regression methods are only one of the many possible ways to train the combination weights. Ueda [395] uses a *probabilistic descent* method to derive the weights for combining neural networks as the base classifiers. Some authors consider using *genetic algorithms* for this task [73, 249].

5.5   A CLASSIFIER AS A COMBINER

Consider the *intermediate feature space* where each point is an expanded version of $DP(\mathbf{x})$ obtained by concatenating its L rows. Any classifier can be applied for labeling this point [189, 384, 395].

5.5.1   The Supra Bayesian Approach

Jacobs [193] reviews methods for combining experts' probability assessments. *Supra Bayesian methods* consider the experts' estimates as data, as many of the combiners do. The problem of estimating $\mu_j(\mathbf{x})$ becomes a problem of Bayesian learning in the intermediate feature space where the decision profile $DP(\mathbf{x})$ provides the $L \times c$ features. Loosely speaking, in supra Bayesian approach for our task, we estimate the probabilities $\mu_j(\mathbf{x}) = P(\omega_j|\mathbf{x})$, $j = 1, \ldots, c$, using the L distributions provided by the ensemble members. Since these distributions are organized in a decision profile $DP(\mathbf{x})$, we have

$$\mu_j(\mathbf{x}) = P(\omega_j|\mathbf{x}) \propto p(DP(\mathbf{x})|\omega_j)P(\omega_j), \quad j = 1, \ldots, c, \qquad (5.57)$$

where $p(DP(\mathbf{x})|\omega_j)$ is the class-conditional likelihood of the decision profile for the given $\mathbf{x}$ and ω_j. We assume that the only prior knowledge that we have is some estimates of the c prior probabilities $P(\omega_j)$.

When the classifier outputs are class labels, the supra Bayesian approach is the theoretical justification of the multinomial combination method, also called BKS (Chapter 4). For continuous-valued outputs, this approach, albeit theoretically well motivated, is impractical [193]. The reason is that the pdf $p(DP(\mathbf{x})|\omega_j)$ is difficult to estimate. In principle, the supra Bayesian approach means that we use the intermediate feature space to build a classifier which is as close as possible to the Bayes classifier thereby guaranteeing the minimum possible classification error rate. Viewed in this light, all combiners that treat the classifier outputs in $DP(\mathbf{x})$ as new features are approximations within the supra Bayesian framework.

DECISION TEMPLATES COMBINER

Training: For $j = 1, \ldots, c$, calculate the mean of the decision profiles of all members of ω_j from the data set $\mathbf{Z}$. Call this mean *decision template DT_j*

$$DT_j = \frac{1}{N_j} \sum_{\substack{y_k = \omega_j \\ \mathbf{z}_k \in \mathbf{Z}}} DP(\mathbf{z}_k),$$

where N_j is the number of elements of $\mathbf{Z}$ from ω_j.

Operation: Given the input $\mathbf{x} \in \mathbb{R}^n$, construct $DP(\mathbf{x})$. Calculate the similarity S between $DP(\mathbf{x})$ and each DT_j,

$$\mu_j(\mathbf{x}) = S(DP(\mathbf{x}), DT_j) \quad j = 1, \ldots, c$$

and label $\mathbf{x}$ to the class with the largest support.

Return the ensemble label of the new object.

FIGURE 5.14   Training and operation algorithm for the decision templates combiner.

5.5.2   Decision Templates

The idea of the decision templates (DT) combiner is to remember the most typical decision profile for each class ω_j, called the *decision template, DT_j*, and then compare it with the current decision profile $DP(\mathbf{x})$ using some similarity measure S. The closest match will label $\mathbf{x}$. Figures 5.14 and 5.15 describe the training and the operation of the decision templates combiner.

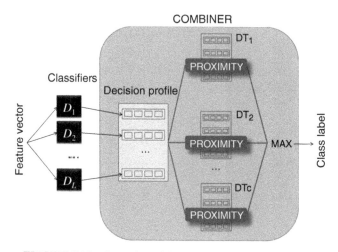

FIGURE 5.15   Operation of the decision templates combiner.

Two typical measures of similarity S are based upon

- *The squared Euclidean distance.* The ensemble support for ω_j is

$$\mu_j(\mathbf{x}) = 1 - \frac{1}{L \times c} \sum_{i=1}^{L} \sum_{k=1}^{c} \left(DT_j(i,k) - d_{i,k}(\mathbf{x})\right)^2, \qquad (5.58)$$

where $DT_j(i,k)$ is the (i,k)th entry in decision template DT_j. The outputs μ_j are within the interval $[0,1]$ but this scaling is not necessary for classification purposes. The class with the maximum support would be the same if we use just

$$\mu_j(\mathbf{x}) = - \sum_{i=1}^{L} \sum_{k=1}^{c} \left(DT_j(i,k) - d_{i,k}(\mathbf{x})\right)^2. \qquad (5.59)$$

This calculation is equivalent to applying the nearest mean classifier in the intermediate feature space. While we use only the Euclidean distance in (5.58), there is no reason to stop at this choice. Any distance could be used, for example, the Minkowski or the Mahalanobis distance.

- *A symmetric difference* coming from the fuzzy set theory [222, 233]. The support for ω_j is

$$\mu_j(\mathbf{x}) = 1 - \frac{1}{L \times c} \sum_{i=1}^{L} \sum_{k=1}^{c} \max\{\min\{DT_j(i,k), (1 - d_{i,k}(\mathbf{x}))\},$$

$$\min\{(1 - DT_j(i,k)), d_{i,k}(\mathbf{x})\}\}. \qquad (5.60)$$

◪ Example 5.6 Decision templates combiner (DT)

Let $c = 3$, $L = 2$, and let the decision templates for ω_1 and ω_2 be, respectively,

$$DT_1 = \begin{bmatrix} 0.6 & 0.4 \\ 0.8 & 0.2 \\ 0.5 & 0.5 \end{bmatrix} \quad \text{and} \quad DT_2 = \begin{bmatrix} 0.3 & 0.7 \\ 0.4 & 0.6 \\ 0.1 & 0.9 \end{bmatrix}.$$

Assume that for an input $\mathbf{x}$, the following decision profile has been obtained:

$$DP(\mathbf{x}) = \begin{bmatrix} 0.3 & 0.7 \\ 0.6 & 0.4 \\ 0.5 & 0.5 \end{bmatrix}.$$

The similarities and the class labels using the Euclidean distance and the symmetric difference are as follows:

DT version	$\mu_1(\mathbf{x})$	$\mu_2(\mathbf{x})$	Label
Euclidean distance	0.9567	0.9333	ω_1
Symmetric difference	0.5000	0.5333	ω_2

The difference in the "opinions" of the two DT versions with respect to the class label is an indication of the flexibility of the combiner.

5.5.3   A Linear Classifier

The LDC seems a good choice for determining the weights of the linear combiner [324, 325, 434]. It has an advantage over the regression method because it minimizes a function directly related to the classification error while regression methods optimize posterior probability approximations. Better still, we can use the SVM classifier with the linear kernel, which is capable of dealing with correlated inputs (the classifier outputs) and small training sets [161]. In fact, any classifier can be applied as the combiner, which brings back the rather philosophical issue raised by Tin Ho [183]: Where do we stop growing the hierarchy of classifiers upon classifiers? Do we even have to?

5.6   AN EXAMPLE OF NINE COMBINERS FOR CONTINUOUS-VALUED OUTPUTS

Consider again the fish data set, generated with 20% label noise. Seventeen random linear classifiers were generated as the base ensemble classifiers. Their classification boundaries are plotted with lines in each data scatterplot in Figure 5.16.

The continuous-valued outputs (posterior probability estimates) were obtained using the MATLAB function `classify`. Each of the nine combiners gives rise to two plots. The left plot contains the grid with the noisy fish data. The region labeled as the fish (black dots) by the ensemble is overlaid. The accuracy displayed under the combiner's name is calculated with respect to the original (noise-free) class labels. The right plot is a gray-scale heat map of the ensemble estimate of $P(\text{fish}|\mathbf{x})$. The contour for $P(\text{fish}|\mathbf{x}) = 0.5$, delineating the classification region for class fish, is plotted with a thick line over the heat map.

In this example, the LDC, the ridge regression ($\lambda = 0.5$, not tuned), and the SVM combiner were the winners, with above 80% correct classification rate, given that the largest prior classifier would give only 64.48%. The worst combiner happened to be the minimum combiner (equal to the maximum combiner for two classes). The average, weighted average, and decision templates combiners were obviously too simplistic for the problem, and gave disappointingly low ensemble accuracies. On the other hand, the "peppery" right plot for the decision tree combiner demonstrates a great deal of over-training. Nonetheless, this combiner achieved over 79% correct

Average (Sum)
69.12

Median
71.44

Min/max
62.96

Weighted average
68.84

LDC combiner
87.96

Ridge regression
81.56

Tree combiner
79.72

Decision Templates
68.32

SVM combiner
80.52

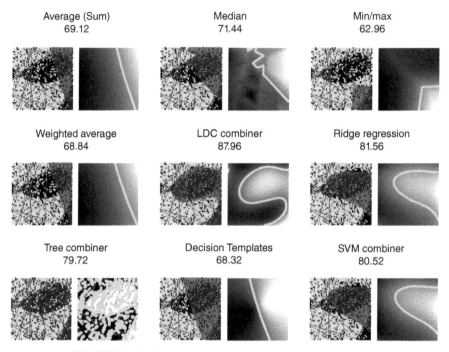

FIGURE 5.16 Comparison of nine combiners on the fish data.

classification, which indicates that, for this problem, even though both alternatives are wrong, over-fitting gives a better pay-off than under-fitting.

As noted before, this example should not be taken to mean that LDC is always the best combiner, and minimum/maximum is the worst. The message is that the choice of a combiner is important, and should not be casually sidelined.

5.7 TO TRAIN OR NOT TO TRAIN?

Some combiners do not need training after the classifiers in the ensemble have been trained individually. An example of this type is the majority vote combiner. Other combiners need additional training, for example, the weighted average combiner. A third class of ensembles develop the combiner during the training of the individual classifiers, an example of which is AdaBoost, discussed later. If a large data set is available, training and testing can be done on large, nonintersecting subsets, which allows for precise tuning while guarding against over-fitting. Small data sets, on the other hand, pose a real challenge. Duin [108] points out the crucial role of the training strategy in these cases and gives the following recommendations:

1. If a single training set is used with a *nontrainable combiner*, then make sure that the base classifiers are not over-trained.

2. If a single training set is used with a *trainable combiner*, then leave the base classifiers under-trained and subsequently complete the training of the combiner on the training set. Here it is assumed that the training set has a certain "training potential." In order to be able to train the combiner reasonably, the base classifiers should not use up all the potential.

3. Use separate training sets for the base classifiers and for the combiners. Then the base classifiers can be over-trained on their training set. The bias will be corrected by training the combiner on the separate training set.

Dietrich et al. [93] suggest that the second training set, on which the ensemble should be trained, may be partly overlapping with the first training set used for the individual classifiers. Let R be the first training set, V be the second training set, and T be the testing set. All three sets are obtained from the available labeled set $\mathbf{Z}$, so $R \cup V \cup T = \mathbf{Z}$. If $\mathbf{Z}$ is small, the three sets might become inadequately small thereby leading to badly trained classifiers and ensemble, and unreliable estimates of their accuracies. To remedy this, the two training sets are allowed to have an overlap controlled by a parameter ρ

$$\rho = \frac{|R \cap V|}{|R|}, \tag{5.61}$$

where $|.|$ denotes cardinality. For $\rho = 0$, R and V are disjoined and for $\rho = 1$, the classifiers and the ensemble are trained on a single set $R = V$. The authors found that better results were obtained for $\rho = 0.5$ compared to the two extreme values. This suggests that a compromise should be sought when the initial data set $\mathbf{Z}$ is relatively small.

Stacked generalization has been defined as a generic methodology for improving generalization in pattern classification [420]. We will present it here through an example, as a protocol for training a classifier ensemble and its combiner.

▣ Example 5.7   Stacked generalization

Let $\mathbf{Z}$ be a data set with N objects partitioned into four parts of approximately equal sizes, denoted A, B, C, and D. Three classifiers, D_1, D_2, and D_3, are trained according to the standard fourfold cross-validation protocol depicted in Figure 5.17. At the end of this training, there will be four versions of each of the classifiers trained on (ABC), (BCD), (ACD), or (ABD), respectively.

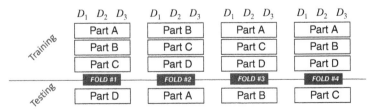

FIGURE 5.17   Standard 4-fold cross-validation set-up.

The combiner is trained on a data set of size N obtained in the following way. For any data point $\mathbf{z}_j$ in subset A, we take the outputs for that point from the versions of D_1, D_2, and D_3 built on (BCD). In this way subset A has not been seen during the training of the individual classifiers. The three outputs together with the label of $\mathbf{z}_j$ form a data point in the training set for the combiner. All the points from subset B are processed by the versions of the three classifiers built on (ACD) and the outputs added to the training set for the combiner, etc. After the combiner has been trained, the four subsets are pooled again into $\mathbf{Z}$ and D_1, D_2, and D_3 are re-trained, this time on the whole of $\mathbf{Z}$. The new classifiers and the combiner are then ready for operation.

Many authors have studied and compared the performance of ensemble combiners [9, 109, 206, 321, 325, 337, 382, 383, 399, 400, 434]. Most of such studies, both empirical and theoretical, do not elect a clear winner. This is to be expected in view of the "no panacea theorem" [187]. The value of such comparative studies is to accumulate knowledge and understanding of the conditions which could guide the choice of a combiner. These conditions may be the type of data or the problem, as well as the ensemble size, homogeneity, diversity, and building strategy.

APPENDIX

5.A.1   THEORETICAL CLASSIFICATION ERROR FOR THE SIMPLE COMBINERS

5.A.1.1   Set-up and Assumptions

We reproduce the scenario and the assumption from the text.

- There are only two classes, $\Omega = \{\omega_1, \omega_2\}$.
- All classifiers produce soft class labels, $d_{j,i}(\mathbf{x}) \in [0, 1]$, $i = 1, 2$, $j = 1, \ldots, L$, where $d_{j,i}(\mathbf{x})$ is an estimate of the posterior probability $P(\omega_i|\mathbf{x})$ by classifier D_j for an input $\mathbf{x} \in \mathbb{R}^n$. We consider the case where for any $\mathbf{x}$, $d_{j,1}(\mathbf{x}) + d_{j,2}(\mathbf{x}) = 1$, $j = 1, \ldots, L$.
- Let $\mathbf{x} \in \mathbb{R}^n$ be a data point to classify. Without loss of generality, we assume that the true posterior probability is $P(\omega_1|\mathbf{x}) = p > 0.5$. Thus, the Bayes-optimal class label for $\mathbf{x}$ is ω_1, and a classification error occurs if label ω_2 is assigned.

Assumption. The classifiers commit independent and identically distributed errors in estimating $P(\omega_1|\mathbf{x})$ such that

$$d_{j,1}(\mathbf{x}) = P(\omega_1|\mathbf{x}) + \eta(\mathbf{x}) = p + \eta(\mathbf{x}), \tag{5.A.1}$$

and respectively $d_{j,2}(\mathbf{x}) = 1 - p - \eta(\mathbf{x})$, where $\eta(\mathbf{x})$ has

(i) a normal distribution with mean 0 and variance σ^2 (we take σ to vary between 0.1 and 1)

(ii) a uniform distribution spanning the interval $[-b, +b]$ (b varies from 0.1 to 1).

We derive the theoretical error rate of an ensemble of L classifiers for a given object $\mathbf{x}$ and the following combiners: majority vote, average (sum), minimum, maximum, and median. For comparison, we include in the list the individual classifier error rate and the so-called "oracle" combiner which outputs the correct class label if at least one of the classifiers produces the correct class label.

Recall that, for the majority vote, we first "harden" the individual decisions by assigning class label ω_1 if $d_{j,1}(\mathbf{x}) > 0.5$, and ω_2 if $d_{j,1}(\mathbf{x}) \leq 0.5, j = 1, \ldots, L$. Then the class label most represented among the L (label) outputs is chosen as the final label for $\mathbf{x}$.

Denote by P_j the output of classifier D_j for class ω_1, that is, $P_j = d_{j,1}(\mathbf{x})$, and let

$$\hat{P}_1 = \mathcal{F}(P_1, \ldots, P_L) \tag{5.A.2}$$

be the fused estimate of $P(\omega_1|\mathbf{x})$. By assumption, the posterior probability estimates for ω_2 are $1 - P_j, j = 1, \ldots, L$. The same fusion method $\mathcal{F}$ is used to find the fused estimate of $P(\omega_2|\mathbf{x})$,

$$\hat{P}_2 = \mathcal{F}(1 - P_1, \ldots, 1 - P_L). \tag{5.A.3}$$

According to the assumptions, we regard the individual estimates P_j as independent, identically distributed random variables, such that $P_j = p + \eta_j$, with pdf $f(y), y \in \mathbb{R}$ and cumulative distribution functions (cdf) $F(t), t \in \mathbb{R}$. Then $\hat{P}_1$ is a random variable with a pdf $f_{\hat{P}_1}(y)$ and cdf $F_{\hat{P}_1}(t)$.

For a single classifier, the average and the median fusion models will result in $\hat{P}_1 + \hat{P}_2 = 1$. The higher of the two estimates determines the class label. The oracle and the majority vote make decisions on the class label outputs, so $\hat{P}_1 = 1, \hat{P}_2 = 0$ for class ω_1, and $\hat{P}_1 = 0, \hat{P}_2 = 1$ for class ω_2. Thus, it is necessary and sufficient to have $\hat{P}_1 > 0.5$ to label $\mathbf{x}$ in ω_1 (the correct label). The probability of error, given $\mathbf{x}$, denoted P_e, is

$$P_e = P(\text{error}|\mathbf{x}) = P(\hat{P}_1 \leq 0.5) = F_{\hat{P}_1}(0.5) = \int_0^{0.5} f_{\hat{P}_1}(y)dy \tag{5.A.4}$$

for the single best classifier, average, median, majority vote, and the oracle.

For the minimum and the maximum rules, however, the sum of the fused estimates is not necessarily one. The class label is then decided by the maximum of $\hat{P}_1$ and $\hat{P}_2$. Thus, an error will occur if $\hat{P}_1 \leq \hat{P}_2$,[7]

$$P_e = P(\text{error}|\mathbf{x}) = P(\hat{P}_1 \leq \hat{P}_2) \tag{5.A.5}$$

for the minimum and the maximum.

---

[7] We note that since $\hat{P}_1$ and $\hat{P}_2$ are continuous-valued random variables, the inequalities can be written with or without the equal sign, that is, $\hat{P}_1 > 0.5$ is equivalent to $\hat{P}_1 \geq 0.5$, and so on.

The two distributions considered are

- Normal distribution, $\hat{P}_1 \sim N(p, \sigma^2)$. We denote by $\Phi(z)$ the cumulative distribution function of $N(0, 1)$. Then

$$F(t) = \Phi\left(\frac{t-p}{\sigma}\right). \tag{5.A.6}$$

- Uniform distribution within $[p - b, p + b]$, that is,

$$f(y) = \begin{cases} \frac{1}{2b}, & y \in [p - b, p + b]; \\ 0, & \text{elsewhere,} \end{cases} \qquad F(t) = \begin{cases} 0, & t \in (-\infty, p - b); \\ \frac{t-p+b}{2b}, & t \in [p - b, p + b]; \\ 1, & t > p + b. \end{cases} \tag{5.A.7}$$

Clearly, using these two distributions, the estimates of the probabilities might fall outside the interval [0,1]. We can accept this, and justify our viewpoint by the following argument. Suppose that p is not a probability but *the amount of support* for ω_1. The support for ω_2 will be again $1 - p$. In estimating p, we do not have to restrict P_js within the interval [0, 1]. For example, a neural network (or *any* classifier for that matter) trained by minimizing the squared error between its output and the zero-one (class label) target function produces an estimate of the posterior probability for that class (cf. [40]). Thus, depending on the parameters and the transition functions, a neural network output (that approximates p) might be greater than 1 or even negative. We take the L values (in $\mathbb{R}$) and fuse them by Equations 5.A.2 and 5.A.3 to get $\hat{P}_1$ and $\hat{P}_2$. The same rule applies: ω_1 is assigned by the ensemble if $\hat{P}_1 > \hat{P}_2$. Then we calculate the *probability* of error P_e as $P(\hat{P}_1 \leq \hat{P}_2)$. This calculation does not require in any way that P_js be probabilities or be within the unit interval.

5.A.1.2   Individual Error

Since $F_{\hat{P}_1}(t) = F(t)$, the error of a single classifier for the normal distribution is

$$P_e = \Phi\left(\frac{0.5 - p}{\sigma}\right), \tag{5.A.8}$$

and for the uniform distribution,

$$P_e = \frac{0.5 - p + b}{2b}. \tag{5.A.9}$$

5.A.1.3   Minimum and Maximum

These two fusion methods are considered together because, as shown in the text, they are identical for $c = 2$ classes and any number of classifiers L.

Substituting $\mathcal{F} = \max$ in Equation 5.A.2, the ensemble's support for ω_1 is $\hat{P}_1 = \max_j\{P_j\}$. The support for ω_2 is therefore $\hat{P}_2 = \max_j\{1 - P_j\}$. A classification error will occur if

$$\max_{j}\{P_{j}\} < \max_{j}\{1 - P_{j}\}, \tag{5.A.10}$$

$$p + \max_{j}\{\eta_{j}\} < 1 - p - \min_{j}\{\eta_{j}\}, \tag{5.A.11}$$

$$\eta_{\max} + \eta_{\min} < 1 - 2p. \tag{5.A.12}$$

The probability of error for the minimum and maximum methods is

$$P_{e} = P\left(\eta_{\max} + \eta_{\min} < 1 - 2p\right) \tag{5.A.13}$$

$$= F_{\eta_{s}}(1 - 2p), \tag{5.A.14}$$

where $F_{\eta_{s}}(t)$ is the cdf of the random variable $s = \eta_{\max} + \eta_{\min}$. For the normally distributed P_{j}s, η_{j} are also normally distributed with mean 0 and variance σ^{2}. However, we cannot assume that $\eta_{\max}$ and $\eta_{\min}$ are independent and analyze their sum as another normally distributed variable because these are *order statistics* and $\eta_{\min} \leq \eta_{\max}$. We have not attempted a solution for the normal distribution case.

For the uniform distribution, we follow an example taken from [285] where the pdf of the midrange $(\eta_{\min} + \eta_{\max})/2$ is calculated for L observations. We derive $F_{\eta_{s}}(t)$ to be

$$F_{\eta_{s}}(t) = \begin{cases} \frac{1}{2}\left(\frac{t}{2b} + 1\right)^{L}, & t \in [-2b, 0]; \\ 1 - \frac{1}{2}\left(1 - \frac{t}{2b}\right)^{L}, & t \in [0, 2b]. \end{cases} \tag{5.A.15}$$

Noting that $t = 1 - 2p$ is always negative,

$$P_{e} = F_{\eta_{s}}(1 - 2p) = \frac{1}{2}\left(\frac{1 - 2p}{2b} + 1\right)^{L}. \tag{5.A.16}$$

5.A.1.4   Average (Sum)

The average combiner gives $\hat{P}_{1} = \frac{1}{L}\sum_{j=1}^{L} P_{j}$. If $P_{1}, \ldots, P_{L}$ are normally distributed and independent, then $\hat{P} \sim N\left(p, \frac{\sigma^{2}}{L}\right)$. The probability of error for this case is

$$P_{e} = P(\hat{P}_{1} < 0.5) = \Phi\left(\frac{\sqrt{L}(0.5 - p)}{\sigma}\right). \tag{5.A.17}$$

The calculation of P_{e} for the case of uniform distribution is not that straightforward. We can assume that the sum of L independent variables will result in a variable of approximately normal distribution. The higher the L, the more accurate the approximation. Knowing that the variance of the uniform distribution for P_j is $\frac{b^2}{3}$, we can assume that $\hat{P} \sim N\left(p, \frac{b^2}{3L}\right)$. Then

$$P_e = P(\hat{P}_1 < 0.5) = \Phi\left(\frac{\sqrt{3L}(0.5 - p)}{b}\right). \qquad (5.A.18)$$

5.A.1.5   Median and Majority Vote

These two fusion methods are pooled because they are identical for the current set-up (see the text). Since only two classes are considered, we restrict our choice of L to odd numbers only. An even L is inconvenient for at least two reasons. First, the majority vote might tie. Second, the theoretical analysis of a median which is calculated as the average of the $(L/2)$ and $(L/2 + 1)$ order statistics is cumbersome.

For the median fusion method

$$\hat{P}_1 = \text{med}\{P_1, \ldots, P_L\} = p + \text{med}\{\eta_1, \ldots, \eta_L\} = p + \eta_m. \qquad (5.A.19)$$

Then the probability of error is

$$P_e = P(p + \eta_m < 0.5) = P(\eta_m < 0.5 - p) = F_{\eta_m}(0.5 - p), \qquad (5.A.20)$$

where F_{η_m} is the cdf of η_m. From the order statistics theory [285],

$$F_{\eta_m}(t) = \sum_{j=\frac{L+1}{2}}^{L} \binom{L}{j} F_\eta(t)^j [1 - F_\eta(t)]^{L-j}, \qquad (5.A.21)$$

where $F_\eta(t)$ is the cdf of η_j, that is, $N(0, \sigma^2)$ or uniform in $[-b, b]$. We can now substitute the two respective cdf, to obtain P_e

- for the normal distribution

$$P_e = \sum_{j=\frac{L+1}{2}}^{L} \binom{L}{j} \Phi\left(\frac{0.5 - p}{\sigma}\right)^j \left[1 - \Phi\left(\frac{0.5 - p}{\sigma}\right)\right]^{L-j}. \qquad (5.A.22)$$

- for the uniform distribution

$$P_e = \begin{cases} 0, & p - b > 0.5; \\ \sum_{j=\frac{L+1}{2}}^{L} \binom{L}{j} \left(\frac{0.5 - p + b}{2b}\right)^j \left[1 - \frac{0.5 - p + b}{2b}\right]^{L-j}, & \text{otherwise.} \end{cases} \qquad (5.A.23)$$

The derivation of these two equations is explained below. The majority vote will assign the wrong class label, ω_2, to $\mathbf{x}$ if at least $\frac{L+1}{2}$ classifiers vote for ω_2. The probability that a single classifier is wrong is given by Equation 5.A.8 for the normal distribution and Equation 5.A.9 for the uniform distribution. Denote this probability by P_s. Since the classifiers are independent, the probability that at least $\frac{L+1}{2}$ are wrong is calculated by the binomial formula

$$P_e = \sum_{j=\frac{L+1}{2}}^{L} \binom{L}{j} P_s^j (1 - P_s)^{L-j}. \qquad (5.A.24)$$

By substituting P_s from Equations 5.A.8 and 5.A.9, we recover Equations 5.A.22 and 5.A.23 for the normal and the uniform distribution, respectively.

5.A.1.6   Oracle

The probability of error for the oracle is

$$P_e = P(\text{all incorrect}) = F(0.5)^L. \qquad (5.A.25)$$

For the normal distribution

$$P_e = \Phi\left(\frac{0.5 - p}{\sigma}\right)^L, \qquad (5.A.26)$$

and for the uniform distribution

$$P_e = \begin{cases} 0, & p - b > 0.5; \\ \left(\frac{0.5 - p + b}{2b}\right)^L, & \text{otherwise.} \end{cases} \qquad (5.A.27)$$

5.A.2   SELECTED MATLAB CODE

Example of the LDC for the Fish Data

The code below generates and plots the data and the 50 linear classification boundaries of the random base classifiers. The LDC combiner is trained on the training data, which consists of all points on the grid, with 20% label noise. The posterior probabilities for class ω_1 are calculated by line 35, using the softmax formula. The points labeled by the ensemble as class fish (black dots) are circled. The code needs the function `fish_data` from Chapter 2, and the statistics toolbox of MATLAB for the `classify` function. An example of the output is shown in Figure 5.A.1

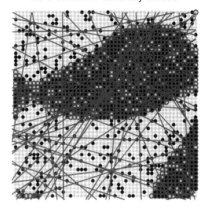

LDC combiner accuracy 0.9508

FIGURE 5.A.1   MATLAB output for the LDC combiner and the fish data.

```
1   %-----------------------------------------------------------%
2   clear all, close all
3   clc
4
5   % Generate and plot the data
6   [~ , ~,labtrue] = fish_data(50,0);
7   % Generate labels with 20% noise
8   [x,y,lb] = fish_data(50,20); figure, hold on
9   plot(x(lb == 1),y(lb == 1),'k.','markers',14)
10  plot(x(lb == 2),y(lb == 2),'k.','markers',14,...
11      'color',[0.87, 0.87, 0.87])
12  axis([0 1 0 1]) % cut the figure to the unit square
13  axis square off % equalize and remove the axes
14
15  % Generate and plot the ensemble of linear classifiers
16  L = 50; % ensemble size
17  N = numel(x); % number of data points
18  [ensemble,P1] = deal(zeros(N,L)); % pre-allocate for speed
19  sc = 1; % scaling constant for the softmax function
20  for i = 1:L
21      p = rand(1,2); % random point in the unit square
22      w = randn(1,2); % random normal vector to the line
23      w0 = p * w'; % the free term (neg)
24      plot([0 1],[w0, (w0-w(1))]/w(2),'r-',...
25          'linewidth',1.4) % plot the linear boundary
26      plot(p(1),p(2),'r.','markersize',15)
27      pause(0.03)
```

```
28      t = 2  - ([x y] * w' - w0 > 0);
29      if mean(t == lb) < 0.5, t = 3-t; end % revert labels
30
31      % Posteriors
32      ou = [x y] * w' - w0;
33      % Store the estimates of the probability for class 1
34      P1(:,i) = 1./(1 + exp(-ou * sc)); % softmax
35
36  end
37
38  % Find and plot the LDC combiner output
39  assigned_labels = classify(P1,P1,lb);
40  % (train with the noisy labels)
41  accuracy_LDC = mean(assigned_labels == labtrue);
42  plot(x(assigned_labels==1),y(assigned_labels==1),...
43      'bo','linewidth',1.5)
44  title(['LDC combiner accuracy ',num2str(accuracy_LDC)])
45  %-----------------------------------------------------------%
```

6

ENSEMBLE METHODS

6.1 BAGGING

6.1.1 The Origins: Bagging Predictors

Breiman introduced the term *bagging* as an acronym for *Bootstrap AGGregatING* [46]. The idea of bagging is simple and appealing: the ensemble is made of classifiers built on bootstrap replicates of the training set. The classifier outputs are combined by the plurality vote [47].

The diversity necessary to make the ensemble work is created by using different training sets. Ideally, the training sets should be generated randomly from the distribution of the problem. In practice, we can only afford one labeled training set, $\mathbf{Z} = \{\mathbf{z}_1, \dots, \mathbf{z}_N\}$, and have to imitate the process or random generation of L training sets. We sample *with replacement* from the original training set (bootstrap sampling [115]) to create a new training set of length N. To make use of the variations of the training set, the base classifier should be *unstable*. In other words, small changes in the training set should lead to large changes in the classifier output. Otherwise, the resultant ensemble will be a collection of almost identical classifiers, therefore unlikely to improve on a single classifier's performance. Figure 6.1 shows the training and operation of bagging.

How large should a bagging ensemble be? Breiman found that 25–50 decision trees are sufficient to get the ensemble error to level off [47]. Practice has shown that this rule of thumb works for a large variety of classifier models and over many different application domains [181, 293].

Combining Pattern Classifiers: Methods and Algorithms, Second Edition. Ludmila I. Kuncheva.
© 2014 John Wiley & Sons, Inc. Published 2014 by John Wiley & Sons, Inc.

BAGGING ENSEMBLE

Training: Given is a labeled data set $\mathbf{Z} = \{\mathbf{z}_1, \dots, \mathbf{z}_N\}$.

1. Choose the ensemble size L and the base classifier model.
2. Take L bootstrap samples from $\mathbf{Z}$ and train classifiers $D_1, \dots, D_L$, one classifier on each sample.

Operation: For each new object

1. Classify the new object $\mathbf{x}$ by all classifiers $D_1, \dots, D_L$.
2. Taking the label assigned by classifier D_i to be a "vote" for the respective class, assign to $\mathbf{x}$ the class with the largest number of votes.

Return the ensemble label of the new object.

FIGURE 6.1   Training and operation algorithm for the bagging ensemble.

Bagging is a parallel algorithm in both its training and operational phases. The L ensemble members can be trained on different processors if needed. A MATLAB example of bagging is given in Appendix 6.A.1.

6.1.2   Why Does Bagging Work?

If the classifier outputs were independent, and the classifiers had the same individual accuracy p, then the majority vote is guaranteed to improve on the individual performance [250]. Bagging aims at developing independent classifiers by taking bootstrap replicates as the training sets. The samples are pseudo-independent because they are taken from the same $\mathbf{Z}$. However, even if they were drawn independently from the distribution of the problem, *the classifiers* built on these training sets might not give independent outputs.

📕 **Example 6.1   Independent and bootstrap samples**
The data for this example was the rotated checker board data (Figure 1.8 in Chapter 1). A training set of 100 points and a testing set of 1000 points were generated 50 times with parameters $a = 0.5$ (side) and $\alpha = -\pi/3$ (rotation angle). Bagging was run with decision trees as the base classifier.[1] The trees were pre-pruned using a fixed threshold $\theta = 3$. To evaluate the effect of bootstrap sampling we ran the same experiment but instead of bootstrap samples of the training set we generated a new training set of 100 objects for each new member of the ensemble.

The purpose of the example is to examine to what extent bootstrap sampling induces dependence between the individual classifiers compared to independent sampling. We also show how this dependence translates into a higher ensemble error.

---

[1] We used the MATLAB code from Chapter 2.

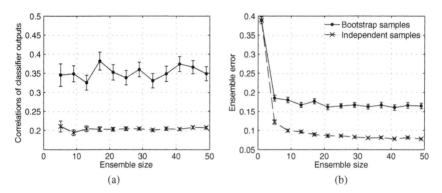

(a)   (b)

FIGURE 6.2   Correlation and error rate for the bagging ensemble versus the ensemble size for the rotated checker board data, using bootstrap samples and independent samples.

The correlation between oracle classifier outputs (correct/wrong labels) is one possible measure of dependence. The correlation between the outputs of classifiers D_i and D_j is calculated as

$$\rho_{i,j} = \frac{N^{11}N^{00} - N^{01}N^{10}}{\sqrt{(N^{11} + N^{10})(N^{01} + N^{00})(N^{11} + N^{01})(N^{10} + N^{00})}}, \tag{6.1}$$

where N^{ab} is the number of objects in the testing set for which classifier D_i gives output a and classifier D_j gives output b, where $a, b \in \{0, 1\}$. To get a single value measuring the correlation for the whole ensemble, $\bar{\rho}$, the pairwise correlations were averaged.

Figure 6.2a shows the ensemble correlation $\bar{\rho}$, averaged over the 50 repetitions, as a function of the ensemble size L. The error bars depict 95% confidence intervals of the estimate ($\frac{1.96}{\sqrt{50}} \times$ standard deviation). Figure 6.2b shows the ensemble error rate averaged over the 50 repetitions.

The plots show that, as expected, the correlation between classifiers built on true independent samples is lower than that produced by bagging. Observe that the correlation for the true independent samples is not zero, which demonstrates that the outputs of classifiers built on independent samples might be dependent.

We can think of the ensembles with lower correlation as more diverse than those with higher correlation. Figure 6.2b shows the benefit of having more diverse ensembles. The base classifiers are the same, the pruning method is the same, the combiner is the same (majority vote) but the error rates are different. The improved error rate can therefore be attributed to higher diversity in the ensemble that uses independently generated training sets. We shall see later that the concept of diversity is not as simple and straightforward as it looks here.

Domingos [101] examines two hypotheses about bagging in the framework of Bayesian learning theory. The first hypothesis is that bagging manages to estimate the posterior probabilities for the classes $\hat{P}(\omega_i|\mathbf{x})$. According to this model, the estimated posterior probability that the class label for the given $\mathbf{x}$ is ω_i, given the training set $\mathbf{Z}$, is averaged across all classifiers. Using $d_{j,i}(\mathbf{x})$ to denote the estimate of $P(\omega_i|\mathbf{x})$ given by classifier D_j, we have

$$\hat{P}(\omega_i|\mathbf{x}, \mathbf{Z}) = \sum_{j=1}^{L} \hat{P}(\omega_i|\mathbf{x}, D_j, \mathbf{Z})\hat{P}(D_j|\mathbf{Z}) = \sum_{j=1}^{L} d_{j,i}(\mathbf{x})\,\hat{P}(D_j|\mathbf{Z}). \qquad (6.2)$$

Take $d_{j,i}(\mathbf{x})$ to be the zero-one output indicating a (hard) class label. For example, in a four-class problem, a classifier output ω_3 corresponds to $d_{j,3}(\mathbf{x}) = 1$ and $d_{j,1}(\mathbf{x}) = d_{j,2}(\mathbf{x}) = d_{j,4}(\mathbf{x}) = 0$. If we set all the model (posterior) probabilities $P(D|\mathbf{Z})$ to $\frac{1}{L}$, we obtain the plurality voting combination which is the traditional combiner for bagging. Domingos also looks at soft class labels and chooses $\hat{P}(D_j|\mathbf{Z})$ so that the combiner is equivalent to simple averaging and weighted averaging. The hypothesis that bagging develops a better estimate of the posterior probabilities within this Bayesian context was not supported by Domingos' experiments [101].

The second hypothesis obtained better empirical support. According to it, bagging shifts the prior distribution of the classifier models toward models that have higher complexity (as the ensemble itself). Such models are assigned a larger likelihood of being the "right" model for the problem. The ensemble is in fact a single (complex) classifier picked from the new distribution.

Domingos' conclusions are matched by an argument in Ref. [352] where the authors challenge the common intuition that voting methods work because they "smooth out" the estimates. They advocate the thesis that voting in fact increases the complexity of the system.

6.1.3   Out-of-bag Estimates

As mentioned in Chapter 1, a bootstrap sample of size N from N data points will leave about 37% of the data out of the sample. These data points are called *out-of-bag*. A proportion of the L classifiers in the ensemble, say $M < L$, will not have seen an object $\mathbf{x}$ in their training set. Then $\mathbf{x}$ can be used to estimate the error of the ensemble of these M classifiers. For each $\mathbf{z}_j$ in the training data set, there would be an individual sub-ensemble of classifiers for which $\mathbf{z}_j$ is an out-of-bag object. If we score the ensemble error on each object in the training data, we can have an estimate of the generalization error of the ensemble. The only caveat here is that the error estimate will refer to a smaller ensemble, and may therefore be slightly pessimistically biased, especially for smaller ensembles [50].

▉ Example 6.2   Out-of-bag estimate of the ensemble error

Using the same experimental set-up as in the previous example with threshold $\theta = 1$ (less pruning), the out-of-bag estimate was calculated for each ensemble. Figure 6.3 shows the independent testing error and the out-of-bag error for the checker board

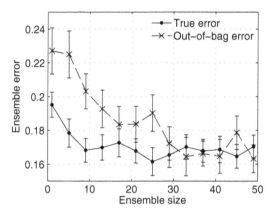

FIGURE 6.3   Independent testing error and out-of-bag error for the checker board data and different ensemble sizes.

data and different ensemble sizes. The error bars depict a 95% confidence interval of the estimate. The higher out-of-bag error curve demonstrates the pessimistic bias of the estimate for small ensemble sizes.

6.1.4   Variants of Bagging

Breiman [49] suggests using the so-called *small votes* whereby the individual classifiers are trained on relatively small subsets of the training set called "bites" [68]. The training sets are sampled from the large data set either randomly, called *Rvotes*, or based on importance, called *Ivotes* [68]. A cascade procedure for filtering the training sets for an ensemble of three neural networks is proposed by Drucker et al. [104]. The small vote ensembles are suitable for large labeled data sets, where the optimization of the computational resources is important.

Skurichina [369] proposes a variant of bagging which she calls *"nice" bagging*. Instead of taking all L classifiers, we only accept classifiers whose training error is smaller than the error made by an individual classifier built on the whole data set. In fact, we may again consider an ensemble of L classifiers by dismissing in the training process the classifiers whose training error is above the threshold and continuing until the ensemble size reaches L. Bagging and "nice" bagging have been found to work for unstable LDC [369]. Many other bagging variants have been proposed, among which are the double bagging [186] and asymmetric bagging [381].

6.2   RANDOM FORESTS

In 2001, Breiman proposed a variant of bagging which he calls a *random forest* [50]. Random forest is a general class of ensemble-building methods using a decision tree as the base classifier.

Definition. [50] A *random forest* is a classifier consisting of a collection of tree-structured classifiers, each tree grown with respect to a random vector Θ_k, where Θ_k, $k = 1, \ldots, L$, are independent and identically distributed. Each tree casts a unit vote for the most popular class at input **x**.

According to this definition, a random forest could be built by sampling from the feature set, from the data set, or just varying randomly some of the parameters of the tree. Any combination of these sources of diversity will also lead to a random forest. For example, we may sample from the feature set *and* from the data set as well [254].

The classical random forest could be thought of as a version of bagging where the base classifier is a random tree, introduced in Chapter 2. Along with selecting bootstrap samples from the training data, random feature selection is carried out *at each node of the tree*. We choose randomly a subset S with M features from the original set of n features, and seek within S the best feature to split the node. A feature subset is selected anew for each node. Breiman suggests to grow a full CART tree (no pruning). The recommended value of M is $\lfloor \log_2 n + 1 \rfloor$ [50].

The random forest heuristic is meant to diversify the ensemble. Alternative approaches include fuzzy trees [43] and random feature weights [278]. Some accuracy of the decision tree may be sacrificed but this pays off through the increased diversity. Indeed, random forest has been the classifier of choice in many application areas such as remote sensing [157, 168, 299], chemistry [378], ecology [79], and medical data analysis [191, 228]. Recent applications include motion and pose recognition from depth images, which is becoming a core component of the Kinect gaming platform [364]. The random forest ensemble method has been implemented for research purposes in R programing language [259] as well as in WEKA [167].

📖 Example 6.3   Bagging and random forest

For now, we will only illustrate the difference between the performance of a bagging ensemble and a random forest ensemble. Measuring the diversity will be discussed in Chapter 8.

Consider the letter data set ($N = 20,000$ objects, $n = 16$ features, $c = 26$ classes). The data was split into 50% training and 50% testing parts. The base classifiers were trained on the whole training set, and majority vote (plurality) was applied to aggregate the individual outputs. The ensemble error was evaluated on the testing data. This procedure was repeated 10 times. Figure 6.4 shows the ensemble error rate, averaged over the 10 repetitions, for the bagging and the random forest ensembles and different ensemble sizes.

The figure shows that, for this data set, random forest outperforms bagging for large ensemble sizes. The inferior performance for small L is likely due to the fact that the individual classifiers may not have had access to all features, hence their error rate would be higher.

For this experiment, we used the MATLAB functions `bagging_train` and `bagging_classify` shown in Appendix 6.A.1.

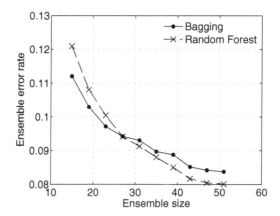

FIGURE 6.4   Ensemble error for bagging and random forest for the letter data set.

6.3   ADABOOST

The boosting family of algorithms have secured a top spot in the rank list of ensemble methods due to their accuracy, robustness, and wide applicability. They have been pronounced the "most accurate available off-the-shelf classifiers on a wide variety of data sets" [48]. AdaBoost was the only ensemble method featured among the "Top 10 algorithms in data mining" by Wu et al. [422]. A lot of enlightening research literature is devoted to these methods, including Schapire and Freund's recent monograph [351]. In 2003, the two authors received the prestigious Goedel Prize (Theoretical Computer Science) for their AdaBoost algorithm. Here we will look at the basic algorithm and several variants, without reproducing the rich and lively theoretical and empirical arguments explaining why boosting works so well.

6.3.1   The AdaBoost Algorithm

Boosting is defined in Ref. [134] as related to the "general problem of produc-ing a very accurate prediction rule by combining rough and moderately inaccurate rules-of-thumb." The general boosting idea is to develop the classifier ensemble D incrementally, adding one classifier at a time. The classifier that joins the ensemble at step k is trained on a data set selectively sampled from the training data set $\mathbf{Z}$. The sampling distribution starts from uniform, and is updated for each new classifier. The likelihood of the objects being misclassified at step $k-1$ is increased so that they have a higher chance of entering the training sample of the next classifier. The algorithm is called *AdaBoost* in Ref. [134] which comes from ADAptive BOOSTing.

There are two implementations of AdaBoost: with *reweighting* and with *resam-pling*. The description above refers to the resampling implementation. For the reweighting implementation we assume that the base classifiers can directly use the probabilities on $\mathbf{Z}$ as weights. No sampling is needed in this case, so the

ADABOOST.M1 (training)

Training: Given is a labeled data set $\mathbf{Z} = \{\mathbf{z}_1, \ldots, \mathbf{z}_N\}$.

1. Choose the ensemble size L and the base classifier model.
2. Set the weights $\mathbf{w}^1 = [w_1^1, \ldots, w_N^1]$, $w_j^1 \in [0, 1]$, $\sum_{j=1}^N w_j^1 = 1$.
   (Usually $w_j^1 = \frac{1}{N}$, $j = 1, \ldots, N$).
3. For $k = 1, \ldots, L$
   (a) Take a sample S_k from $\mathbf{Z}$ using distribution $\mathbf{w}^k$.
   (b) Build a classifier D_k using S_k as the training set.
   (c) Calculate the weighted ensemble error at step k by

   $$\epsilon_k = \sum_{j=1}^N w_j^k l_k^j, \tag{6.3}$$

   $\left(l_k^j = 1 \text{ if } D_k \text{ misclassifies } \mathbf{z}_j \text{ and } l_k^j = 0 \text{ otherwise} \right)$.
   (d) If $\epsilon_k = 0$, reinitialize the weights w_j^k to $\frac{1}{N}$ and continue.
      i. Else if $\epsilon_k \geq 0.5$, ignore D_k, reinitialize the weights w_j^k to $\frac{1}{N}$ and continue.
      ii. else, calculate

   $$\beta_k = \frac{\epsilon_k}{1 - \epsilon_k}, \quad \text{where} \quad \epsilon_k \in (0, 0.5), \tag{6.4}$$

   and update the individual weights

   $$w_j^{k+1} = \frac{w_j^k \beta_k^{(1 - l_k^j)}}{\sum_{i=1}^N w_i^k \beta_k^{(1 - l_k^j)}}, \quad j = 1, \ldots, N. \tag{6.5}$$

4. Return $D = \{D_1, \ldots, D_L\}$ and $\beta_1, \ldots, \beta_L$.

FIGURE 6.5   Training algorithm for AdaBoost.M1.

algorithm becomes completely *deterministic*. There is no strong evidence favoring one of the versions over the other [48, 134, 135].

AdaBoost was proposed initially for two classes and then extended for multiple classes. Figures 6.5 and 6.6 show, respectively, the training and operation of AdaBoost.M1, which is the most straightforward multi-class extension of AdaBoost [134]. Here we give the resampling implementation.

■ **Example 6.4   Illustration of AdaBoost**
The performance of the AdaBoost algorithm is illustrated on the fish data. The experimental set-up was: 50 runs with a training set with 20% label noise and the

---

ADABOOST.M1 (operation)

Operation: For each new object

1. Classify the new object **x** by all classifiers $D_1, \ldots, D_L$.
2. Calculate the support for class ω_t by

$$\mu_t(\mathbf{x}) = \sum_{D_k(\mathbf{x})=\omega_t} \ln\left(\frac{1}{\beta_k}\right). \qquad (6.6)$$

3. The class with the maximum support is chosen as the label for **x**.

Return the ensemble label of the new object.

FIGURE 6.6   Operation algorithm for AdaBoost.M1.

nondistorted data as the testing set. The base classifier was the nonpruned decision tree. The testing error averaged across the 50 runs is shown in Figure 6.7. For comparison we also show the bagging ensemble error rate.

6.3.2   The arc-x4 Algorithm

Breiman studies bagging and boosting from various curious angles in Ref. [48]. He calls the class of boosting algorithms *arcing* algorithms as an acronym for "Adaptive Resample and Combining." His *arc-fs* algorithm is AdaBoost (named "fs" after its authors Freund and Schapire [134]). Breiman proposed a boosting algorithm called *arc-x4* to investigate whether the success of AdaBoost roots in its technical details

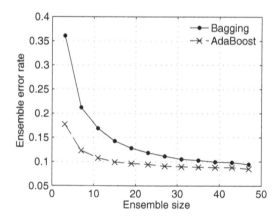

FIGURE 6.7   Testing error of AdaBoost and bagging versus the ensemble size for the fish data set.

ARC-X4

Training: Given is a labeled data set $\mathbf{Z} = \{\mathbf{z}_1, \ldots, \mathbf{z}_N\}$.

1. Choose the ensemble size L and the base classifier model.
2. Set the weights $\mathbf{w}^1 = [w_1^1, \ldots, w_N^1]$, $w_j^1 \in [0, 1]$, $\sum_{j=1}^N w_j^1 = 1$. $\left(\text{Usually } w_j^1 = \frac{1}{N}\right)$.
3. For $k = 1, \ldots, L$
   (a) Take a sample S_k from $\mathbf{Z}$ using distribution $\mathbf{w}^k$.
   (b) Build a classifier D_k using S_k as the training set.
   (c) Find m_j as the proportion of classifiers currently in the ensemble which misclassify $\mathbf{z}_j$. Update the individual weights

$$
w_j^{k+1} = \frac{1 + m_j^4}{\sum_{i=1}^N 1 + m_i^4} , \quad j = 1, \ldots, N. \tag{6.7}
$$

Operation: For each new object

1. Classify the new object $\mathbf{x}$ by all classifiers $D_1, \ldots, D_L$.
2. Taking the label assigned by classifier D_i to be a "vote" for the respective class, assign to $\mathbf{x}$ the class with the largest number of votes.

Return the ensemble label of the new object.

FIGURE 6.8   Training and operation algorithm for arc-x4.

or in the resampling scheme it uses. The difference between AdaBoost and arc-x4 is twofold. First, the weight for object $\mathbf{z}_j$ at step k is calculated as the proportion of times $\mathbf{z}_j$ has been misclassified by the $k - 1$ classifiers built so far. Second, the final decision is made by plurality voting rather than weighted majority voting. The arc-x4 algorithm is described in Figure 6.8.

The parameter of the algorithm, the power of m_j (Figure 6.8), has been fixed to the constant 4 (hence the name) by a small experiment. Breiman compared the behaviors of AdaBoost and arc-x4 and found that AdaBoost makes more abrupt moves while arc-x4 has a more gradual behavior, both showing a similar overall performance. This is reflected, for example, in the standard deviations of the weights assigned to a single data point. This was found to be much larger for AdaBoost than for arc-x4. Breiman concludes that the two boosting variants could be on the two edges of a scale, and new even more successful algorithms could be found in-between [48].

6.3.3   Why Does AdaBoost Work?

One of the explanations for the success of AdaBoost comes from the algorithm's property to drive the training error to zero very quickly, practically in the first few iterations.

6.3.3.1   The Upper Bound on the Training Error   Freund and Schapire prove an upper bound on the training error of AdaBoost [134]. The following theorem gives their result:

Theorem 6.1   Let $\Omega = \{\omega_1, \dots, \omega_c\}$. Let ϵ be the ensemble training error and let $\epsilon_i, i = 1, \dots, L$ be the weighted training errors of the classifiers in D as in Equation (6.3) where $\epsilon_i < 0.5$. Then

$$\epsilon \; < \; 2^L \prod_{i=1}^{L} \sqrt{\epsilon_i(1 - \epsilon_i)}. \tag{6.8}$$

This theorem indicates that, by adding more classifiers with individual error smaller than 0.5, the training error of the ensemble approaches zero. Freund and Schapire argue that the 0.5 threshold is too strict a demand for a multiple-class *weak learner*. They proceed to propose another version of AdaBoost, called AdaBoost.M2 which does not require $\epsilon_i < 0.5$ [134]. Note that ϵ_i is *not* the error of classifier D_i; it is a *weighted* error. This means that if we applied D_i on a data set drawn from the problem in question, its (conventional) error could be different from ϵ_i, it could be larger or smaller.

6.3.3.2   The Margin Theory   Experiments with AdaBoost showed an unexpected phenomenon: the testing error continues to decrease with adding more classifiers even after the training error reaches zero. This prompted another look into the possible explanations and brought forward the margin theory [135, 352].

    The concept of *margins* (see Chapter 2) comes from the statistical learning theory [402] in relation to the Vapnik–Chervonenkis dimension (VC dimension). In layman's terms, the VC dimension gives an upper bound on the classification ability of classifier models. Although the bound is loose, it has proven to be an important theoretical accessory in pattern recognition and machine learning. Intuitively, the margin for an object is related to the certainty of its classification. Objects for which the assigned label is correct and highly certain will have large margins. Negative margins signify incorrect classification. Objects with uncertain classification are likely to have small margins. A small margin will cause instability of the classification label, that is, the object might be assigned to different classes by two similar classifiers.

    For c classes, the classification margin (also ensemble margin or voting margin) of object $\mathbf{x}$ is calculated using the degrees of support $\mu_j(\mathbf{x}), j = 1, \dots, c$, as

$$m(\mathbf{x}) = \mu_k(\mathbf{x}) - \max_{j \neq k}\{\mu_j(\mathbf{x})\}, \tag{6.9}$$

where ω_k is the (known) class label of $\mathbf{x}$ and $\sum_{j=1}^{c} \mu_j(\mathbf{x}) = 1$.

    Thus all objects which are misclassified will have negative margins, and those correctly classified will have positive margins. Trying to maximize the margins (called boosting the margins) will intuitively lead to "more confident" classifiers.

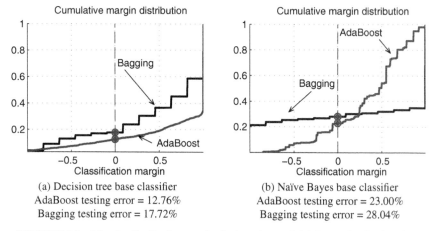

FIGURE 6.9   Margin distribution graphs for bagging and AdaBoost for the fish data.

Schapire et al. [352] prove upper bounds (loose) on the testing error which depend on the classification margin.

Example 6.5   AdaBoost and ensemble margins

To illustrate the effect of bagging and AdaBoost on the margins, we ran both algorithms for the fish data with 20% label noise and ensembles with $L = 11$ classifiers. The margins were calculated according to Equation 6.9. Figure 6.9 shows the margin distribution graphs for AdaBoost and bagging for two base classifier models. The x-axis is the margin m, and the y-axis is the number of points whose margin is less than or equal to m. If all training points are classified correctly, there will be only positive margins. Ideally, all points should be classified correctly and with the maximum possible certainty, that is, the cumulative graph should be a single vertical line at $m = 1$. Figure 6.9 shows the margin distribution graphs for bagging and AdaBoost.[2] AdaBoost gives lower classification error for both classifiers. Its margin curve is entirely underneath the curve for the bagging ensemble with the decision tree base classifier, demonstrating the superiority of the algorithm. However, bagging shows a better curve in the positive margin's range for the Naïve Bayes classifier.

The difference between AdaBoost and bagging can be illustrated further by the distribution from which the samples are drawn. While bagging is expected to have a uniform coverage of the available data, AdaBoost is likely to favor difficult objects, as illustrated by the example below.

Example 6.6   Heat map of AdaBoost sampling distribution

The fish data was used again with 10% label noise. The ensemble size was set at $L = 101$ in order to have a reasonable estimate of the sampling distribution of

---

[2] In bagging, the support for class ω_j, $\mu_j(\mathbf{x})$ was calculated as the proportion of votes for that class.

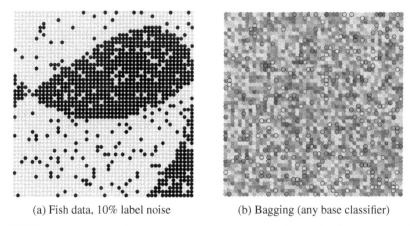

(a) Fish data, 10% label noise   (b) Bagging (any base classifier)

FIGURE 6.10   Heat map of the sampling distribution for bagging for the fish data ($L = 101$). Error with the Naïve Bayes base classifier was 27.76% and with the decision tree classifier, 2.52%.

AdaBoost and bagging. Figures 6.10 and 6.11 show the heat maps for the two ensemble methods, for the Naïve Bayes classifier, and the (nonpruned) decision tree classifier. The noise points are circled. Dark color signifies an often selected point.

The resampling for AdaBoost depends on the base classifier because the subsequent resampling distributions are determined by the previously misclassified objects. Hence there are two distribution plots in Figure 6.11. The bagging resampling, on the other hand, does not depend on the outcome of the previous classifiers, therefore there is only one distribution pattern as shown in Figure 6.10b.

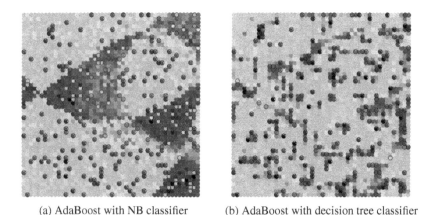

(a) AdaBoost with NB classifier   (b) AdaBoost with decision tree classifier

FIGURE 6.11   Heat map of the sampling distribution for AdaBoost for the fish data ($L = 101$). Error with the Naïve Bayes base classifier was 23.04% and with the decision tree classifier, 5.28%.

As expected, bagging covers the data space uniformly. AdaBoost, on the other hand, demonstrates a clear nonuniform pattern. In both plots in Figure 6.11, the label noise is among the most often selected points. This illustrates the tendency of AdaBoost to focus on noise. While this tendency has been regarded as a drawback of AdaBoost, in some cases such isolated points are valid islands of data which should not be ignored.

There is a notable difference between the patterns in the two subplots in Figure 6.11. The Naïve Bayes classifier persistently mislabels parts of the classification regions of the two classes, forcing AdaBoost to keep sampling from these regions, as shown by the darker patches. This parallel gives an insight into stable and unstable base classifiers. Being a stable classifier, NB fails on the same training objects, and these are repeatedly selected by AdaBoost in the subsequent training samples. Such a pattern does not exist for the decision tree classifier. The ensemble error rate for both bagging and AdaBoost is by an order of magnitude smaller for the unstable base classifier—the decision tree.

6.3.3.3   Bias and Variance Reduction   The results of extensive experimental studies can be summarized as follows: bagging is assumed to reduce variance without changing the bias. However, it has been found to reduce the bias as well, for some high-bias classifiers. AdaBoost has different effects at its different stages. At its early iterations boosting primarily reduces bias while at the later iterations it has been found to reduce mainly variance [29, 48, 96, 102, 135, 352, 396].

Freund and Schapire [135] argue that bias–variance decomposition is not the appropriate analysis tool for boosting, especially boosting with reweighting. Their explanation of the success of AdaBoost is based on the margin theory. Schapire et al. [352] state that it is unlikely that a "perfect" theory about voting methods exists, that is, a theory applicable to any base classifier and any source of independent identically distributed labeled data. Friedman et al. note that in practice, boosting achieves results far more impressive than the theoretical bounds and analyses would imply [136].

Many authors have compared bagging and AdaBoost including some large-scale experiments [29, 48, 94, 95, 319]. The general consensus is that boosting reaches lower testing error in spite of its sensitivity to noise and outliers, especially for small data sets [29, 48, 319].

6.3.4   Variants of Boosting

There is a great variety of methods drawn upon the basic idea of boosting. The majority of these address the sensitivity of AdaBoost to noise or alter the optimization criterion. Table 6.1 gives a chronological list of some AdaBoost variants.

For further reading, consider the monographs by Schapire and Freund [351] and Zhou [439].

6.3.5   A Famous Application: AdaBoost for Face Detection

Viola and Jones's algorithm for detecting a face in a still image [405] has been recognized as one of the most exciting breakthroughs in computer vision. The detector

TABLE 6.1 AdaBoost Variants (Classification Task)

Year	Name
1997	AdaBoost [134]
1998	Arc-x4 [48]
1999	AdaBoost.MH [350], AdaCost [66], DynaBoost [283]
2000	Real AdaBoost, LogitBoost, Gentle AdaBoost [136], MultiBoost [412], DOOM [274], MadaBoost [100]
2001	Agnostic boosting [32], BrownBoost [131], ATA, AdaBoost-reg [323], FloatBoost [257]
2002	AdaBoost.M1W [117], LPBoost [88], Stochastic gradient boosting [138]
2003	Smooth boosting [356], L2Boost [59], SMOTEBoost [69], AdaBoost-VC [267], AveBoost [296]
2004	
2005	AdaBoost.M2 [118, 133], Martingale boosting [266]
2006	TotalBoost [411], SAMME [441], MPBoost [125], MutualBoost [362]
2007	One-pass boosting, Picky AdaBoost [27]
2008	RotBoost [435], FilterBoost [45]
2009	RobustBoost [132]
2010	RUSBoost [354], Twin boosting [60]
2011	
2012	SampleBoost [2]
2013	POEBoost [114]

scans the image, sliding a square window and classifying it as positive (face) or negative (nonface). The procedure is run at several different magnifications of the image. The detector is a cascade of AdaBoost ensembles. Each ensemble is trained using decision stumps, selecting from a large set of Haar-like features. The beauty of the Viola–Jones detector lies in the simple way of calculating the image features and the quick labeling of negative windows in the first steps of the classification cascade. This makes the method significantly faster than other similar face detection methods.

The method uses three types of features, shown diagrammatically in Figure 6.12. The features are calculated from the gray level intensities of the image. The value of a "two-rectangle" feature is the difference between the sum of the pixels within two rectangular regions. The regions have the same size and shape and are horizontally or vertically adjacent. If the sums of the gray level intensities are denoted by a and b,

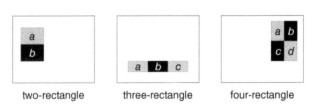

two-rectangle   three-rectangle   four-rectangle

FIGURE 6.12 The three types of Haar-like features used in Viola–Jones face detector.

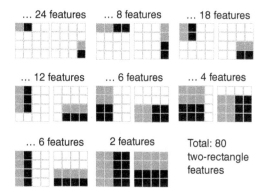

... 24 features   ... 8 features   ... 18 features

... 12 features   ... 6 features   ... 4 features

... 6 features   2 features   Total: 80
two-rectangle
features

FIGURE 6.13   The two-rectangle Haar-like features for a window of size 4×4.

respectively, the value of the feature is $a - b$. The value of a "three-rectangle" feature, shown in the middle plot, is $b - (a + c)$. The value of the "four-rectangle" feature is $(a + d) - (b + c)$.

The detector works with a window of size 24×24 pixels, which gives rise to over 160,000 features. As an example, Figure 6.13 shows the 80 possible two-rectangle features for a window of size 4×4. The authors propose a clever way for a fast calculation of these features using an *integral image*.

Each feature is regarded as a classifier itself. The class label is obtained by thresholding the value. Thus, AdaBoost with decision stumps serves as a method for feature selection [405]. The cascade detector is designed to take advantage of the fact that the overwhelming majority of windows in an image are negative. It operates as shown in Figure 6.14.

The training process goes through the following steps:

1. Choose the target false positive rate F_{target} and the acceptable false positive and false negative rates for the cascade stages.

2. Obtain a data set with positive and negative examples, and set aside a validation set.

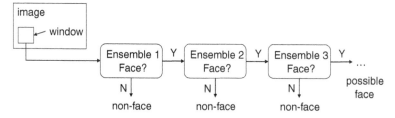

FIGURE 6.14   Operation of the cascade face detector.

FIGURE 6.15 An example of the output from the Viola–Jones face detector.

3. Add a new AdaBoost ensemble trained on the current set of positive and negative examples.
4. Test the current cascade classifier on the validation set.
5. If F_{target} has not been reached, discard the current negative examples and assemble a new set of negative examples from the false positives obtained at the previous step. Continue from step 3. Otherwise, return the trained cascade of ensembles.

Can the detector work without AdaBoost? Probably, but we would like to believe that AdaBoost shares a large portion of the credit for the remarkable performance of the Viola and Jones' face detector.

The detector has been trained on 5000 manually labeled positive examples (face windows) and 300 million negative examples (nonface windows). The final cascade consists of 38 ensembles (stages). The first ensemble uses only two features, rejects about 50% of the windows, and detects nearly 100% of the face windows. The second ensemble uses 10 features, and again detects nearly 100% of the face windows, while discarding a further 30% of nonfaces. The total number of features used by the cascade is 6060. However, on average, 10 features are needed to label a window.

According to the authors, training of the detector took weeks of CPU time but the operation is remarkably fast. The trained classifier is implemented in many programing languages, and is also encoded within OpenCV library.[3] Finally, an example of the result from the Viola–Jones face detector is shown in Figure 6.15.[4] Six positive windows were returned, all of which are suitably positioned on the image.

---

[3]OpenCV (Open Source Computer Vision) is a library of programming functions for real-time computer vision. http://opencv.org/

[4]We used a MATLAB interface to OpenCV written by Dirk-Jan Kroon. http://www.mathworks.com/matlabcentral/fileexchange/29437-viola-jones-object-detection

---

RANDOM SUBSPACE

Training: Given is a labeled data set $\mathbf{Z} = \{\mathbf{z}_1, \dots, \mathbf{z}_N\}$, where each object is described by the features in the feature set $X = \{X_1, \dots, X_n\}$.

1. Choose the ensemble size L, the base classifier model, and the number of features d ($d < n$) to be sampled for each classifier.
2. Take L samples of size d from X, and train classifiers $D_1, \dots, D_L$. The samples are usually taken without replacement.

Operation: For each new object:

1. Classify the new object $\mathbf{x}$ by all classifiers $D_1, \dots, D_L$. For classifier D_i, use only the features that this classifier was trained on.
2. Taking the label assigned by classifier D_i to be a "vote" for the respective class, assign to $\mathbf{x}$ the class with the largest number of votes.

Return the ensemble label of the new object.

FIGURE 6.16   Training and operation algorithm for the random subspace ensemble.

6.4   RANDOM SUBSPACE ENSEMBLES

In problems with a large number of features, a natural ensemble-building heuristic is to use different feature subsets to train the ensemble members. Since formally introduced and named *Random Subspace Ensemble* by Ho [182], this method has taken the lead in many application domains characterized by high-dimensional data. Examples include, but are not limited to, phenotype recognition [433], cancer diagnosis [19, 36, 37], functional magnetic resonance imaging (fMRI) data analysis [227, 228], face recognition [70, 410, 442], credit scoring [407], and bankruptcy prediction [288]. The random subspace heuristic can be used together with any other compatible heuristic for building the ensemble, for example, taking bootstrap samples together with taking a feature sample [58, 254, 407].

The random subspace method is detailed in Figure 6.16.

MATLAB code for the random subspace ensemble is given in Appendix 6.A.3.

▣ Example 6.7   Random subspace ensemble
For this example we used the "mfeat" data set from the UCI Machine Learning Repository [22].[5] This data set was donated by Bob Duin, Delft University of Technology. It consists of features of handwritten numerals from 0 to 9 extracted from a collection of Dutch utility maps. Examples from all classes are shown in Figure 6.17. There are

---

[5]http://archive.ics.uci.edu/ml/

0 1 2 3 4 5 6 7 8 9

FIGURE 6.17 Examples of the handwritten digits in the mfeat UCI data set.

2000 data points in total, 200 from each class. Six feature sets were extracted and stored:

- 76 Fourier coefficients of the character shapes;
- 216 profile correlations;
- 64 Karhunen–Loève coefficients;
- 240 pixel averages in 2×3 windows;
- 47 Zernike moments; and
- 6 morphological features.

Using all features makes the problem too easy, hence we chose only the 216 profile correlations as the feature set X. We ran a 10-fold cross-validation for the random subspace ensemble method, splitting the data in a systematic way. Out of the 200 objects for a given class, the testing folds were formed from objects 1–20, 21–40, 41–60, and so on. The remaining objects were the training data. For comparison, we ran AdaBoost and bagging for the same splits. In all cases we used decision tree as the base classifier and an ensemble size of $L = 10$ classifiers. The average testing error rates and the standard deviations were as follows:

$$
\begin{array}{rl}
\text{Random subspace:} & 4.80\% \pm 1.38 \\
\text{Bagging:} & 6.55\% \pm 1.76 \\
\text{AdaBoost:} & 5.00\% \pm 0.94
\end{array}
$$

The random subspace ensemble outperformed the two competitors for these settings. This does not mean that it is universally better for high-dimensional spaces, and for any ensemble size. The example was meant to demonstrate that there are cases where the RS ensemble has an edge.

6.5 ROTATION FOREST

The rotation forest ensemble method [226, 331] relies on the instability of the decision tree classifier with respect to rotation of the space, an illustration of which was shown in Section 2.2.7. The algorithm is unashamedly heuristic, but is gaining popularity due to its empirical success, often outperforming Bagging, AdaBoost, random forest, and random subspace. Recent application areas include (but are not limited to) cancer classification [264], diagnosis of coronary artery disease [202] and Parkinson's disease [298], segmentation of computed tomography angiography (CTA) images

[21], genomic and proteomic research [377, 423], remote sensing [169, 203], image processing [432], and finance [85, 272].

The intuition behind rotation forest is that diversity can be enforced without sacrificing data, either objects or features. All other ensemble methods sample from the data or the feature set, potentially losing accuracy of the individual classifiers, which is then compensated by the increased diversity. The rotation forest method does not throw away data; it only presents it to the classifier from a different angle.

The rotation forest method prepares the training set for each individual classifier in the following way. Start with a labeled data set $\mathbf{Z} = \{\mathbf{z}_1, \ldots, \mathbf{z}_N\}$ described by feature set $X = \{X_1, \ldots, X_n\}$. Suppose that $\mathbf{Z}$ is organized as a numerical matrix of size $N \times n$. Choose the ensemble size L and an integer K. First, split randomly the feature set into K (disjoint) subsets of approximately equal size. For simplicity, suppose that K is a factor of the number of features n so that each feature subset contains $M = n/K$ features.

Denote by $S_{i,j}$ the jth subset of features for the training set of classifier D_i. For every such subset, select randomly a nonempty subset of classes and then draw a bootstrap sample of objects, of size 75% of the data count. Run principal component analysis (PCA) using only the M features in $S_{i,j}$ and the selected subset of classes. Store the coefficients of the principal components, $\mathbf{a}_{i,j}^{(1)}, \ldots, \mathbf{a}_{i,j}^{(M)}$. Note that it is possible that some of the eigenvalues are zero, therefore we may not have all M vectors but have instead $M_j \leq M$ vectors. Running PCA on a subset of classes is done in a bid to avoid identical coefficients if the same feature subset is chosen for different classifiers.

Next, organize the obtained vectors with coefficients in a sparse rotation matrix R_i

$$R_i = \begin{bmatrix} \mathbf{a}_{i,1}^{(1)}, \mathbf{a}_{i,1}^{(2)}, \ldots, \mathbf{a}_{i,1}^{(M_1)}, & [\mathbf{0}] & \cdots & [\mathbf{0}] \\ [\mathbf{0}] & \mathbf{a}_{i,2}^{(1)}, \mathbf{a}_{i,2}^{(2)}, \ldots, \mathbf{a}_{i,2}^{(M_2)}, & \cdots & [\mathbf{0}] \\ \vdots & \vdots & \ddots & \vdots \\ [\mathbf{0}] & [\mathbf{0}] & \cdots & \mathbf{a}_{i,K}^{(1)}, \mathbf{a}_{i,K}^{(2)}, \ldots, \mathbf{a}_{i,K}^{(M_K)} \end{bmatrix}. \quad (6.10)$$

(The rotation matrix will have dimensionality $n \times \sum_j M_j$.) To calculate the training set for classifier D_i we first rearrange the rows of R_i so that they correspond to the original features. Denote the rearranged rotation matrix R_i^a. Then, the training set for classifier D_i is $\mathbf{Z}R_i^a$.

For the operation of the rotation forest, each new object $\mathbf{x}$ (vector row) must be transformed using the rotation matrix for the specific classifier, D_i,

$$\mathbf{y} = \mathbf{x}R_i^a,$$

and subsequently classified. After obtaining the decisions of all L classifiers, $d_{i,j}$ (as the decision profile), we calculate the final degree of support using the average

---

ROTATION FOREST

Training: Given is a labeled data set $\mathbf{Z} = \{\mathbf{z}_1, \ldots, \mathbf{z}_N\}$ described by feature set $X = \{X_1, \ldots, X_n\}$.

1. Choose the ensemble size L, the base classifier model (decision tree is recommended) and K, the number of feature subsets.
2. For $i = 1 \ldots L$
   - (a) Prepare the rotation matrix R_i^a:
      - i. Split the feature set X into K subsets:
         $S_{i,j}$ (for $j = 1 \ldots K$).
      - ii. For $j = 1 \ldots K$
         - Let $Z_{i,j}$ be the data set for the features in $S_{i,j}$ (an $N \times |S_{i,j}|$ matrix).
         - Eliminate a random subset of classes from $Z_{i,j}$, resulting in a data set $Z'_{i,j}$.
         - Select a bootstrap sample from $Z'_{i,j}$, of size 75% of the number of objects in $Z'_{i,j}$. Denote the new set by $Z''_{i,j}$.
         - Apply PCA on $Z''_{i,j}$ to obtain and store the coefficients in a matrix $C_{i,j}$.
      - iii. Arrange the $C_{i,j}$, for $j = 1 \ldots K$ in a rotation matrix R_i as in Equation 6.10.
      - iv. Construct R_i^a by rearranging the the rows of R_i so as to match the order of features in X.
   - (b) Build classifier D_i using $\mathbf{Z}R_i^a$ as the training set, with the given class labels.

Operation: For each new object $\mathbf{x}$

1. For $i = 1 \ldots L$, calculate the transformed object $y = \mathbf{x}R_i^a$ and run it through classifier D_i. Denote by $d_{i,j}(\mathbf{y})$ the support assigned by classifier D_i to the hypothesis that $\mathbf{x}$ comes from class ω_j.
2. Calculate the confidence for each class, ω_j, by the average combination method.
3. Assign $\mathbf{x}$ to the class with the largest confidence.

Return the ensemble label of the new object.

FIGURE 6.18   Training and operation algorithm for the rotation forest ensemble.

combination method (Chapter 5). The training and operation of the rotation forest method are shown in Figure 6.18.

MATLAB code for the rotation forest ensemble method is given in Appendix 6.A.4. Continuing the example with the mfeat data set and the 216 profile correlation features, the ensemble error for the rotation forest ensemble was

$$\text{Rotation forest:} \quad 4.50\% \pm 1.63.$$

◼ Example 6.8   An example of rotation forest training
Here we present a worked numerical example for calculating the rotation matrix for a hypothetical classifier D_i. The data set Z consists of 10 objects described by $n = 9$ features, as shown in Table 6.2.

TABLE 6.2   Data for the Numerical Example with Rotation Forest

Data									Labels
1	5	2	−2	−1	1	−4	5	−4	1
−2	3	−1	−1	5	−2	−2	1	4	1
3	0	4	0	0	2	4	1	5	1
−3	0	1	−2	−3	3	−4	−2	3	2
2	0	−1	4	5	−2	5	0	−4	2
−3	−1	5	−3	5	−3	3	2	−2	2
−1	1	4	−2	0	−2	0	2	−1	3
2	1	1	−3	−3	−1	1	−1	2	3
3	4	2	−2	−2	0	−2	−1	−3	3
−4	3	1	0	0	1	0	5	3	3

Let the random permutation that shuffled the features be

$$1 \quad 3 \quad 8 \quad 2 \quad 4 \quad 5 \quad 6 \quad 7 \quad 9$$

Splitting into $K = 3$ subsets, we get $S_{i,1} = \{1,3,8\}$, $S_{i,2} = \{2,4,5\}$ and $S_{i,3} = \{6,7,9\}$. Suppose that the subset of classes for $S_{i,1}$ was ω_1 and ω_3. The data set is reduced to columns 1, 3, and 8 and rows 1–3 and 6–10. Taking a 75% bootstrap sample with indices $[1,6,6,2,7]$, the data submitted to the PCA calculation is as follows:

$$
\begin{array}{rrr}
1 & 2 & 5 \\
3 & 2 & -1 \\
3 & 2 & -1 \\
-2 & -1 & 1 \\
-4 & 1 & 5
\end{array}
$$

The three vectors with the principal component coefficients are:

$$
\begin{array}{rrr}
-0.7270 & 0.5283 & 0.4385 \\
-0.1470 & 0.5041 & -0.8510 \\
0.6707 & 0.6832 & 0.2889
\end{array}
$$

These vectors are inset as the top 3×3 entries of the rotation matrix R_i. The second 3×3 block striding the leading diagonal corresponds to feature subset $S_{i,2} = \{2,4,5\}$. The data set is reduced again, this time using another random subset of classes, say ω_2 and ω_3. A 75% bootstrap sample is taken from the reduced set and the principal components are calculated. In the same way, we calculate the last 3×3 block. The rotation matrix R_i is shown as a heat map in Figure 6.19a. To make this matrix usable with the original data set, the rows must be rearranged to correspond to the features. Thus the first row stays the same, the second row is replaced by the current fourth

(a) Unarranged rotation matrix R_i (b) Arranged rotation matrix R_i^a

FIGURE 6.19   Illustration of the rotation matrices R_i and R_i^a for classifier D_i.

row, and so on. The arranged matrix, R_i^a is shown in Figure 6.19b. The training data for classifier D_i is $\mathbf{Z}R_i^a$.

An experimental study [226] reported that

1. The splitting of the feature set X into subsets is essential for the success of rotation forest. This was demonstrated by comparing sparse with non-sparse random projections; the results were favourable to sparse random projections.
2. No pattern of dependency was found between $K(M)$ and the ensemble accuracy. As $M = 3$ worked well in previous experiments, this value was recommended as the default parameter of the algorithm.
3. In the experiments with 35 benchmark data sets, rotation forest was found to be better than bagging, AdaBoost, and random forest for all ensemble sizes, more noticeably so for smaller ensembles sizes.
4. Interestingly, PCA was found to be the best method for rotating the feature space. It was compared with nonparametric discriminant analysis (NDA), which takes into account the class labels, and also with sparse and nonsparse random projections. This issue was recently revisited, and supervised projections were advocated instead [148].

6.6   RANDOM LINEAR ORACLE

The Random Linear oracle (RLO) [239] is an ensemble method which combines the classifier fusion and classifier selection approaches. Each classifier in the ensemble is replaced by a mini-ensemble consisting of two classifiers. A linear function (oracle) decides which of the two classifiers gets to label the object. This approach encourages extra diversity in the ensemble while allowing for high accuracy of the individual ensemble members. The RLO algorithm is shown in Figure 6.20.

The RLO is a "wrapping" approach, which means that it can be used with nearly any ensemble method, as well as on its own. It was found that the RLO framework improves upon the standard versions of bagging, AdaBoost, Multiboost, random forest, random subspace, and rotation forest [239].

RANDOM LINEAR ORACLE

Training: Given is a labeled data set $\mathbf{Z} = \{\mathbf{z}_1, \ldots, \mathbf{z}_N\}$.

1. Choose the ensemble size L and the base classifier model.
2. For $i = 1, \ldots, L$
   (a) Generate a random hyperplane (the ith oracle) through the feature space so that there is at least one object on each side of the hyperplane. Store the coefficients of the plane equation.
   (b) Split the data into ($+$) and ($-$) subsets, depending on which side of the hyperplane the data points lie. Train classifiers $D_{i(+)}$ and $D_{i(-)}$, respectively.
3. Return the $2L$ trained classifiers and the L oracles.

Operation: For each new object $\mathbf{x}$

1. For $i = 1, \ldots, L$, apply ith oracle to $\mathbf{x}$ and subsequently classify $\mathbf{x}$ with $D_{i(+)}$ or $D_{i(-)}$. Take the output to be the vote of the ith ensemble member (label or continuous valued).
2. Assign to $\mathbf{x}$ the class with the largest number of votes (or largest aggregated degree of support).

Return the ensemble label of the new object.

FIGURE 6.20   Training and operation algorithm for the random linear oracle ensemble.

If the classifier for the two parts of the RLO is a decision tree, each member of the RLO ensemble can be viewed as the so-called "omnivariate tree" [258, 426]. In omnivariate decision trees, the split at each node is done through a function which is selected or tuned during the induction of the tree. The base classifiers in the RLO ensemble are, in essence, omnivariate trees with a random linear node at the root and standard univariate subtrees to the left and to the right. The similarity between the RLO base classifier and the omnivariate tree is only structural because the RLO function at the root node is not trained but is chosen at random. The randomness of this choice is the most important heuristic of the RLO ensemble.

▣ Example 6.9   Random linear oracle for the fish data

We generated the fish data set with 35% label noise as the training data, and used again the whole grid with the noise-free labels for testing. Naïve Bayes was chosen as the base classifier. Figure 6.21a shows the training data and the classification regions of the single Naïve Bayes classifier. Figure 6.21b shows the regions for a mini-ensemble classifier consisting of two NB classifiers. The oracle was constructed in the following way. Pick two random points in the data set, and find the equation of the hyperplane (line in this case) orthogonal to the segment and passing through the middle of it. Let the points be A and B with position vectors $\mathbf{a}$ and $\mathbf{b}$, respectively. The equation of

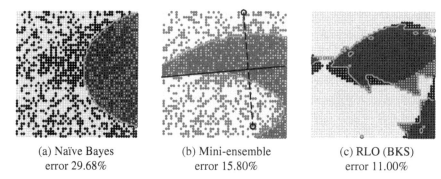

| (a) Naïve Bayes | (b) Mini-ensemble | (c) RLO (BKS) |
| error 29.68% | error 15.80% | error 11.00% |

FIGURE 6.21 Classification regions for the NB classifier, the mini-ensemble of two NB classifier, and the RLO ensemble combined by BKS ($L = 15$). The error rate is calculated with respect to the data with noise-free labels.

the oracle hyperplane is $\mathbf{w}^T\mathbf{x} + w_0 = 0$, where $\mathbf{w} = \mathbf{a} - \mathbf{b}$, and $w_0 = (\mathbf{b}^T\mathbf{b} - \mathbf{a}^T\mathbf{a})/2$. The two points and the oracle line are plotted in Figure 6.21b.

Figure 6.21c shows the data with the noise-free labels and the classification regions of an RLO ensemble. The ensemble size was $L = 15$, and the combination rule was the BKS (we used the code from Chapter 5). The error rates of the classifiers with respect to the noise-free data labels are also shown.

Figure 6.22 shows the classification regions for RLO with majority vote and with BKS for several values of L. For this configuration of the classes, majority vote combiner happened to be consistently worse than the BKS combiner except for the smallest ensemble size of $L = 5$. BKS shows its over-fitting tendency for larger L, which is demonstrated by the increasing level of noise incorporated into the boundary (top row).

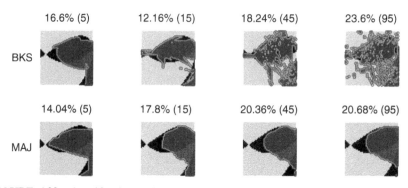

FIGURE 6.22 Classification regions for the RLO and the NB classifier. Two combination methods were tried: BKS (top row) and majority vote (bottom row), for different ensemble sizes as shown in the brackets. The ensemble error rate is shown in the plot titles.

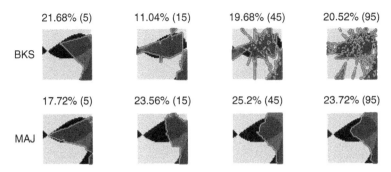

FIGURE 6.23   Classification regions for the RLO and the LDC. Two combination methods were tried: BKS (top row) and majority vote (bottom row), for different ensemble sizes as shown in the brackets.

The results are not much different if we use the linear discriminant classifier as the base classifier for the RLO ensembles. The results for several values of L are shown in Figure 6.23.

The example shows that the random oracle can be a useful diversifying heuristic, which can convert a stable classifier, such as the parametric Naïve Bayes or LDC, into a nonstable base classifier. The RLO mini-ensemble nearly halves the error of the NB classifier. This supports the idea of classifier selection, where the regions of competence can be random. A simple classifier trained separately in each region may lead to an excellent ensemble.

Observe that the ensemble error depends on the ensemble size L, the type of combiner, and the type of base classifier. This suggest that it is important to train all the components of the ensemble together rather than try to perfect an isolated heuristic in a global context. That heuristic may have its niche in relation to other ensemble parameters.

6.7   ERROR CORRECTING OUTPUT CODES (ECOC)

The ECOC ensemble strategy was developed for problems with multiple classes [4, 14, 97, 98, 217, 275, 276, 358]. The idea is to avoid solving the multi-class problem directly and to break it into dichotomies instead. Each classifier in the ensemble discriminates between two possibly compound classes. For example, let $\Omega = \{\omega_1, \ldots, \omega_{10}\}$. We can break Ω into $\Omega = \{\Omega^{(1)}, \Omega^{(2)}\}$, where $\Omega^{(1)} = \{\omega_1, \ldots, \omega_5\}$ and $\Omega^{(2)} = \{\omega_6, \ldots, \omega_{10}\}$. The ECOC strategy is the natural multi-class extension for classification methods specifically designed for two-class problems. Examples of such methods are SVM and AdaBoost [10, 209, 304, 416].

There are two main issues in ECOC ensembles: how to design the two-class sets (the dichotomies) to be solved by the ensemble members, and how to decode the output. Many solutions have been proposed for both problems [10, 122, 416].

6.7.1 Code Designs

6.7.1.1 The Code Matrix   We can represent each split of the set of c classes as a binary vector of length c with 1s for the classes in $\Omega^{(1)}$ and 0s for the classes in $\Omega^{(2)}$. The corresponding vector for the above example is

$$[1, 1, 1, 1, 1, 0, 0, 0, 0, 0]^T.$$

The set of all such vectors has 2^c elements. However, not all of them correspond to different splits. Consider $[0, 0, 0, 0, 0, 1, 1, 1, 1, 1]^T$. Even if the Hamming distance between the two binary vectors is equal to maximum possible value 10, the two subsets are again $\Omega^{(1)}$ and $\Omega^{(2)}$, only with swapped labels. Since there are two copies of each split within the total of 2^c splits, the number of different splits is $2^{(c-1)}$. The splits $\{\Omega, \emptyset\}$ and the corresponding $\{\emptyset, \Omega\}$ are of no use because they do not represent any discrimination task. Therefore the number of possible *different* splits of a set of c class labels into two nonempty disjoint subsets (dichotomies) is

$$2^{(c-1)} - 1. \tag{6.11}$$

We can choose L dichotomies to be the classifier assignments. These can be represented as a binary *code matrix* C of size $c \times L$. The (i,j)th entry of C, denoted $C(i,j)$ is 1 if class ω_i is in $\Omega_j^{(1)}$ (the compound class with label "1" for classifier D_j) or 0, if class ω_i is in $\Omega_j^{(2)}$. Thus each row of the code matrix, called a *codeword*, corresponds to a class and each column corresponds to a classifier.

The two main categories of the ECOC design methods are problem independent and problem specific. Recently, the problem-specific methods have flourished, including methods based on class separability [316, 317], sub-class structures [123], and feature selection [23]. Below we explain the more generic approach; that of generating problem-independent ECOC.

To make the most of an ECOC ensemble, the code matrix should be built according to two main criteria.

Row separation. In order to avoid misclassifications, the codewords should be as far apart from one another as possible. We can still recover the correct label for $\mathbf{x}$ even if several classifiers have guessed wrongly. A measure of the quality of an error correcting code is the minimum Hamming distance between any pair of codewords. If this distance is denoted by H_c, the number of errors that the code is guaranteed to be able to correct is

$$\left\lfloor \frac{H_c - 1}{2} \right\rfloor. \tag{6.12}$$

Column separation. It is important that the dichotomies given as the assignments to the ensemble members are as different from each other as possible too. This will encourage low correlation between the classification errors and will increase the ensemble accuracy [98]. The distance between the columns must be maximized keeping in mind that the complement of a column gives the same split of the set of classes. Therefore, the column separation should be sought by maximizing

$$H_L = \min_{i,j,i\neq j} \; \min \left\{ \sum_{k=1}^{c} |C(k,i) - C(k,j)| \,,\, \sum_{k=1}^{c} |1 - C(k,i) - C(k,j)| \right\},$$

$$i,j \in \{1, 2, \dots, L\}. \tag{6.13}$$

6.7.1.2 ECOC Generation Methods   Below we explain five simple ECOC generation methods which have been used as benchmark in the literature [10].

1. *One per class* (also called "one versus all" or "one against all") [329]. The target function for class ω_j is a codeword containing 1 at position j and 0s elsewhere. Thus, the code matrix is the identity matrix of size c, and we only build $L = c$ classifiers. This encoding is of low quality because the Hamming distance between any two rows is 2, and so the error correcting power is $\lfloor \frac{2-1}{2} \rfloor = 0$.

2. *All pairs* (also called "one versus one") [80, 200, 275]. The code matrix for this class is called "ternary" because there must be a third element in the code indicating that the respective class is not included in either the positive or the negative group. The conventional notation is 1 for the positive class, -1 for the negative class, and 0 for excluded classes. Then the code matrix for the all-pairs code for a c-class problem will require $c(c-1)/2$ columns, which will also be the ensemble size L. For problems with a large number of classes, for example in face recognition (say, 100 employees of a company), this coding scheme is impractical.

3. *Exhaustive code* (also called "complete code"). Dietterich and Bakiri [98] give the following procedure for generating all possible $2^{(c-1)} - 1$ different classifier assignments for c classes.

   (a) Row 1 is all ones.

   (b) Row 2 consists of $2^{(c-2)}$ zeros followed by $2^{(c-2)} - 1$ ones.

   (c) Row 3 consists of $2^{(c-3)}$ zeros, followed by $2^{(c-3)}$ ones, followed by $2^{(c-3)}$ zeros, followed by $2^{(c-3)} - 1$ ones.

   (d) In row i, there are alternating $2^{(c-i)}$ zeros and ones.

   (e) The last row is 0,1,0,1,0,1,...,0.

   The exhaustive code for $c = 4$ obtained through this procedure is as follows:

	D_1	D_2	D_3	D_4	D_5	D_6	D_7
ω_1	1	1	1	1	1	1	1
ω_2	0	0	0	0	1	1	1
ω_3	0	0	1	1	0	0	1
ω_4	0	1	0	1	0	1	0

The same class splits can be obtained by taking the integers from 1 to $2^{(c-1)}$ and converting them to binary numbers with c binary digits. MATLAB code for the exhaustive code is given in Appendix 6.A.6.

Dieterich and Bakiri [98] suggest that exhaustive codes should be used for $3 \leq c \leq 7$. For $8 \leq c \leq 11$, an optimization procedure can be used to select columns from the exhaustive code matrix. For values of c larger than 11, random code generation is recommended.

4. *Random generation—dense code.* Authors of studies on ECOC ensembles share the opinion that random generation of the codewords is a reasonably good method [98, 358]. Although these studies admit that more sophisticated procedures might lead to better codes, they also state that the improvement in the code might have only marginal effect on the ensemble accuracy. The example below illustrates the random ECOC generation. In the dense code, all classes are present in each dichotomy. In the ternary notation, the code matrix contains only 1s and -1s. Each bit of the matrix takes one of these values with probability 0.5. The recommended ensemble size (codeword length) in this case is $\lceil 10 \log_2(c) \rceil$ [10], where c is the number of classes.

5. *Random generation—sparse code.* In this case, the randomly sampled entries in the code matrix are -1, 0, and 1. Zero is sampled with probability 0.5, and any of the other two values, with probability 0.25. The recommended ensemble size (codeword length) in this case is $\lceil 15 \log_2(c) \rceil$ [10].

To create a random code, a predefined large number of code matrices are sampled and evaluated with respect to a given criterion. The best code matrix is returned. The criterion could be the maximum row and column separation

$$\max_{C}\{H_c + H_L\}, \tag{6.14}$$

where C denotes the code matrix.

Evolutionary algorithms, and specifically genetic algorithms, have been favoured for designing ECOC matrices because they can handle large binary spaces, and can optimize any fitness function [23, 30, 224].

6.7.2 Decoding

Suppose that the classifiers in the ensemble produce binary labels $(s_1, \dots, s_L)$ for a given input $\mathbf{x}$. The Hamming distance between the classifier outputs and the codewords for the classes is calculated, and the class with the shortest Hamming distance is chosen as the label of $\mathbf{x}$. The support for class ω_j can be expressed as

$$\mu_j(\mathbf{x}) = - \sum_{i=1}^{L} |s_i - C(j, i)|. \tag{6.15}$$

6.7.2.1 ECOC, Voting and Decision Templates   Suppose that the combiner of an ECOC ensemble is the minimum Hamming distance. This can be viewed as majority voting as follows. Suppose that classifier D_i solves the dichotomy $\{\Omega_i^{(1)}, \Omega_i^{(2)}\}$. Let the decision of D_i be $s_i = 1$, that is, compound class $\Omega_i^{(1)}$ is chosen. Each individual class within $\Omega_i^{(1)}$ will obtain one vote from D_i. Since each dichotomy contains all the classes, each class will obtain a vote from each classifier. Selecting the class with the largest sum of these votes is equivalent to making a decision in favor of the class whose codeword has the lowest Hamming distance to the binary word of the L outputs $s_1, \ldots, s_L$. If the classifiers are made to learn very different boundaries, then there is a good chance that their errors will be unrelated.

■ **Example 6.10   ECOC, majority voting, and decision templates**
Table 6.3 shows the code matrix for $c = 8$ classes and codeword length $L = 15$ found with the random generation method ($H_c = 5$, $H_L = 1$).

TABLE 6.3   Code matrix for $c = 8$ classes and $L = 15$ dichotomizers

ω_1	1	1	1	1	1	0	0	0	1	1	0	0	0	0	0
ω_2	1	0	0	1	1	1	0	0	1	1	1	1	0	1	0
ω_3	1	0	0	0	1	1	0	1	1	0	0	1	0	0	0
ω_4	0	1	0	0	1	1	0	1	1	0	1	0	1	0	0
ω_5	1	1	1	0	1	1	0	1	0	1	1	0	0	0	1
ω_6	0	0	0	0	1	0	0	0	0	1	0	0	1	1	0
ω_7	1	0	0	1	1	0	0	0	1	0	0	0	0	1	1
ω_8	1	0	1	1	1	0	1	1	0	0	0	0	1	0	0

Suppose that the ensemble $\mathcal{D} = \{D_1, \ldots, D_{15}\}$ produces the following set of outcomes for some input **x** :

$$[1, 1, 1, 1, 1, 1, 1, 1, 1, 1, 1, 1, 1, 1, 1]. \tag{6.16}$$

The Hamming distances to the eight codewords are as follows: $8, 6, 9, 8, 6, 11, 9, 8$. There is a tie between ω_2 and ω_5, so any of the two labels can be assigned.

By giving label $s_1 = 1$, classifier D_1 votes for classes $\omega_1, \omega_2, \omega_3, \omega_5, \omega_7$, and ω_8. Thus, the number of votes that class ω_1 will get is exactly the Hamming distance between the class codeword and the classifiers' output.

The Hamming distance between binary vectors is equal to the Euclidean distance between them. If we regard the codewords as the templates for the classes, then by labeling **x** according to the minimal Hamming distance, we implement the decision template combiner.

6.7.2.2 Soft ECOC Labels   Kong and Dietterich [217] use soft labels $d_{i,j}(\mathbf{x}) = \hat{P}(\Omega_i^{(j)}|\mathbf{x})$ instead of the 0/1 labels from the L classifiers in the ensemble. Since there are only two compound classes, we have $d_{i,2}(\mathbf{x}) = 1 - d_{i,1}(\mathbf{x})$. The Hamming distance between the classifier outputs and the codeword for class ω_j is

$$\mu_j(\mathbf{x}) = \sum_{i=1}^{L} |d_{i,1}(\mathbf{x}) - C(j, i)|. \qquad (6.17)$$

A thriving research area of ECOC ensemble are the decoding strategies [14, 122, 417]. Examples of such strategies are loss-based decoding [10], probabilistic decoding [188, 304], reject rules [366], and classifier-based decoding, treating the ECOC outputs as features [438]. Escalera et al. [121] offer a MATLAB library for ECOC ensembles containing a variety of coding and decoding functions.

6.7.3   Ensembles of Nested Dichotomies

An alternative approach to solving multi-class problems is the *ensemble of nested dichotomies* (ENDs) [130, 332]. Although ENDs are not exactly examples of ECOC ensembles, they exploit a similar idea. Each classifier in the ensemble solves a cascade of binary classification sub-problems. Such a classifier has a tree-wise structure and can be built in a top–down or a bottom–up approach. In the top–down approach [221, 317], the set of classes reaching a node is progressively split into two parts based on a criterion related to the separability between the two meta-classes. Pujol et al. [317] use mutual information as the separability criterion, and Kumar et al. [221] use a probability estimate.

▉ Example 6.11   Building a class-hierarchy tree

To illustrate the idea of a nested dichotomy, we used the mfeat UCI data set, an example of which is shown in Figure 6.17. Without loss of generality, only the "pix" subset of features was used. To generate the tree in a top–down manner, we checked all possible splits into two nonempty sets at each node. For example, at the root, there are 10 classes, therefore there are $2^{(10-1)} - 1 = 511$ possible splits. For each split, we calculated the resubstitution error rate of the linear classifier (function `classify` from Statistics Toolbox of MATLAB). The split with the smallest error happened to be classes (27) versus classes (01345689). Continuing in the same way, the whole tree was built (Figure 6.24).

Pujol et al. [317] translate the tree into a ternary ECOC matrix, arguing that the class assignments are more meaningful and economical than those in the random codes. Each split of the tree defines one classifier. The 10×8 ECOC code matrix for the tree in Figure 6.24 is shown in Figure 6.25.

The columns of the matrix are the assignments for the classifiers. For example, the first column defines the task for the classifier at the root of the tree. It splits the class set into positive (01346789) and negative (27) subsets. The second column ignores all classes but 2 (positive) and 7 (negative), and so on.

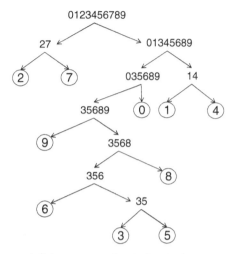

FIGURE 6.24   The nested dichotomy tree for the handwritten digits data set mfeat (UCI). The numbers at the nodes and the leaves of the tree are the class labels (digits) from 0 to 9.

Instead of using one tree to generate the ECOC matrix, the ENDs use a random tree structure as the ensemble diversifying heuristic [130]. The number of different nested dichotomies for a set of c classes can be calculated through the following recursive equation [130]:

$$T(c) = (2c - 3) \times T(c - 1), \tag{6.18}$$

where $T(1) = 1$. A random subset of the possible nested dichotomies makes an ensemble. An extensive experimental study by Rodríguez et al. [332] found that the nested

	Classifiers									Classifiers
0	1	0	1	0	-1	0	0	0	0	
1	1	0	-1	-1	0	0	0	0	0	
2	-1	1	0	0	0	0	0	0	0	
3	1	0	1	0	1	1	1	1	1	
4	1	0	-1	1	0	0	0	0	0	
5	1	0	1	0	1	1	1	1	-1	
6	1	0	1	0	1	1	1	-1	0	
7	-1	-1	0	0	0	0	0	0	0	
8	1	0	1	0	1	1	-1	0	0	
9	1	0	1	0	1	-1	0	0	0	

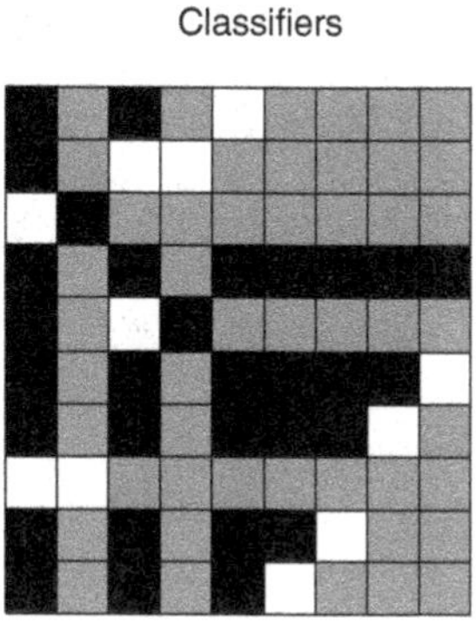

(Classes — row labels 0 through 9)

FIGURE 6.25   Numerical and color-based representation of the ternary code matrix for the nested dichotomy in Figure 6.24.

dichotomies, while not spectacularly accurate on their own, make a valuable ensemble accessory, which improves upon the accuracies of the classical configurations.

APPENDIX

6.A.1   BAGGING

Function `bagging_train` trains an ensemble, and `bagging_classify` calculates the ensemble output and error. A base classifier model needs to be specified as two function handles, one for the training and one for the classification. For this example, the base classifier was Naïve Bayes (code given in Chapter 2). The example shows how a bagging ensemble fares on the Fisher's iris data (introduced in Chapter 1), when the training data is a noise-distorted version of the original.

Figure 6.A.1 plots classes "versicolor" and "virginica" as the original data (Figure 6.A.1a) and the noise-contaminated data (Figure 6.A.1b). The noise amplitude and the ensemble size are parameters that can be re-set or varied in a loop, should we want to examine the behavior of bagging.

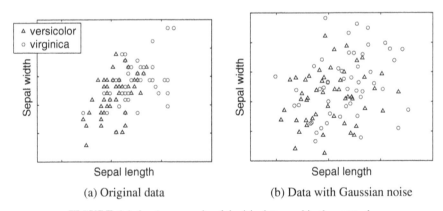

(a) Original data   (b) Data with Gaussian noise

FIGURE 6.A.1   An example of the iris data used in the example.

```
1  %------------------------------------------------------------%
2  load fisheriris % load the iris data
3
4  % create numerical labels
5  labels = [ones(50,1);2*ones(50,1);3*ones(50,1)];
6
7  % standardize the data for easier manipulation
8  data = zscore(meas); % mean 0, std 1
```

```
 9  % inject noise to make a new training data
10  na = 0.8; % noise amplitude
11  data_noise = data + randn(size(data)) * na;
12
13  L = 5; % ensemble size
14  E = bagging_train(data_noise,labels,L,@naive_bayes_train);
15  [~,e] = bagging_classify(E,@naive_bayes_classify,data,labels);
16  fprintf('Iris data\nNoise level = %.2f\n',na)
17  fprintf('Ensemble size = %i\nEnsemble error = %.4f\n',L,e)
18  %-----------------------------------------------------------%
```

```
 1  %-----------------------------------------------------------%
 2  function E = bagging_train(data, labels, L, train_D)
 3  % --- train a bagging ensemble
 4  %
 5  % Input: -------------------------------------
 6  %      data:  N-by-n array with data
 7  %    labels:  N-by-1 numerical labels: 1, 2, 3, ...
 8  %         L:  ensemble size
 9  %   train_D:  classifier model (a function handle)
10  %
11  % Output:  -------------------------------------
12  %         E:  a cell array containing the L classifiers
13
14  N = size(data,1); % number of objects
15  E = cell(1,L); % pre-allocate the ensemble array for speed
16  for i = 1:L
17      bi = randi(N,N,1); % bootstrap index
18      bd = data(bi,:); % bootstrap data
19      bl = labels(bi); % bootstrap labels
20      E{i} = train_D(bd,bl); % train the i-th classifier
21  end
22  %-----------------------------------------------------------%
```

```
 1  %-----------------------------------------------------------%
 2  function [al,e] = bagging_classify(E, class_D, data, labels)
 3  % --- classify with a trained bagging ensemble
 4  %
 5  % Input: -------------------------------------
 6  %         E:  ensemble trained with bagging_train
 7  %   class_D:  a function handle for classification
 8  %             with a trained classifier D
 9  %      data:  N-by-n array with data to classify
10  %    labels:  N-by-1 numerical labels (optional)
11  %
```

```
12  % Output:   -------------------------------------
13  %       al:  N-by-1 labels assigned by the ensemble
14  %        e:  the ensemble classification error, if labels
15  %             are provided
16
17  L = numel(E); % ensemble size
18
19  % pre-allocate the ensemble output array for speed
20  ens = zeros(size(data,1),L);
21
22  for i = 1:L
23      ens(:,i) = class_D(E{i},data);
24      % take the labels from the the i-th classifier
25  end
26  al = mode(ens')'; % plurality vote labels
27
28  e = Inf; % pre-set ensemble error
29  if nargin > 3
30      e = mean(al ~ = labels); % the ensemble error
31  end
32  %--------------------------------------------------------%
```

Example 6.3 in the text uses the above functions. To call the base classifier, a handle to the `classregtree` function from the Statistics toolbox of MAT-LAB was created. The pruning was turned off in order to destabilize the base classifiers. For the random forest ensemble, the line calling the bagging training function was

```
1  bagging_train(tr,trl,1,...
2      @(x,y) classregtree(x,y,'nvartosample',5,'prune','off'));
```

For the bagging ensemble,

```
1  bagging_train(tr,trl,1,...
2      @(x,y) classregtree(x,y,'prune','off'));
```

For the classification part, we used

```
1  bagging_classify(ER,@eval,ts,tsl);
```

6.A.2   ADABOOST

Function `adaboost_train` trains an ensemble, and `adaboost_classify` calculates the ensemble output and error. A base classifier model needs to be specified as two function handles, one for the training and one for the classification. For this example, the base classifier was Naïve Bayes.

```
1  %-------------------------------------------------------------%
2  function [E,be] = adaboost_train(data, labels, L, train_D,
       class_D)
3  % --- train an adaboost ensemble
4  %
5  % Input: ---------------------------------------
6  %     data:   N-by-n array with data
7  %   labels:   N-by-1 numerical labels: 1, 2, 3, ...
8  %        L:   ensemble size
9  % train_D:   classifier model (a function handle)
10 % class_D:   a function handle for classification
11 %               with a trained classifier D
12 % Output:   ---------------------------------------
13 %        E:   a cell array containing the L classifiers
14
15 N = size(data,1); % number of objects
16 w = ones(1,N)/N; % initialize the weights
17 E = cell(1,L); % pre-allocate the ensemble array for speed
18 i = 1; % ensemble size
19 be = zeros(1,L);
20 while i <= L
21     ai = p_sample(w); % indices of the objects in the sample
22     ad = data(ai,:); % data
23     al = labels(ai); % labels
24     E{i} = train_D(ad,al); % train the i-th classifier
25     lo = (class_D(E{i},data) ~ = labels); % loss
26     ep  = w * lo; % weighted loss
27     if ep < 0.5 && ep ~ = 0
28         be(i) = ep / (1 - ep); % classifier's weight
29         w = w .* be(i).^(1 - lo'); % objects' weights
30         w = w / sum(w); % normalize
31         i = i + 1; % add the classifier to the ensemble
32     else
33         w = ones(1,N)/N; % re-initialize the weights
34     end
35 end
36 end
37
38 function ind = p_sample(w)
39 % sample from distribution w
40 N = length(w);
41 cdf = cumsum(w);
```

```
42  r = rand(N,1); % uniform random numbers
43  d1 = repmat(r(:),1,N);
44  d2 = repmat(cdf(:)',N,1);
45  [~, ind] = max((d1 < d2)');
46  end
47  %-----------------------------------------------------------%


 1  %-----------------------------------------------------------%
 2  function [al,e] = adaboost_classify(E, be, class_D, data,
        labels)
 3  % --- classify with a trained adaboost ensemble
 4  %
 5  % Input: ----------------------------------------
 6  %        E:  ensemble trained with adaboost_train
 7  %       be:  ensemble weights
 8  % class_D:  a function handle for classification
 9  %            with a trained classifier D
10  %    data:  N-by-n array with data to classify
11  %  labels:  N-by-1 numerical labels (optional)
12  %
13  % Output:  ----------------------------------------
14  %       al:  N-by-1 labels assigned by the ensemble
15  %        e:  the ensemble classification error, if labels
16  %            are provided
17
18  L = numel(E); % ensemble size
19  W = log(1./be); % classifiers' weights
20
21  % pre-allocate the ensemble output array for speed
22  ens = zeros(size(data,1),L);
23
24  for i = 1:L
25      ens(:,i) = class_D(E{i},data);
26      % take the labels from the the i-th classifier
27  end
28
29  class_scores = [];
30  for i = 1:max(ens(:)) % for all classes
31      class_scores = [class_scores; W * (ens == i)'];
32  end
33
34  [~ ,al] = max(class_scores); % weighted vote labels
35  al = al';
36
37  e = Inf; % pre-set ensemble error
```

```
38  if nargin > 4
39      e = mean(al ~ = labels); % the ensemble error
40  end
41  %------------------------------------------------------------%
```

The script that illustrates the operation of these two functions can be adapted from the script for the bagging ensemble above. Lines 14 and 15 should be replaced by

```
1  %------------------------------------------------------------%
2  [E,be] = adaboost_train(data_noise,labels,L,...
3      @naive_bayes_train,@naive_bayes_classify);
4  [~ ,e] = adaboost_classify(E,be,@naive_bayes_classify,data,
     labels);
5  %------------------------------------------------------------%
```

6.A.3 RANDOM SUBSPACE

Function `rs_train` trains an ensemble, and `rs_classify` calculates the ensemble output and error. A base classifier model needs to be specified as two function handles, one for the training and one for the classification. For this example, the base classifier is the classification tree encoded in the Statistics Toolbox of MATLAB, function `classregtree`.

The script uploads the file mfeat-fac with the 216 profile correlations. It then creates the class labels and the indices of the cross-validation folds. The loop performs the 10-fold cross-validation of the random subspace ensemble. The average error rate and the standard deviation thereof are printed at the end. The code in this section is standalone and requires internet access to the data and the Statistics Toolbox of MATLAB.

```
1  %------------------------------------------------------------%
2  clear all, close all;  clc
3
4  p1 = 'http://archive.ics.uci.edu/ml/machine-learning-
          databases';
5  p2 = '/mfeat/mfeat-fac'; % data set name
6  data = str2num(urlread([p1,p2])); % load up the data
7  labels = repmat(1:10,200,1); labels = labels(:);
8  L = 10; % ensemble size
9
10  % Form the 10-fold cross-validation indices
11  I = 1:2000; II = reshape(I,200,10); III = reshape(II',200,10);
12
13  for i = 1:10
14      tsI = III(:,i); % testing data indiced
15      trI = setxor(tsI,I); % training data indices
```

```
16      [E,F] = rs_train(data(trI,:),labels(trI),L,...
17          floor(size(data,2)/3),...
18          @(x,y) classregtree(x,y,'prune','off'));
19      [~, e(i)] = rs_classify(E,F,@eval,data(tsI,:),labels(tsI));
20      fprintf('RS error = %.4f\n',e(i))
21 end
22
23 % Print the average RS ensemble error and the standard devia-
   tion
24 fprintf('\nAverage RS testing error %.4f +- %.4f\n\n',...
25      mean(e),std(e))
26 %------------------------------------------------------------%
```

```
1  %------------------------------------------------------------%
2  function [E,F] = rs_train(data, labels, L, d, train_D)
3  % --- train a random subspace ensemble
4  %
5  % Input: ------------------------------------------
6  %     data:  N-by-n array with data
7  %   labels:  N-by-1 numerical labels: 1, 2, 3, ...
8  %        L:  ensemble size
9  %        d:  number of features to select
10 % train_D:  classifier model (a function handle)
11 %
12 % Output:  ------------------------------------------
13 %        E:  a cell array containing the L classifiers
14
15 n = size(data,2); % number of features
16 E = cell(1,L); % pre-allocate the ensemble array for speed
17 F = cell(1,L); % pre-allocate the selected features array
18 for i = 1:L
19     rp = randperm(n); % sample without replacement
20     F{i} = rp(1:d); % sampled features
21     % train the i-th classifier
22     E{i} = train_D(data(:,F{i}),labels);
23 end
24
25 %------------------------------------------------------------%
```

```
1  %------------------------------------------------------------%
2  function [al,e] = rs_classify(E, F, class_D, data, labels)
3  % --- classify with a trained random subspace ensemble
4  %
5  % Input: ------------------------------------------
6  %        E:  ensemble trained with rs_train
```

```
 7 %         F:   cell array with the selected features
 8 % class_D:   a function handle for classification
 9 %              with a trained classifier D
10 %    data:   N-by-n array with data to classify
11 %  labels:   N-by-1 numerical labels (optional)
12 %
13 % Output:   -------------------------------------
14 %      al:   N-by-1 labels assigned by the ensemble
15 %       e:   the ensemble classification error, if labels
16 %              are provided
17
18 L = numel(E); % ensemble size
19
20 % pre-allocate the ensemble output array for speed
21 ens = zeros(size(data,1),L);
22
23 for i = 1:L
24     ens(:,i) = class_D(E{i},data(:,F{i}));
25     % take the labels from the the i-th classifier
26 end
27 al = mode(ens')'; % plurality vote labels
28
29 e = Inf; % pre-set ensemble error
30 if nargin > 4
31     e = mean(al ~ = labels); % the ensemble error
32 end
33
34 %-----------------------------------------------------------%
```

6.A.4   ROTATION FOREST

The training and classification functions are respectively rotation_forest_train and rotation_forest_classify. A base classifier model needs to be specified as two function handles, one for the training and one for the classification. The recommended base classifier is the decision tree, for example, the classification tree encoded in the Statistics Toolbox of MATLAB (function classregtree).

There are two issues with this code. First, the original rotation forest algorithm uses the average combiner for obtaining the ensemble decision. Here we used the plurality vote, which does not seem to adversely affect the method.

Second, the implementation of the PCA in MATLAB centers the data before applying the PCA by subtracting the mean from each feature. This will rotate and center the space. The code below does not take the centering into account when training the classifier and then for classifying **x**. This does not do any harm because the centering only shifts the data in the PCA space in addition to the rotation. Decision trees are invariant with respect to shift of the features, therefore the training will result in the same tree, with or without shift. But even if the shift mattered, since the classifier is trained and is operating on data coming from the same distribution, the rotation forest ensemble is not likely to suffer any loss of accuracy or diversity.

```
1  %------------------------------------------------------------%
2  function [E,R] = rotation_forest_train(data, labels, L, K,...
       train_D)
3  % --- train an adaboost ensemble
4  %
5  % Input: ------------------------------------------
6  %     data:  N-by-n array with data
7  %   labels:  N-by-1 numerical labels: 1, 2, 3, ...
8  %        L:  ensemble size
9  %        K:  number of feature subsets
10 % train_D:  a function handle for training a
11 %           base classifier D
12 % Output:   ------------------------------------
13 %        E:  a cell array containing the L classifiers
14 %        R:  a cell array with L rotation matrices
15
16 n = size(data,2); % number of features
17 c = max(labels); % number of classes
18 E = cell(1,L); % pre-allocate the ensemble array for speed
19 R = cell(1,L); % pre-allocate the matrices array for speed
20
21 for i = 1:L
22     rp = randperm(n); % shuffle the features
23     si = linspace(1,n+1,K+1); % split indices
24     ro = zeros(n); % rotation matrix for classifier i
25     for j = 1:K
26         % eliminate a random subset of classes
27         ctr = zeros(1,c); % classes to remain
28         while sum(ctr) == 0, ctr = rand(1,c) > 0.5; end
29         itr = ismember(labels,find(ctr)); % index to remain
30         d = data(itr,rp(si(j):si(j+1)-1)); % new data
31
32         % take 75% bootstrap sample
33         Nd = size(d,1); % number of objects in the new data
34         bi = randi(Nd,round(0.75*Nd),1); % bootstrap 75% index
35
36         % apply PCA and inset the columns in ro
37         co = princomp(d(bi,:));
38         ro(si(j):si(j+1)-1,si(j):si(j+1)-1) = co;
39     end
```

```
40        [~,rrp] = sort(rp); % reverse the shuffle index
41        R{i} = ro(rrp,:); % store the rearranged rotation matrix
42        E{i} = train_D(data * R{i}, labels); % store classifier
43  end
44  %----------------------------------------------------------%
```

```
1   %----------------------------------------------------------%
2   function [al,e] = rotation_forest_classify(E, R, class_D,
      data, ... labels)
3   % --- classify with a trained rotation forest ensemble
4   %
5   % Input: ----------------------------------------
6   %        E:  ensemble trained with rotation_forest_train
7   %        R:  cell array with the rotaion matrices
8   % class_D:  a function handle for classification
9   %            with a trained classifier D
10  %    data:  N-by-n array with data to classify
11  %  labels:  N-by-1 numerical labels (optional)
12  %
13  % Output:  --------------------------------------
14  %       al:  N-by-1 labels assigned by the ensemble
15  %        e:  the ensemble classification error, if labels
16  %            are provided
17
18  L = numel(E); % ensemble size
19
20  % pre-allocate the ensemble output array for speed
21  ens = zeros(size(data,1),L);
22
23  for i = 1:L
24      ens(:,i) = class_D(E{i},data * R{i});
25      % take the labels from the the i-th classifier
26  end
27  al = mode(ens')'; % plurality vote labels *
28  % * Note: The original Rotation Forest algorithm
29  % assumes that the classifier outputs are
30  % continuous-valued and applies the average combiner
31
32  e = Inf; % pre-set ensemble error
33  if nargin > 4
34      e = mean(al ~ = labels); % the ensemble error
35  end
36  %----------------------------------------------------------%
```

6.A.5   RANDOM LINEAR ORACLE

Function `rlo_train` trains an ensemble, and `rlo_classify` calculates the ensemble output and error. A base classifier model needs to be specified as two function handles, one for the training and one for the classification.

```
 1  %-------------------------------------------------------------%
 2  function [E,H] = rlo_train(data, labels, L, train_D)
 3  % --- train a random linear oracle ensemble
 4  %
 5  % Input: ------------------------------------------
 6  %    data:  N-by-n array with data
 7  %  labels:  N-by-1 numerical labels: 1, 2, 3, ...
 8  %       L:  ensemble size
 9  % train_D:  classifier model (a function handle)
10  %
11  % Output:  ------------------------------------
12  %       E:  a cell array containing the L classifiers
13
14  N = size(data,1); % number of objects
15  E = cell(L,2); % pre-allocate the ensemble array for speed
16  H = cell(L,1); % hyperplane array
17  for i = 1:L
18      rp = randperm(N); % choose 2 points
19      A = data(rp(1),:); B = data(rp(2),:);
20      H{i} = [A-B, (B*B' - A*A')/2]; % hyperplane/line
21      left = [data ones(N,1)] * H{i}' > 0; % split the data
22      E{i,1} = train_D(data(left,:),labels(left));
23      E{i,2} = train_D(data(~ left,:),labels(~ left));
24  end
25  %-------------------------------------------------------------%
```

```
 1  %-------------------------------------------------------------%
 2  function [al,e] = rlo_classify(E, H, class_D, data, labels)
 3  % --- classify with a trained rlo ensemble
 4  %
 5  % Input: ------------------------------------------
 6  %       E:  ensemble trained with rlo_train
 7  %       H:  hyperplane array
 8  % class_D:  a function handle for classification
 9  %           with a trained classifier D
10  %    data:  N-by-n array with data to classify
11  %  labels:  N-by-1 numerical labels (optional)
12  %
13  % Output:  ------------------------------------
```

```
14  %        al:  N-by-1 labels assigned by the ensemble
15  %         e:  the ensemble classification error, if labels
16  %             are provided
17
18  L = size(E,1); % ensemble size
19
20  % pre-allocate the ensemble output array for speed
21  ens = zeros(size(data,1),L);
22
23  for i = 1:L
24      left = ([data ones(size(data,1),1)] * H{i}') > 0; % split
                  the ... data
25      ens(left,i) = class_D(E{i,1},data(left,:));
26      ens(~ left,i) = class_D(E{i,2},data(~ left,:));
27      % take the labels from the the i-th classifier
28  end
29  al = mode(ens')'; % plurality vote labels
30
31  e = Inf; % pre-set ensemble error
32  if nargin > 4
33      e = mean(al ~ = labels); % the ensemble error
34  end
35  %------------------------------------------------------------%
```

6.A.6   ECOC

```
1  %------------------------------------------------------------%
2  c = 6; % number of classes
3
4  % Exhaustive code
5  C_s = dec2bin(1:2^(c-1)-1,c)'; % string code matrix
6  C_n = reshape(str2num(C_s(:)),size(C_s)); % numerical
7  %------------------------------------------------------------%
```

7

CLASSIFIER SELECTION

7.1 PRELIMINARIES

The presumption in classifier selection is that there is an oracle that can identify the best expert for a particular input **x**. This expert's decision is accepted as the decision of the ensemble for **x**. The operation of a classifier selection example is shown in Figure 7.1.

We note that classifier selection is different from creating a large number of classifiers and then selecting an ensemble among these. We call the latter approach *overproduce and select* and discuss it in Chapter 8.

The idea of using different classifiers for different input objects was suggested by Dasarathy and Sheela back in 1979 [82]. They combine a linear classifier and a *k*-nearest neighbor (*k*-nn) classifier. The composite classifier identifies a conflict domain in the feature space and uses *k*-nn in that domain while the linear classifier is used elsewhere. In 1981, Rastrigin and Erenstein [322] gave a methodology for classifier selection almost in the form it is used now.

We may assume that the classifier "realizes" its competence for labeling **x**. For example, if the 10-nearest neighbor is used, and 9 of the 10 neighbors suggest the same class label, then the confidence in the decision is high. If the classifier outputs are reasonably well calibrated estimates of the posterior probabilities, then the confidence of classifier $D_i \in D$ for object **x** can be measured as

$$C(D_i|\mathbf{x}) = \max_{j=1}^{c} \hat{P}(\omega_j|\mathbf{x}, D_i), \tag{7.1}$$

where c is the number of classes.

Combining Pattern Classifiers: Methods and Algorithms, Second Edition. Ludmila I. Kuncheva.
© 2014 John Wiley & Sons, Inc. Published 2014 by John Wiley & Sons, Inc.

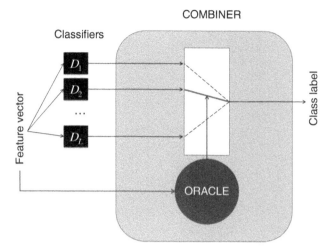

FIGURE 7.1   Operation of the classifier selection ensemble.

To construct a classifier selection ensemble, the following questions need to be answered:

- How do we build the individual classifiers? Should they be stable or unstable? Homogeneous or heterogeneous? Woods et al. [421] considered heterogeneous selection ensembles but found little improvement on the best classifier (which often happens to be the k-nearest neighbor classifier). Some authors suggest using bagging or boosting to develop the ensemble and use a selection strategy for combining the outputs [20, 255, 363].
- How do we evaluate the competence of the classifiers for a given $\mathbf{x}$?
- Once the competences are found, what selection strategy shall we use? The standard strategy is to select one most competent classifier and take its decision. However, if there are several classifiers of equally high competence, do we take one decision or shall we fuse the decisions of the most competent classifiers? When is it beneficial to select one classifier to label $\mathbf{x}$ and when should we be looking for a fused decision?

7.2   WHY CLASSIFIER SELECTION WORKS

Consider an ensemble $D = \{D_1, \ldots, D_L\}$. Let the feature space $\mathbb{R}^n$ be divided into K *selection regions* (called also *regions of competence*), where $K > 1$. Denote the regions $R_1, \ldots, R_K$. These regions are not associated with specific classes, nor do they need to be of a certain shape or size. The following example demonstrates the rationale for classifier selection ensembles.

▣ Example 7.1   Selection regions

An example of partitioning of the feature space $\mathbb{R}^2$ into three *selection* regions is shown in Figure 7.2. Depicted is a banana data set with 2000 data points. There are two classes and therefore two *classification* regions.

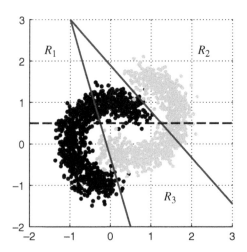

FIGURE 7.2   An example of partitioning the feature space into three selection regions.

Suppose that we have three classifiers: D_1, which always predicts class "black"; D_2, which always predicts class "gray"; and a linear classifier D_3 whose discriminant function is shown as the horizontal dashed line in Figure 7.2. This classifier predicts class "black" above the line and class "gray" underneath. The individual accuracy of each of the three classifiers is approximately 0.5. A majority vote between them is also useless as it will always match the decision of the arbiter D_3 and lead to 50% error as well. However, if we use the three selection regions and nominate one classifier for each region (D_1 in R_1, D_2 in R_2, and D_3 in R_3), the error of the ensemble will be negligible.

This example only shows the potential of the classifier selection approach. In practice, we will hardly be so fortunate to find regions that will have such a dramatic effect on the ensemble performance.

Let $D^* \in D$ be the ensemble member with the highest average accuracy over the whole feature space $\mathbb{R}^n$. Denote by $P(D_i|R_j)$ the probability of correct classification by D_i in region R_j. Let $D_{i(j)} \in D$ be the classifier responsible for region $R_j, j = 1, \ldots, K$. The overall probability of correct classification of our classifier selection system is

$$P(\text{correct}) = \sum_{j=1}^{K} P(R_j)P(D_{i(j)}|R_j), \tag{7.2}$$

where $P(R_j)$ is the probability that an input $\mathbf{x}$ drawn from the distribution of the problem falls in R_j. To maximize $P(\text{correct})$, we assign $D_{i(j)}$ so that

$$P(D_{i(j)}|R_j) \geq P(D_t|R_j), \quad \forall\, t = 1, \ldots, L. \tag{7.3}$$

Ties are broken randomly. From Equations 7.2 and 7.3,

$$P(\text{correct}) \geq \sum_{j=1}^{K} P(R_j)P(D^*|R_j) = P(D^*). \tag{7.4}$$

The above equation shows that the combined scheme is at least as good as the best classifier D^* in the pool $\mathcal{D}$, *regardless of the way the feature space has been partitioned* into selection regions. The only condition (and, of course, the trickiest one) is to ensure that $D_{i(j)}$ is indeed the best amongst the L classifiers from $\mathcal{D}$ for region R_j. The extent to which this is satisfied determines the success of the classifier selection model.

7.3 ESTIMATING LOCAL COMPETENCE DYNAMICALLY

Let $\mathbf{x}$ be the object to be labeled. In *dynamic classifier selection*, the "competences" of the ensemble members are estimated in the vicinity of $\mathbf{x}$, and the most competent classifier is chosen to label $\mathbf{x}$ [92, 153, 421].

7.3.1 Decision-Independent Estimates

Giacinto and Roli [153] call this approach "a priori" because the competence is determined based only on the location of $\mathbf{x}$, prior to finding out what labels the classifiers suggest for $\mathbf{x}$.

7.3.1.1 Direct k-nn Estimate
One way to estimate the competence is to identify the K nearest neighbors of $\mathbf{x}$ from either the training set or a validation set, and find out how accurate the classifiers are on these K objects [322, 421]. K is a parameter of the algorithm, which needs to be tuned prior to the operational phase.

■ Example 7.2 Decision-independent competence
Consider the fish data and two classifier models: the linear discriminant classifier (LDC) and the Naïve Bayes classifier (NB). The data was generated with 10% label noise. The two classifiers were trained on the noisy data and evaluated on the noise-free data. The competence values were calculated for each point on the grid using the direct k-nn approach with $K = 10$ neighbors. Figures 7.3a and 7.3b show the competence heat map for the two classifiers. Dark color indicates low competence, and light color, high competence. The classification error with respect to the noise-free data is shown in the subplot caption.

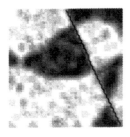 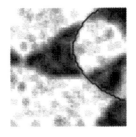 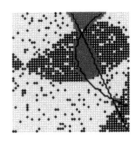

(a) Linear 30.52%   (b) Naïve Bayes 27.64%   (c) Selection ensembles 24.64%

FIGURE 7.3 (a) and (b): competence heat map (direct k-nn) for the LDC and NB classifiers for the fish data with 10% label noise and $K = 10$ neighbors (high competence is shown with lighter color); (c) the selection regions for NB (shaded) and LDC (the remaining part of the space). The classification error with respect to the noise-free data is also given.

The competence patterns are as expected. For example, LDC classifies all points in the left part of the space as gray dots, and the right part as the black dots (the fish). Thus, the part of the fish in the gray class region appears in dark color because the classifier is mistaken there (not competent). By the same argument, the fish "head" in the right region is correctly recognized, hence the light color. The variation of the gray level intensity is due to the label noise.

Figure 7.3c shows the data set and the classification boundaries of the two classifiers. The regions where NB is deemed to have a higher competence than the linear classifier are shaded. In a classifier selection ensemble, the oracle will authorise NB to make the decision in the shaded regions and LDC to make the decision elsewhere. As shown in Figure 7.3c, the error in the ensemble is smaller than the error of either of the two individual classifiers.

The competence estimate depends on several factors, among which is the size of the training or validation set on which the estimate is calculated. The following example illustrates this point.

Example 7.3 Effect of sample size on the competence maps

The LDC and the Naïve Bayes classifier (NB) were applied on the banana data set. (The MATLAB code for generating the two classes is shown in Appendix 7.A.1.) The two "bananas" are aligned, and can be generated with different degrees of noise. Figure 7.4 shows the competence heat maps for LDC and NB for two sizes of the training set: $N = 100$ objects (50 per class) and $N = 800$ (400 per class). In all cases, the direct k-nn method was applied with $K = 10$.

The above example suggests that, for large N, the confidence pattern is generally more pronounced, in that the regions of high and low confidence are more consistent with the expected patterns. Observe that the competence maps for the linear classifier are similar for the two values of N, whereas the maps for NB are quite different.

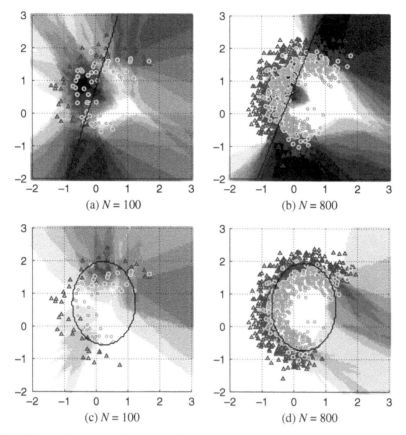

FIGURE 7.4 Competence heat map (direct k-nn) for the aligned banana data for two sizes of the training set. (a) and (b): linear classifier; (c) and (d): NB classifier. High competence is shown with lighter color ($K = 10$).

This raises the question about the size of the reference set (training or validation) on which the competence is dynamically estimated. The choice of N may have a much more dramatic effect on the results compared to the choices of other parameters of the algorithm, for example, the number of neighbors K. Besides, the choice of N could have different effects for different base classifiers.

7.3.1.2 *Distance-based k-nn Estimate*   If the classifiers produce soft outputs, these can be used in the estimate. Giacinto and Roli [153] propose to estimate the competence of classifier D_i as a weighted average. Denote by $P_i(y_j|\mathbf{z}_j)$ the estimate given by D_i of the probability for the true class label y_j of object $\mathbf{z}_j$. For example, suppose that the output of D_i for $\mathbf{z}_j$ in a five-class problem is $[0.1, 0.4, 0.1, 0.1, 0.3]$. Let the true class label of $\mathbf{z}_j$ be $y_j = \omega_5$. Although the decision of D_i would be for class ω_2, the estimated probability for the correct class label is $P_i(y_j|\mathbf{z}_j) = 0.3$. The probabilities

are weighted by the distances to the K neighbors [153,372]. Let $N_\mathbf{x}$ denote the set of the K nearest neighbors of $\mathbf{x}$. The competence of classifier D_i for $\mathbf{x}$ is

$$C(D_i|\mathbf{x}) = \frac{\sum_{\mathbf{z}_j \in N_\mathbf{x}} P_i(y_j|\mathbf{z}_j)\frac{1}{d(\mathbf{x},\mathbf{z}_j)}}{\sum_{\mathbf{z}_j \in N_\mathbf{x}} \frac{1}{d(\mathbf{x},\mathbf{z}_j)}}, \tag{7.5}$$

where $d(\mathbf{x}, \mathbf{z}_j)$ is the distance between $\mathbf{x}$ and its nearest neighbor $\mathbf{z}_j \in N_\mathbf{x}$ according to an appropriate distance measure chosen in advance.

▣ Example 7.4 Competence maps for the distance-based k-nn method

We used the banana data set from the previous example and LDC for the same two data sizes. This time, the competences were calculated with the distance-based k-nn method for two values of the number of neighbors K, $K = 2$ (Figure 7.5) and $K = 10$ (Figure 7.6).

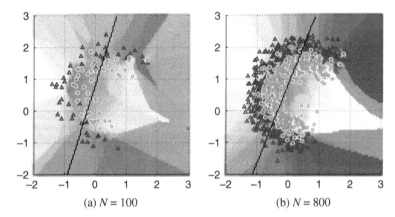

(a) $N = 100$ (b) $N = 800$

FIGURE 7.5 Competence heat map (distance-based k-nn) for the aligned banana data for two sizes of the training set and the linear classifier. High competence is shown with lighter gray color ($K = 2$).

The figures show much smoother competence maps even for $K = 2$. There is also an indication that for a combination of smaller N and larger K (Figure 7.6a), the competence map could be over-smoothed into a flat blur. This defeats the purpose of the competence estimation approach.

The examples illustrate the sensitivity of competence estimates to the choices of parameter values and data. Typically, such choices are not based on a rigorous pre-study, hence the competence maps may turn out to be somewhat spurious. Even so, the classifier selection approach may still work well. If the ensemble is based on strong classifiers, a strong classifier will be elected to make a decision even if it is not the most competent judge for the object submitted for labeling. Such situations will happen in parts of the feature space where the estimates of the classifiers' competence

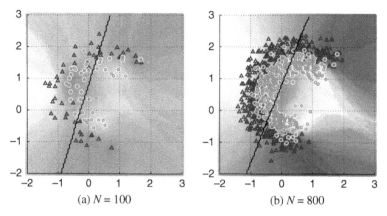

(a) $N = 100$ (b) $N = 800$

FIGURE 7.6 Competence heat map (distance-based k-nn) for the aligned banana data for two sizes of the training set and the linear classifier. High competence is shown with lighter gray color ($K = 10$).

are not very precise. Assuming that the competences will be estimated correctly in most of the space, the classifier selection strategy will work. Note that the competence estimates do not have to be precise as long as they are, what we can call, "top correct". This means that the true most competent classifier has the largest competence value compared to the other classifiers. The exact order of the remaining competences does not matter, nor do the exact values.

7.3.1.3 Potential Functions Estimate

7.3.1.3 Potential Functions Estimate   Rastrigin and Erenstein [322] also consider a distance-based competence coming from the so-called *potential functions* model. We regard the feature space as a field and assume that each point in the data set contributes to the potential in **x**. The potential for classifier D_i at **x** corresponds to its competence. The higher the potential, the higher the competence. The individual contribution of $\mathbf{z}_j \in \mathbf{Z}$ to $C(D_i|\mathbf{x})$ is

$$\phi(\mathbf{x}, \mathbf{z}_j) = \frac{g_{ij}}{1 + \alpha_{ij}(d(\mathbf{x}, \mathbf{z}_j))^2}, \qquad (7.6)$$

where

$$g_{ij} = \begin{cases} 1, & \text{if } D_i \text{ recognizes correctly } \mathbf{z}_j \in \mathbf{z} \\ -1, & \text{otherwise,} \end{cases} \qquad (7.7)$$

and α_{ij} is a parameter which weights the contribution of $\mathbf{z}_j$. In the simplest case, $\alpha_{ij} = \alpha$ for all $i = 1, \dots, L$, and all $j = 1, \dots, N$. The competence is calculated as

$$C(D_i|\mathbf{x}) = \sum_{\mathbf{z}_j \in \mathbf{Z}} \phi(\mathbf{x}, \mathbf{z}_j). \qquad (7.8)$$

It is not clear whether the distance-based versions are better than the direct k-nn estimate of the competence. One advantage though is that ties are less likely to occur for the distance-based estimates.

7.3.2 Decision-Dependent Estimates

This approach is termed "a posteriori" in Ref. [153]. The class predictions of all the classifiers are known.

7.3.2.1 Direct k-nn Estimate
Let $s_i \in \Omega$ be the class label assigned to $\mathbf{x}$ by classifier D_i. Denote by $N_{\mathbf{x}}^{(s_i)}$ the set of K nearest neighbors of $\mathbf{x}$ from $\mathbf{Z}$, which classifier D_i labeled as s_i. The competence of classifier D_i for the given $\mathbf{x}$, $C(D_i|\mathbf{x})$, is calculated as the proportion of elements of $N_{\mathbf{x}}^{(s_i)}$ whose true class label was s_i. This estimate is called, in Ref. [421], the *local class accuracy*.

7.3.2.2 Distance-based k-nn Estimate
Denote by $P_i(s_i|\mathbf{z}_j)$ the estimate given by D_i of the probability that the true class label of $\mathbf{z}_j$ is s_i. The competence of D_i can be measured by $P_i(s_i|\mathbf{z}_j)$ averaged across the data points in the vicinity of $\mathbf{x}$ whose true labels were s_i. Using the distances to $\mathbf{x}$ as weights, we calculate the competence of D_i as

$$C(D_i|\mathbf{x}) = \frac{\sum_{\mathbf{z}_j} P_i(s_i|\mathbf{z}_j)\frac{1}{d(\mathbf{x},\mathbf{z}_j)}}{\sum_{\mathbf{z}_j}\frac{1}{d(\mathbf{x},\mathbf{z}_j)}}, \tag{7.9}$$

where the summation is on $\mathbf{z}_j \in N_{\mathbf{x}}$ such that $y_j = s_i$. A different number of neighbors K can be considered for $N_{\mathbf{x}}$ and $N_{\mathbf{x}}^{(s_i)}$. Woods et al. [421] found the direct decision-dependent k-nn estimate of competence superior to that provided by the decision-independent estimate. They recommend $K = 10$ for determining the set $N_{\mathbf{x}}^{(s_i)}$.

▣ **Example 7.5   Estimation of local competence of classifiers**
Table 7.1 gives the true class labels and the guessed class labels using classifier D_i for a hypothetical set of 15 nearest neighbors of $\mathbf{x}$, $N_{\mathbf{x}}$. The indices of $\mathbf{z}_j$ in the first row in the table are the actual values of $\mathbf{z}_j \in \mathbb{R}$ and $\mathbf{x} \in \mathbb{R}$ is located at 0. Euclidean distance is used, for example, $d(\mathbf{x}, \mathbf{z}_3) = 3$. We assume that D_i provides only label outputs.

TABLE 7.1   True and guessed class labels for the 15 hypothetical nearest neighbors of x sorted by proximity to x

Object (z_j)	1	2	3	4	5	6	7	8	9	10	11	12	13	14	15
True label (y_j)	2	1	2	2	3	1	2	1	3	3	2	1	2	2	1
Guessed label (s_i)	2	3	2	2	1	1	2	2	3	3	1	2	2	2	1

The probability for the chosen output is $P_i(s_i|\mathbf{z}_j) = 1$ and for any other $s \in \Omega$, $s \neq s_i$, we have $P_i(s|\mathbf{z}_j) = 0$. Thus, if the suggested class label is ω_2, the soft vector with the suggested probabilities will be $[0, 1, 0]$.

The decision-independent direct k-nn estimate of competence of D_i would be the accuracy of D_i calculated on $N_\mathbf{x}$, that is, $C(D_i|\mathbf{x}) = \frac{10}{15} \approx 0.33$. Suppose that the output suggested by D_i for $\mathbf{x}$ is ω_2. The decision-dependent direct k-nn estimate of competence of D_i for $K = 5$ is the accuracy of D_i calculated on $N_\mathbf{x}^{(\omega_2)}$ where only the five nearest neighbors labeled in ω_2 are considered. Then $N_\mathbf{x}^{(\omega_2)}$ consists of objects $\{1, 3, 4, 7, 8\}$. As four of the elements of $N_\mathbf{x}^{(\omega_2)}$ have true labels ω_2, the local competence of D_i is $C(D_i|\mathbf{x}) = \frac{4}{5} = 0.80$.

The decision-independent distance-based k-nn estimate of the competence of D_i (using the whole of $N_\mathbf{x}$) will be

$$C(D_i|\mathbf{x}) = \frac{1 + \frac{1}{3} + \frac{1}{4} + \frac{1}{6} + \frac{1}{7} + \frac{1}{9} + \frac{1}{10} + \frac{1}{13} + \frac{1}{14} + \frac{1}{15}}{1 + \frac{1}{2} + \frac{1}{3} + \cdots + \frac{1}{15}} \approx 0.70.$$

The potential function estimate is calculated in a similar way.

Once again, let the output suggested by D_i for $\mathbf{x}$ be ω_2. For the decision-dependent estimate, using the whole of $N_\mathbf{x}$ and taking only the elements whose true label was ω_2,

$$C(D_i|\mathbf{x}) = \frac{1 + \frac{1}{3} + \frac{1}{4} + \frac{1}{7} + \frac{1}{13} + \frac{1}{14}}{1 + \frac{1}{3} + \frac{1}{4} + \frac{1}{7} + \frac{1}{11} + \frac{1}{13} + \frac{1}{14}} \approx 0.95.$$

Woods et al. [421] give a procedure for tie break in case of equal competences. In view of the potential instability of the competence estimates, simple tie-break procedures should suffice.

The search for better competence estimation methods continues with

- probabilistic competence measures [418, 419];
- distribution match between $\mathbf{x}$ and the data used to train the base classifiers, called "Bank of classifiers" [367];
- more elaborate but accurate and adaptive estimates of the neighborhood of $\mathbf{x}$ [91, 156]; and
- using diversity and accuracy together in the selection criterion [363, 424].

7.4 PRE-ESTIMATION OF THE COMPETENCE REGIONS

Estimating the competence dynamically might be too computationally demanding. First, the K nearest neighbors of $\mathbf{x}$ have to be found. For the decision-dependent estimates of competence, $L \times K$ neighbors might be needed. Second, the estimates of the competence have to be calculated. To decrease the computational complexity, competences can be pre-calculated across *regions of competence*. When **x** is submitted for classification, the region of competence in which **x** falls is identified and **x** is labeled by the classifier responsible for that region. The problem becomes one of identifying the regions of competence and the corresponding classifiers.

The regions of competence can be chosen arbitrarily as long as we have reliable estimates of the competences within these regions. Denote by K the number of regions of competence. This number does not have to be equal to the number of classifiers L or the number of classes c. Next, we decide which classifier from $D = \{D_1, \ldots, D_L\}$ should be picked for each region $R_j, j = 1, \ldots, K$. Some classifiers might never be nominated and therefore they are not needed in the operation of the combination scheme. Even the classifier with the highest accuracy over the whole feature space might be dropped from the final set of classifiers. On the other hand, one classifier might be nominated for more than one region.

7.4.1   Bespoke Classifiers

This is perhaps the easiest way of building a classifier selection ensemble. The regions $R_1, \ldots, R_K$ can be chosen prior to training any classifier. A classifier can be trained on the data for each region R_j to become the expert there, denoted $D_{i(j)}$. The advantage of this approach is that the classifiers are trained exclusively on their own regions, which could give them extra competence. The drawback, however, is that the regions may not contain enough data to ensure sound training. This can be addressed by using a simple classifier model.

▉ **Example 7.6   Classifier selection ensembles of bespoke classifiers**
Two competence regions were formed on the fish data by a random linear split. Figure 7.7 shows:

- (a) The classification regions of LDC trained on the noise-free data. The shaded area is what LDC classifies as black dots (the fish).
- (b) The regions of the selection ensemble of LDCs trained on the respective regions of competence.

(a) LDC 30.52%   (b) 2 LDC 26.32%   (c) NB 26.92%   (d) 2 NB 13.24%

FIGURE 7.7   Classification regions of LDC and NB, and for two selection ensembles with random regions of competence delineated by the thick line in (b) and (d). The classification error is given in the captions.

- (c) The classification regions of NB trained on the noise-free data. The shaded area is what NB classifies as black dots (the fish).
- (d) The regions of the selection ensemble of NB classifiers trained on the respective regions of competence.

The classification errors on the noise-free data are shown in the subplot captions.

The example shows that the using regions of competence may lead to a dramatic improvement on the individual classifier's error. However, this is not guaranteed but it is likely, and more so if the regions of competence are determined by some analysis rather than randomly.

7.4.2   Clustering and Selection

The feature space can be split into regular regions but in this case some of them might contain only a small amount of data points and lead to spurious estimates of the competences. To ensure that the regions contain a sufficient amount of points we can use clustering, and regard each cluster as a region of competence. The classifier whose accuracy is estimated to be the highest within a region R will be assigned to make the decisions for any future $\mathbf{x} \in R$.

A classifier selection method called *clustering and selection* is proposed in Refs. [232, 234]. Figure 7.8 shows the training and operation of clustering and selection.

The clustering approach makes sure that there is sufficient data to train or evaluate the classifier. However, this approach has a serious drawback. If the classes are

---

CLUSTERING AND SELECTION

Training: Given is a labeled data set $\mathbf{Z} = \{\mathbf{z}_1, \dots, \mathbf{z}_N\}$.

1. Choose the ensemble size L and the base classifier model.
2. Design the individual classifiers $D_1, \dots, D_L$ using $\mathbf{Z}$.
3. Pick the number of regions K.
4. Disregarding the class labels, cluster $\mathbf{Z}$ into K clusters, $C_1, \dots, C_K$, using, for example, the K-means clustering procedure [106]. Find the cluster centroids $\mathbf{v}_1, \dots, \mathbf{v}_K$ as the arithmetic means of the points in the respective clusters.
5. For each cluster C_j, (defining region R_j), estimate the classification accuracy of $D_1, \dots, D_L$. Pick the most competent classifier for R_j and denote it by $D_{i(j)}$.
6. Return $\mathbf{v}_1, \dots, \mathbf{v}_K$ and $D_{i(1)}, \dots, D_{i(K)}$.

Operation: For each new object

1. Given the input $\mathbf{x} \in \mathbb{R}^n$, find the nearest cluster center from $\mathbf{v}_1, \dots, \mathbf{v}_K$, say, $\mathbf{v}_j$.
2. Use $D_{i(j)}$ to label $\mathbf{x}$, that is, $\mu_k(\mathbf{x}) = d_{i(j),k}(\mathbf{x}), \quad k = 1, \dots, c$.

Return the ensemble label of the new object.

FIGURE 7.8   Training and operation algorithm for clustering and selection.

compact and fairly separable, each class will appear as a cluster and will be designated as a competence region. Then, only a small fraction of the data will be labeled in the alternative classes. This presents a difficult *imbalanced* classification problem in the region of competence.

7.5   SIMULTANEOUS TRAINING OF REGIONS AND CLASSIFIERS

An interesting ensemble method which belongs to the classifier selection group is the so-called *mixture of experts* (ME) [194, 198, 199, 291]. As illustrated in Figure 7.9, in this model the selector makes use of a separate classifier which determines the participation of the experts in the final decision for an input **x**. The ME architecture has been proposed for neural networks. The "experts" are neural networks (NNs), which are trained so that each NN is responsible for a part of the feature space. The selector uses the output of another neural network called the *gating network*. The input to the gating network is **x** and the output is a set of coefficients $P_1, \ldots, P_L$, all depending on **x**. Typically, $\sum_{i=1}^{L} P_i = 1$, and P_i is interpreted as the probability that expert D_i is the most competent expert to label the particular input **x**. The probabilities are used together with the classifier outputs in one of the following ways:

- *Stochastic selection.* The classifier to label **x** is chosen by sampling from $D = \{D_1, \ldots, D_L\}$ according to the distribution $P_1, \ldots, P_L$.
- *Winner-takes-all.* The classifier to label **x** is chosen by the maximum of P_i.
- *Weights.* The probabilities are used as the weighting coefficients to the classifier outputs. For example, suppose that the classifiers produce soft outputs for the c

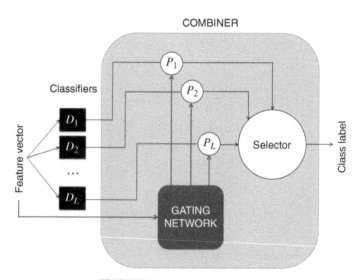

FIGURE 7.9   Mixture of experts.

classes, $d_{i,j}(\mathbf{x}) \in [0, 1]$, $i = 1, \ldots, L, j = 1, \ldots, c$. Then the final (soft) output for class ω_j is

$$\mu_j(\mathbf{x}) = \sum_{i=1}^{L} P_i \, d_{i,j}(\mathbf{x}). \qquad (7.10)$$

If $d_{i,j}(\mathbf{x})$ are interpreted as probabilities, then P_i can be thought of as mixing proportions in estimating $\mu_j(\mathbf{x})$ as a mixture of distributions.

The most important question is how to train the ME model. Two training procedures are suggested in the literature. The first procedure is the standard error backpropagation implementing a *gradient descent*. According to this procedure, the training of an ME is no different in principle to training a single neural network with a complicated structure. The second training approach is based on the *expectation maximization* method which appears to be faster than the gradient descent [199]. ME has been designed mainly for function approximation rather than classification. Its key importance for multiple classifier systems is the idea of training the gating network (therefore the selector) together with the individual classifiers through a standardized training protocol. A similar approach for training a neural network ensemble was proposed by Smieja [371], called a "pandemonium system of reflective agents."

The limitation of the ME training algorithm is that all base classifiers, as well as the gating classifier, must be neural networks. A possible solution for other base classifiers are the evolutionary algorithms (EA) [192]. Using EA, the regions of competence and the classifiers can be trained together.

◼ Example 7.7   A toy evolutionary algorithm for the banana data

Consider a classifier selection ensemble of three linear classifiers for the banana data. A simple evolutionary algorithm was tried, aiming at splitting the space into three regions of competence. The algorithm evolved three points as the centroids of the regions of competence. The regions themselves were the Voronoi cells corresponding to the centroids.

A population of chromosomes was evolved for 150 generations. Each chromosome contained the coordinates of the three centroids (six values in $\mathbb{R}$). At each generation, only the mutation operator was applied by adding small Gaussian noise to the centroids. The set of offspring was created by mutating the whole parent population. The fitness function was the accuracy of the ensemble, where a linear classifier was trained in each cell. The classifiers in the cells were trained and tested on the same training set. After the fitness of all chromosomes in the offspring set was evaluated, the combined set of parents and children was arranged by fitness, and cut to the population size. Thus, the best subset continued as the next generation.

Figure 7.10 shows the classification regions of the evolved ensemble. The centroids and the boundaries of the regions of competence are also plotted. The testing accuracy of the evolved ensembles was typically between 88% and 90%.

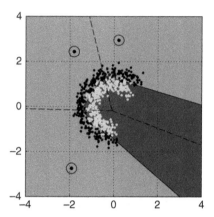

FIGURE 7.10 Classification regions of the evolved ensemble. The centroids and boundaries of the regions of competence are also shown (dashed lines).

Standalone MATLAB code for this example, including the plotting, is given in Appendix 7.A.2. A more detailed description of a genetic algorithm is given in Section 9.4.3.

7.6 CASCADE CLASSIFIERS

Cascade classifiers can be thought of as a version of a classifier selection ensemble. They are extremely useful for real-time systems where the majority of the data will only need to be processed by a few classifiers [13, 57, 116, 146, 230, 247]. When the classifier's confidence is high, we take the class suggested by this classifier as the label for **x**. Alternatively, if the confidence is low, **x** is passed on to the next classifier in the cascade, which processes it in the same fashion. Finally, if the last classifier is also uncertain, the cascade might refrain from making a decision or may select the most likely class anyway. The worth of cascade classifiers was proved beyond doubt by the face detection algorithm of Viola and Jones [405] which is a cascade of AdaBoost ensembles of decision stumps.

APPENDIX: SELECTED MATLAB CODE

7.A.1 BANANA DATA

```
1  %-------------------------------------------------------------------%
2  function [d, labd] = samplebanana(N,na)
3  % N = number of points per class, na = noise amplitude
4  t = -linspace(-pi/4,pi,N)'; z = [sin(t),cos(t)];
5  d = [randn(N,2)*na + 1.5*z; randn(N,2)*na + z];
6  labd = [ones(N,1);2*ones(N,1)];
7  %-------------------------------------------------------------------%
```

7.A.2   EVOLUTIONARY ALGORITHM FOR A SELECTION ENSEMBLE FOR THE BANANA DATA

```
1   %-----------------------------------------------------------------------%
2   function EA_SelectionEnsemble
3   close all
4   global tr trl M N
5   %% Set the parameters
6   N = 600; na = 0.17; % # objects and noise amplitude
7   %......................................................
8   %% Generate the data
9   [tr,trl] = samplebanana(N/2,na); % training data
10  [ts,tsl] = samplebanana(N/2,na); % testing data
11  %......................................................
12  %% Evolve Voronoi cells
13  ps = 20; % population size
14  M = 3; % # of regions of competence
15  T = 150; % generations
16  Pm = 0.8; % mutation shift
17  P = rand(ps,M*2)*8 - 4; % initial population
18  f = fitness(P); % evaluate initial parents
19  for i = 1:T
20      O = P + Pm*randn(size(P)); % mutation
21      O(O<-4) = 0; O(O>4) = 0; % bring outliers to center
22      F = [f,fitness(O)]; % fitness of parents & offspring
23      [t1,t2] = sort(F,'descend');
24      G =[P;O]; P = G(t2(1:ps),:); % survivors
25      f = t1(1:ps); % fitness of the survivors
26  end
27  %......................................................
28  %% Estimate the final ensemble accuracy
29  % retrieve the centroids from the best chromosome
30  cen = reshape(P(1,:),M,2);
31  [xxs,yys] = meshgrid(-4:0.01:4,-4:0.01:4);
32  xx = xxs(:); yy = yys(:); % grid points
33  d = pdist2(cen,tr); [~,reg] = min(d);
34  d2 = pdist2(cen,[xx,yy]);[~,rege] = min(d2);
35  d3 = pdist2(cen,ts); [~,regt] = min(d3);
36  BigLab = zeros(size(xx)); TestLab = zeros(size(tsl));
37  for j = 1:M
38      if sum(reg==j) > 3
39          la2 = classify([xx(rege==j),yy(rege==j)],...
40              tr(reg==j,:),trl(reg==j));
41          BigLab(rege == j) = la2;
42          la3 = classify([ts(regt==j,1),...
43              ts(regt==j,2)],tr(reg==j,:),trl(reg==j));
```

```
44              TestLab(regt == j) = la3;
45        end
46  end
47  fprintf('Ensemble testing error = %.2f %% \n',...
48        mean(TestLab == tsl)*100)
49  %...............................................
50  %% Plot the data
51  figure('color','w'), hold on
52  plot(xx(BigLab == 1),yy(BigLab == 1),'g.','markers',10)
53  plot(xx(BigLab == 2),yy(BigLab == 2),'r.','markers',10)
54  plot(tr(trl == 1,1),tr(trl == 1,2),'k.','markers',10)
55  plot(tr(trl == 2,1),tr(trl == 2,2),'k.','markers',10,...
56        'color',[0.87, 0.87, 0.87])
57  voronoi(cen(:,1),cen(:,2),'k')
58  plot(cen(:,1),cen(:,2),'ko','markersize',15)
59  set(gca,'FontName','Calibri','FontSize',16,'layer','top')
60  axis([-4 4 -4 4]),axis square, grid on
61  %...............................................
62  end
63
64  function f = fitness(P)
65  global tr trl M N
66  for i = 1:size(P,1)
67        cen = reshape(P(i,:),M,2); d = pdist2(cen,tr);
68        [~,reg] = min(d); % regions of competence
69        for j = 1:M
70              z(j) = 0;
71              if sum(reg==j) > 5
72                    la = classify(tr(reg==j,:),...
73                        tr(reg==j,:),trl(reg==j));
74                  z(j) = sum(la == trl(reg==j));
75              end
76        end
77        f(i) = sum(z)/N;
78  end
79  end
80  %-----------------------------------------------------------------%
```

8

DIVERSITY IN CLASSIFIER ENSEMBLES

Common sense suggests that the classifiers in the ensemble should be as accurate as possible and should not make coincident errors. Ensemble-creating methods which rely on inducing diversity in an intuitive manner have proven their value. Even weakening the individual classifiers for the sake of better diversity appears to be an excellent ensemble building strategy, unequivocally demonstrated by the iconic AdaBoost. Ironically, trying to measure diversity and using it explicitly in the process of building the ensemble does not share the success of the implicit methodologies.

8.1 WHAT IS DIVERSITY?

If we have a perfect classifier which makes no errors, then we do not need an ensemble. If however, the classifier does make errors, then we seek to complement it with another classifier which makes errors on different objects. The diversity of the classifier outputs is therefore a vital requirement for the success of the ensemble. However, diversity alone is not responsible for the ensemble performance. It is intricately related with other characteristics of the ensemble. For example, individual classifier accuracy can be sacrificed in order to make the classifiers more diverse. This pays off by a more accurate ensemble compared to that employing the more accurate classifiers. Then how do we strike a compromise between diversity and individual accuracy? How far can we "shake" the ensemble members without destroying the ensemble performance? How can we implement this process, how can we control the degree of this compromise?

Combining Pattern Classifiers: Methods and Algorithms, Second Edition. Ludmila I. Kuncheva.
© 2014 John Wiley & Sons, Inc. Published 2014 by John Wiley & Sons, Inc.

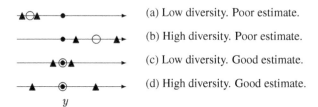

(a) Low diversity. Poor estimate.

(b) High diversity. Poor estimate.

(c) Low diversity. Good estimate.

(d) High diversity. Good estimate.

FIGURE 8.1   Examples of low and high diversity for a point-value estimation. •—true value y; o—estimate $\hat{y}$; ▲—estimator outputs.

8.1.1   Diversity for a Point-Value Estimate

When the task is to approximate a numerical value by an ensemble (called a regression ensemble), the outputs may compensate for each other's discrepancies. The concept of diversity for this case lends itself to mathematical analyses, and so the area has been well researched [56, 220].

Consider a simple example which illustrates the problems associated with the concept of ensemble diversity. The task is to estimate an unknown value y. The ensemble consists of just two estimators, and the estimate $\hat{y}$ is calculated as their average.

Four cases are shown in Figure 8.1. The true y is depicted as a dot, the estimators as triangles, and the estimate $\hat{y}$, as an open circle. The ensemble diversity for this example can be associated with the distance between the two triangles. The closer the triangles, the lower the diversity.

This example gives a good insight into the difficulty in unequivocally relating large diversity to better ensemble performance. The accuracy of the individual estimators is measured by how close the triangles are to the dot y. If both estimators are all accurate (Figure 8.1c), diversity is necessarily small and not particularly important. The outcome is a good estimate. On the other hand, if the estimators are inaccurate (Figures 8.1a, 8.1b, and 8.1d), diversity may or may not be beneficial. High diversity does not necessarily entail good estimate $\hat{y}$, as demonstrated by (Figure 8.1b). Figure 8.1d shows that good estimate can be obtained from arbitrarily inaccurate estimators as long as their deviations cancel one another. Unfortunately, engineering such ensembles is not a trivial matter.

8.1.2   Diversity in Software Engineering

A major issue in software engineering is the reliability of software. Multiple programs (called versions) can be run in parallel in the hope that if one or more fail, the others will compensate for it by producing correct outputs. It is tempting to assume that the errors of the versions will be independent if the versions are created independently. It appears however that independently created versions fail together on difficult assignments and run correctly together on easy assignments [113, 262, 303]. The programs (versions) correspond to the classifiers in the ensemble and the inputs correspond to the points in the feature space.

The model proposed by Littlewood and Miller [262] and developed further by Partridge and Krzanowski [302, 303] considers a set of programs and a set of inputs. The quantity of interest, which underpins several measures of diversity (discussed later) is the probability that two randomly selected programs will fail simultaneously on a randomly chosen input.

8.1.3 Statistical Measures of Relationship

Recall the types of classifier outputs detailed in Section 4.1, and in particular the oracle output. For a given data set $\mathbf{Z}$, classifier D_i produces an output vector $\mathbf{y}_i$ such that

$$y_{ij} = \begin{cases} 1, & \text{if } D_i \text{ classifies object } \mathbf{z}_j \text{ correctly,} \\ 0, & \text{otherwise.} \end{cases} \tag{8.1}$$

Clearly, oracle outputs are only possible for a labeled data set and their use is limited to the design stage of the classifiers and the ensemble.

Various measures of relationship between two variables can be found in statistical literature [65, 74, 373].

8.1.3.1 Correlation Correlation coefficients can be calculated for pairs of continuous-valued (soft) outputs. Every pair of classifiers will give rise to c coefficients, one per class. To get a single measure of diversity between the two classifiers, the correlations can be averaged across classes.

Correlation can be calculated also for a pair of oracle outputs because we can treat the two values (0 and 1) numerically. To illustrate the calculation, consider a table of the joined (oracle) outputs of classifiers D_i and D_j as shown in Table 8.1. The entries in the table are the probabilities for the respective pair of correct/incorrect outputs.

The correlation between two binary classifier outputs is

$$\rho_{i,j} = \frac{ad - bc}{\sqrt{(a + b)(c + d)(a + c)(b + d)}}. \tag{8.2}$$

The derivation of ρ is given in Appendix 8.A.1.

8.1.3.2 The Q Statistic Using Table 8.1, Yule's Q statistic [429] for classifiers D_i and D_j, is

$$Q_{i,j} = \frac{ad - bc}{ad + bc}. \tag{8.3}$$

TABLE 8.1   The 2 × 2 Relationship Table with Probabilities

	D_j correct (1)	D_j wrong (0)
D_i correct (1)	a	b
D_i wrong (0)	c	d

Total, $a + b + c + d = 1$

For statistically *independent* classifiers, $Q_{i,j} = 0$. Q varies between -1 and 1. Classifiers that tend to recognize *the same* objects correctly will have positive values of Q. For any two classifiers, Q and ρ have the same sign, and it can be proved that $|\rho| \leq |Q|$.

8.1.3.3 Interrater Agreement, κ   A statistic developed as a measure of interrater reliability, called κ, can be used when different raters (here classifiers) assess subjects (here $\mathbf{z}_j$) to measure the level of agreement while correcting for chance [128]. For c class labels, κ is defined on the $c \times c$ coincidence matrix M of the two classifiers. The entry $m_{k,s}$ of M is the proportion of the data set (used currently for testing of both D_i and D_j) which D_i labels as ω_k and D_j labels as ω_s. The agreement between D_i and D_j is given by

$$\kappa_{i,j} = \frac{\sum_k m_{kk} - \text{ABC}}{1 - \text{ABC}}, \tag{8.4}$$

where $\sum_k m_{kk}$ is the observed agreement between the classifiers and "ABC" is agreement by chance

$$\text{ABC} = \sum_k \left(\sum_s m_{k,s} \right) \left(\sum_s m_{s,k} \right). \tag{8.5}$$

Low values of κ signify higher disagreement and hence higher diversity. Using the oracle output in Table 8.1

$$\kappa_{i,j} = \frac{2(ad - bc)}{(a+b)(b+d) + (a+c)(c+d)}. \tag{8.6}$$

The derivation of κ is shown in Appendix 8.A.1.

8.2 MEASURING DIVERSITY IN CLASSIFIER ENSEMBLES

8.2.1 Pairwise Measures

These measures and the ones discussed hitherto consider a pair of classifiers at a time. An ensemble of L classifiers will produce $\frac{L(L-1)}{2}$ pairwise diversity values. To get a single value we average across all pairs.

8.2.1.1 The Disagreement Measure   This is probably the most intuitive measure of diversity between a pair of classifiers. For the oracle outputs, this measure is equal to the probability that the two classifiers will disagree on their decisions, that is,

$$D_{i,j} = b + c. \tag{8.7}$$

Without calling it a disagreement measure, this statistic has been used in the literature for analyzing classifier ensembles [182, 368].

8.2.1.2 The Double-Fault Measure   The double fault measure is another intuitive choice, as it gives the probability of classifiers D_i and D_j both being wrong,

$$DF_{i,j} = d. \tag{8.8}$$

Ruta and Gabrys [344] note that DF is a *nonsymmetrical* diversity measure. In other words, if we swap the 0s and the 1s, DF will no longer have the same value. This measure is based on the concept that it is more important to know when *simultaneous errors* are committed than when both classifiers are correct. Thus, the measure is related by design to the ensemble performance.

All diversity measures introduced so far are *pairwise*. To get an overall value for the ensemble we can average across all pairs. Choi and coauthors [74] list 76 (!) binary similarity and distance measures, each of which can be adopted as a diversity measure.

8.2.2 Nonpairwise Measures

The measures of diversity introduced below consider all the classifiers together and calculate directly one diversity value for the ensemble. Oracle classifier outputs are assumed again, where 1 means that the object is correctly labeled, and 0, that the object is misclassified.

8.2.2.1 The Entropy Measure E   Intuitively, the ensemble is most diverse for a particular $z_j \in Z$ when $\lfloor L/2 \rfloor$ of the votes for this object are 0s and the other $L - \lfloor L/2 \rfloor$ votes are 1s.[1] If they all were 0s or all were 1s, there is no disagreement, and the classifiers cannot be deemed diverse. One possible measure of diversity based on this concept is

$$E = \frac{1}{N} \frac{2}{L-1} \sum_{j=1}^{N} \min \left\{ \left(\sum_{i=1}^{L} y_{ij} \right), \left(L - \sum_{i=1}^{L} y_{ij} \right) \right\}, \tag{8.9}$$

where y_{ij} is the oracle output of classifier D_i for object z_j. E varies between 0 and 1, where 0 indicates no difference and 1 indicates the highest possible diversity. Let all classifiers have the same individual accuracy p. While the value 0 is achievable for any number of classifiers L and any p, the value 1 can only be attained for $p \in \left[\frac{L-1}{2L}, \frac{L+1}{2L} \right]$.

It should be noted that E is not a standard entropy measure because it does not use the logarithm function. A classical version of this measure is proposed by Cunningham and Carney [78] (we denote it here as E_{CC}). Taking the expectation over the whole feature space, letting the number of classifiers tend to infinity ($L \to \infty$), and denoting by a the proportion of 1s (correct outputs) in the ensemble, the two

---

[1] $\lfloor a \rfloor$ is the "floor" function. It returns the largest integer smaller than a. $\lceil a \rceil$ is the "ceiling" function. It returns the smallest integer greater than a.

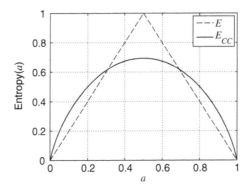

FIGURE 8.2   The two entropies $E(a)$ and $E_{CC}(a)$ plotted versus a.

entropies are

$$E(a) = 2 \min\{a, 1 - a\} \tag{8.10}$$

$$E_{CC}(a) = -a \log(a) - (1 - a) \log(1 - a). \tag{8.11}$$

Figure 8.2 plots the two entropies versus a.

The two measures are equivalent up to a (nonlinear) monotonic transformation. This means that they will have a similar pattern of relationship with the ensemble accuracy.

8.2.2.2 *Kohavi–Wolpert Variance*   Kohavi and Wolpert derived a decomposition formula for the error rate of a classifier [215]. Consider a classifier model. Let y be the predicted class label for $\mathbf{x}$. The variance of y across different training sets which were used to build the classifier is defined to be

$$\text{variance}_{\mathbf{x}} = \frac{1}{2} \left(1 - \sum_{i=1}^{c} P(y = \omega_i | \mathbf{x})^2 \right). \tag{8.12}$$

The variance (8.12) is the *Gini index* for the distribution of the classifier output regarded as a set of probabilities, $P(y = \omega_1 | \mathbf{x}), \dots, P(y = \omega_c | \mathbf{x})$.

We use the general idea for calculating the variance for each $\mathbf{z}_j$ in the following way. We look at the variability of the predicted class label for $\mathbf{z}_j$ for the given training set using the classifier models $D_1, \dots, D_L$. Instead of the whole of Ω, here we consider just the two possible outputs: correct and incorrect. In the Kohavi–Wolpert framework, $P(y = \omega_i | \mathbf{z}_j)$ is estimated as an average across different data sets. In our case, $P(y = 1 | \mathbf{z}_j)$ and $P(y = 0 | \mathbf{z}_j)$ will be obtained as an average across the set of classifiers D, that is,

$$\hat{P}(y = 1 | \mathbf{z}_j) = \frac{l(\mathbf{z}_j)}{L} \quad \text{and} \quad \hat{P}(y = 0 | \mathbf{z}_j) = \frac{L - l(\mathbf{z}_j)}{L}, \tag{8.13}$$

where $l(\mathbf{z}_j)$ is the number of correct votes for $\mathbf{z}_j$ among the L classifiers, that is,

$$l(\mathbf{z}_j) = \sum_{i=1}^{L} y_{i,j}.$$

Substituting Equation 8.13 in Equation 8.12,

$$\text{variance}_{\mathbf{z}_j} = \frac{1}{2}\left(1 - \hat{P}(y=1|\mathbf{z}_j)^2 - \hat{P}(y=0|\mathbf{z}_j)^2\right), \qquad (8.14)$$

and averaging over the whole of $\mathbf{Z}$, we set the *KW* measure of diversity to be

$$KW = \frac{1}{NL^2}\sum_{j=1}^{N} l(\mathbf{z}_j)(L - l(\mathbf{z}_j)). \qquad (8.15)$$

Interestingly, *KW* differs from the averaged disagreement measure (Equation 8.7), denoted D_{av}, by a coefficient, that is,

$$KW = \frac{L-1}{2L}D_{av}. \qquad (8.16)$$

(The proof of the equivalence is given in Appendix 8.A.2).

8.2.2.3   *Measurement of Interrater Agreement, κ, for $L > 2$*   If we denote $\bar{p}$ to be the average individual classification accuracy, then [128]

$$\kappa = 1 - \frac{\frac{1}{L}\sum_{j=1}^{N} l(\mathbf{z}_j)(L - l(\mathbf{z}_j))}{N(L-1)\bar{p}(1-\bar{p})}. \qquad (8.17)$$

It is easy to see that κ is related to *KW* and D_{av} as follows:

$$\kappa = 1 - \frac{L}{(L-1)\bar{p}(1-\bar{p})}KW = 1 - \frac{1}{2\bar{p}(1-\bar{p})}D_{av}. \qquad (8.18)$$

Note that κ is not equal to the averaged pairwise kappa, $\kappa_{i,j}$, in Equation 8.6.

8.2.2.4   *The Measure of Difficulty, θ*   The idea for this measure came from a study by Hansen and Salamon [174]. We define a discrete random variable X taking values in $\{\frac{0}{L}, \frac{1}{L}, \dots, 1\}$ and denoting the proportion of classifiers in D that correctly classify an input $\mathbf{x}$ drawn randomly from the distribution of the problem. To estimate the probability mass function (pmf) of X, the L classifiers in D are run on the data set $\mathbf{Z}$.

Figure 8.3 shows three possible histograms of X for $L = 7$ and $N = 100$ data points. We assumed that all classifiers have individual accuracy $p = 0.6$. The leftmost plot shows the histogram if the seven classifiers were independent. In this case the

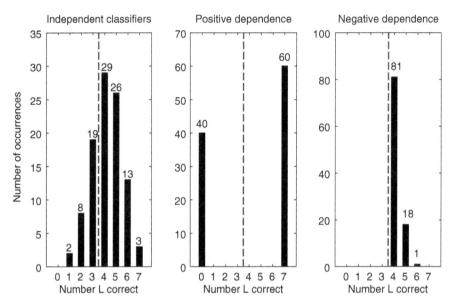

FIGURE 8.3   Patterns of difficulty for three classifier ensembles with $L = 7$, $p = 0.6$, and $N = 100$. The dashed line is the majority vote border. The histograms show the number of points (out of 100) which are correctly labeled by i of the L classifiers. The x-axis shows the number of correct classifiers.

discrete random variable LX has a binomial distribution ($p = 0.6, n = L$). The middle plot shows seven identical classifiers. They all recognize correctly *the same* 60 points and misclassify the remaining 40 points in **Z**. The rightmost plot corresponds to the case of negatively dependent classifiers. They recognize different subsets of **Z**. The distributions are calculated so that each classifier recognizes correctly exactly 60 of the 100 data points.

Hansen and Salamon [174] talk about a pattern of difficulty of the points in the feature space, which in our case is represented by the histogram over **Z**. If the same points have been *difficult* for all classifiers, and the other points have been *easy* for all classifiers, we obtain a plot similar to the middle one (no diversity in the ensemble). If the points that were difficult for some classifiers were easy for other classifiers, the distribution of X is as the one on the right. Finally, if each point is equally difficult for all classifiers, the distribution on the left is the most likely one. Diverse ensembles $\mathcal{D}$ will have smaller variance of X (right plot). Ensembles of similar classifiers will have high variance, as the pattern in the middle plot. Let the three variables X in Figure 8.3 be X_a (left), X_b (middle), and X_c (right). The three variances are

$$\theta_a = Var(X_a) = 0.034 \quad \theta_b = Var(X_b) = 0.240, \quad \theta_c = Var(X_c) = 0.004.$$

Therefore we define the measure of *difficulty* θ to be $Var(X)$. For convenience we can scale θ linearly into $[0, 1]$, taking $p(1 - p)$ as the highest possible value. The higher the value of θ, the worse the classifier ensemble. Ideally, $\theta = 0$, but this is an unrealistic scenario. More often, real classifiers are positively dependent and will exhibit patterns similar to Figure 8.3b.

8.2.2.5 Generalized Diversity Partridge and Krzanowski [303] consider a random variable Y, expressing the proportion of classifiers (out of L) that are incorrect (or *fail*) on a randomly drawn object $\mathbf{x} \in \mathbb{R}^n$. Denote by p_i the probability that $Y = \frac{i}{L}$, that is, the probability that *exactly* i out of the L classifiers fail on a randomly chosen input. (Note that $Y = 1 - X$, where X was introduced for θ.) Denote by $p(i)$ the probability that i *randomly chosen* classifiers will fail on a randomly chosen $\mathbf{x}$. Suppose that two classifiers are randomly picked from $\mathcal{D}$. Partridge and Krzanowski argue that maximum diversity occurs when failure of one of these classifiers is accompanied by correct labeling by the other classifier. In this case, the probability of both classifiers failing is $p(2) = 0$. Minimum diversity occurs when a failure of one classifier is always accompanied by a failure of the other classifier. Then the probability of both classifiers failing is the same as the probability of one randomly picked classifier failing, $p(1)$. Using

$$p(1) = \sum_{i=1}^{L} \frac{i}{L} p_i, \quad \text{and} \quad p(2) = \sum_{i=1}^{L} \frac{i}{L} \frac{(i-1)}{(L-1)} p_i, \tag{8.19}$$

the generalization diversity measure GD is defined as

$$GD = 1 - \frac{p(2)}{p(1)}. \tag{8.20}$$

GD varies between 0 (minimum diversity when $p(2) = p(1)$) and 1 (maximum diversity when $p(2) = 0$).

8.2.2.6 Coincident Failure Diversity Coincident failure diversity (CFD) is a modification of GD also proposed by Partridge and Krzanowski [303]. Again we use the notation p_i as the probability that exactly i out of the L classifiers fail on a randomly chosen input. The diversity measure is defined as

$$CFD = \begin{cases} 0, & p_0 = 1.0; \\ \frac{1}{1-p_0} \sum_{i=1}^{L} \frac{L-i}{L-1} p_i, & p_0 < 1. \end{cases} \tag{8.21}$$

This measure is designed such that it has a minimum value of 0 when all classifiers are always correct or when all classifiers are simultaneously either correct or wrong. Its maximum value 1 is achieved when all misclassifications are unique. In other words, maximum CFD implies that at most one classifier will fail on any randomly chosen object.

Various other diversity measures, frameworks and summaries have been proposed [6, 25, 52, 166, 265, 342, 415], which illustrates the diversity of diversity.

TABLE 8.2   A Distribution of the Votes of Three Classifiers (Row 27 from Table 4.3)

D_1, D_2, D_3	111	101	011	001	110	100	010	000
Frequency	3	0	0	3	2	1	1	0

8.3   RELATIONSHIP BETWEEN DIVERSITY AND ACCURACY

The general anticipation is that diversity measures will be helpful in designing the individual classifiers, the ensemble, and choosing the combination method. For this to be possible, there should be a relationship between diversity and the ensemble performance.

8.3.1   An Example

To investigate the hypothetical relationship between diversity and the ensemble accuracy we recall the example in Section 4.3.3. We generated all possible distributions of correct/incorrect votes for 10 objects and 3 classifiers, such that each classifier recognizes exactly 6 of the 10 objects (individual accuracy $p = 0.6$). The 28 possible vote distributions are displayed in Table 4.3. The accuracy of the ensemble of three classifiers, each of accuracy 0.6, varied between 0.4 and 0.9. The two limit distributions were called the "pattern of success" and the "pattern of failure," respectively.

◪ Example 8.1   Calculation of diversity measures
We take as our example row 27 from Table 4.3. The 10 objects are distributed in such a way that even though all three classifiers have accuracy $p = 0.6$, the ensemble accuracy is low; $P_{\text{maj}} = 0.5$. For an easier reference, the distribution of the votes (correct/wrong) of row 27 of Table 4.3 is duplicated in Table 8.2.

Tables 8.3a, 8.3b, and 8.3c show the probability estimates for the classifier pairs.
The pairwise measures of diversity are calculated as follows:

$$Q_{1,2} = \frac{5 \times 3 - 1 \times 1}{5 \times 3 + 1 \times 1} = \frac{7}{8}$$

$$Q_{1,3} = Q_{2,3} = \frac{3 \times 1 - 3 \times 3}{3 \times 1 + 3 \times 3} = -\frac{1}{2}$$

$$Q_{\text{av}} = \frac{1}{3}\left(\frac{7}{8} - \frac{1}{2} - \frac{1}{2}\right) = -\frac{1}{24} \approx -0.04; \qquad (8.22)$$

TABLE 8.3   The Three Pairwise Tables for the Distribution in Table 8.2

		D_1	D_2			D_1	D_3			D_2	D_3
(a)	D_1	0.5	0.1	(b)	D_1	0.3	0.3	(c)	D_2	0.3	0.3
	D_2	0.1	0.3		D_3	0.3	0.1		D_3	0.3	0.1

$$\rho_{1,2} = \frac{5 \times 3 - 1 \times 1}{\sqrt{(5+1)(1+3)(5+1)(1+3)}} = \frac{7}{12}$$

$$\rho_{1,3} = \rho_{2,3} = \frac{3 \times 1 - 3 \times 3}{(5+1)(1+3)} = -\frac{1}{4}$$

$$\rho_{av} = \frac{1}{3}\left(\frac{7}{12} - \frac{1}{4} - \frac{1}{4}\right) = \frac{1}{36} \approx \mathbf{0.03}; \tag{8.23}$$

$$D_{av} = \frac{1}{3}((0.1 + 0.1) + (0.3 + 0.3) + (0.3 + 0.3)) = \frac{7}{15} \approx \mathbf{0.47}; \tag{8.24}$$

$$DF_{av} = \frac{1}{3}(0.3 + 0.3 + 0.1) = \frac{1}{6} \approx \mathbf{0.17}. \tag{8.25}$$

The nonpairwise measures KW, κ, and E are calculated by

$$KW = \frac{1}{10 \times 3^2}(3 \times (1 \times 2) + 2 \times (2 \times 1) + 1 \times (1 \times 2) + 1 \times (1 \times 2))$$

$$= \frac{7}{45} \approx \mathbf{0.16}; \tag{8.26}$$

$$\kappa = 1 - \frac{D}{2 \times 0.6 \times (1 - 0.6)} = 1 - \frac{7/15}{12/25}$$

$$= \frac{1}{36} \approx \mathbf{0.03}; \tag{8.27}$$

$$E = \frac{1}{10} \times \frac{2}{(3-1)} \times (3 \times \min\{3, 0\} + 3 \times \min\{1, 2\}$$

$$+ 2 \times \min\{2, 1\} + 1 \times \min\{1, 2\} + 1 \times \min\{1, 2\})$$

$$= \frac{7}{10} = \mathbf{0.70}. \tag{8.28}$$

The distribution of the random variables X and Y needed for θ, GD, and CFD are depicted in Figure 8.4.

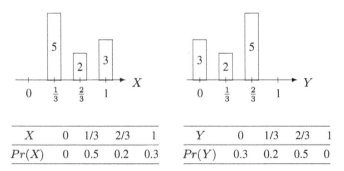

X	0	1/3	2/3	1
$Pr(X)$	0	0.5	0.2	0.3

Y	0	1/3	2/3	1
$Pr(Y)$	0.3	0.2	0.5	0

FIGURE 8.4 The frequencies and the probability mass functions of variables X and Y needed for calculating diversity measures θ, GD, and CFD.

The mean of X is 0.6, and the mean of Y ($p(1)$) is 0.4. The respective measures are calculated as follows:

$$\theta = Var(X) = (1/3 - 0.6)^2 \times 0.5 + (2/3 - 0.6)^2 \times 0.2 + (1 - 0.6)^2 \times 0.3$$

$$= \frac{19}{225} \approx 0.08; \tag{8.29}$$

$$p(2) = \frac{2}{3} \times \frac{(2-1)}{(3-1)} \times 0.5 = \frac{1}{6};$$

$$GD = 1 - \frac{1/6}{0.4} = \frac{7}{12} \approx 0.58; \tag{8.30}$$

$$CFD = \frac{1}{1 - 0.3} \left(\frac{(3-1)}{(3-1)} \times 0.2 + \frac{(3-2)}{(3-1)} \times 0.5 \right)$$

$$= \frac{9}{14} \approx 0.64. \tag{8.31}$$

Calculated in this way, the values of the 10 diversity measures for all distributions of classifier votes from Table 4.3 are shown in Table 8.4. To enable cross-referencing, the last column of the table shows the majority vote accuracy of the ensemble, P_{maj}. The rows are arranged in the same order as in Table 4.3.

With 10 objects, it is not possible to model pairwise independence. The table of probabilities for this case will contain $a = 0.36$, $b = c = 0.24$, and $d = 0.16$. To use 10 objects, we have to round so that $a = 0.4$, $b = c = 0.2$, and $d = 0.2$, but instead of 0, this gives a value of Q

$$Q = \frac{0.08 - 0.04}{0.08 + 0.04} = \frac{1}{3}.$$

In this sense, closest to independence are rows 2, 17, and 23.

8.3.2   Relationship Patterns

It is not easy to spot by eye in Table 8.4 any relationship between diversity and accuracy for any of the diversity measures. To visualize a possible relationship we give a scatterplot of diversity Q_{av} versus improvement in Figure 8.5.

Each point in the figure corresponds to a classifier ensemble. The x-coordinate is the diversity value, Q_{av}, averaged for the three pairs of classifiers, (D_1, D_2), (D_1, D_3), and (D_2, D_3). The y-value is the improvement $P_{maj} - p$. Since each classifier has individual accuracy 0.6, the y-value is simply $P_{maj} - 0.6$. The scatter of the points does not support the intuition about the relationship between diversity and accuracy. If there was a relationship, the points would be distributed along a straight or a curved line with a backward slop, indicating that the lower the values of Q (high diversity), the greater the improvement. What the figure shows is that the best ensemble is not found for the minimum Q_{av} and the worst ensemble is not found for the maximum

TABLE 8.4   The 10 Diversity Measures and the Majority Vote Accuracy, P_{maj}, for the 28 Distributions of Classifier Votes in Table 4.3.

No	Q	ρ	D	DF	KW	κ	E	θ	GD	CFD	P_{maj}
1	−0.50	−0.25	0.60	0.10	0.20	−0.25	0.90	0.04	0.75	0.90	0.9
2	0.33	0.17	0.40	0.20	0.13	0.17	0.60	0.11	0.50	0.75	0.8
3	−0.22	−0.11	0.53	0.13	0.18	−0.11	0.80	0.06	0.67	0.83	0.8
4	−0.67	−0.39	0.67	0.07	0.22	−0.39	1.0	0.02	0.83	0.90	0.8
5	−0.56	−0.39	0.67	0.07	0.22	−0.39	1.0	0.02	0.83	0.90	0.8
6	0.88	0.58	0.20	0.30	0.07	0.58	0.30	0.17	0.25	0.50	0.7
7	0.51	0.31	0.33	0.23	0.11	0.31	0.50	0.13	0.42	0.64	0.7
8	0.06	0.03	0.47	0.17	0.16	0.03	0.70	0.08	0.58	0.75	0.7
9	−0.04	0.03	0.47	0.17	0.16	0.03	0.70	0.08	0.58	0.75	0.7
10	−0.50	−0.25	0.60	0.10	0.20	−0.25	0.90	0.04	0.75	0.83	0.7
11	−0.39	−0.25	0.60	0.10	0.20	−0.25	0.90	0.04	0.75	0.83	0.7
12	−0.38	−0.25	0.60	0.10	0.20	−0.25	0.90	0.04	0.75	0.83	0.7
13	1.0	1.0	0.00	0.40	0.00	1.0	0.00	0.24	0.00	0.00	0.6
14	0.92	0.72	0.13	0.33	0.04	0.72	0.20	0.20	0.17	0.30	0.6
15	0.69	0.44	0.27	0.27	0.09	0.44	0.40	0.15	0.33	0.50	0.6
16	0.56	0.44	0.27	0.27	0.09	0.44	0.40	0.15	0.33	0.50	0.6
17	0.33	0.17	0.40	0.20	0.13	0.17	0.60	0.11	0.50	0.64	0.6
18	0.24	0.17	0.40	0.20	0.13	0.17	0.60	0.11	0.50	0.64	0.6
19	0.00	0.17	0.40	0.20	0.13	0.17	0.60	0.11	0.50	0.64	0.6
20	−0.22	−0.11	0.53	0.13	0.18	−0.11	0.80	0.06	0.67	0.75	0.6
21	−0.11	−0.11	0.53	0.13	0.18	−0.11	0.80	0.06	0.67	0.75	0.6
22	−0.21	−0.11	0.53	0.13	0.18	−0.11	0.80	0.06	0.67	0.75	0.6
23	−0.33	−0.11	0.53	0.13	0.18	−0.11	0.80	0.06	0.67	0.75	0.6
24	0.88	0.58	0.20	0.30	0.07	0.58	0.30	0.17	0.25	0.30	0.5
25	0.51	0.31	0.33	0.23	0.11	0.31	0.50	0.13	0.42	0.50	0.5
26	0.06	0.03	0.47	0.17	0.16	0.03	0.70	0.08	0.58	0.64	0.5
27	−0.04	0.03	0.47	0.17	0.16	0.03	0.70	0.08	0.58	0.64	0.5
28	0.33	0.17	0.40	0.20	0.13	0.17	0.60	0.11	0.50	0.50	0.4

The ensembles separated with lines are: (row 1) pattern of success, (row 13) identical classifiers, and (row 28) pattern of failure.

Q_{av}. The patterns of success and failure occur for values of Q_{av} within the span of possible values for this experiment.

The hypothetical independence point shows only mild improvement of about 0.05 on the individual accuracy p, much smaller than the improvement of 0.30 corresponding to the pattern of success. Note, however, that values of $Q_{av} = 0$ are associated with improvement of 0.10 and also with a decline of the performance by 0.10. For the hypothetical independence point, all three pairwise Q are 0, that is, $Q_{1,2} = Q_{1,3} = Q_{2,3} = 0$ while for the other points at the same Q_{av}, the individual diversities just add up to 0. This suggests that a single measure of diversity might not be accurate enough to capture all the relevant diversities in the ensemble.

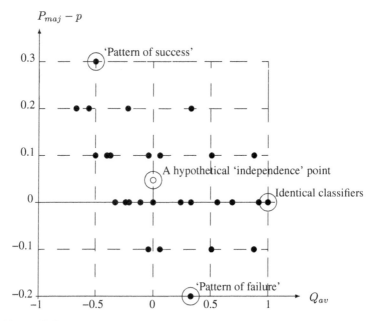

FIGURE 8.5   Improvement on the individual accuracy ($P_{maj} - p$) versus Q_{av}.

The relationship patterns were not substantially different for the other measures. To populate the scatterplots with points, we repeated the simulation but took $N = 30$ objects ($L = 3$ classifiers, each of accuracy $p = 0.6$), which gave a total of 563 ensembles. Figure 8.6 shows the scatterplots of the improvement versus diversity for the 10 measures. As the figure shows, diversity and accuracy are not strongly related.

So far, we assumed that all distributions of the classifier votes are equally likely. The picture is quite different if we assume approximately equal pairwise dependencies $Q_{i,j}$. To illustrate this, we generated 300 ensembles with $L = 3$ classifiers and 300 with $L = 5$ classifiers. The pairwise accuracies within each ensemble were kept approximately equal. Let $P_{max}(D_i)$ be the observed maximum individual accuracy of ensemble D_i, $P_{maj}(D_i)$ be the majority vote accuracy, and $Q_{av}(D_i)$ be the average diversity Q. The ensembles were generated in such a way that all $Q_{k,s}, k, s = 1, \dots, L$, $k \neq s$ were approximately the same for each ensemble, hence approximately equal to $Q_{av}(D_i)$. The relationship between Q_{av} and $P_{maj} - P_{max}$ is illustrated in Figure 8.7. Each point in the scatterplot corresponds to a classifier ensemble.

This time the relationship between diversity and improvement over the single best ensemble member is clearly visible. Smaller Q (more diverse classifiers) leads to higher improvement over the single best classifier. Negative Q (negative dependency) is better than independence ($Q = 0$) as it leads to an even bigger improvement. The zero improvement is marked with a horizontal line in the figure. The points below the line correspond to classifier ensembles that fared worse than the single best classifier. In all these cases, the ensembles consist of positively related but not identical classifiers.

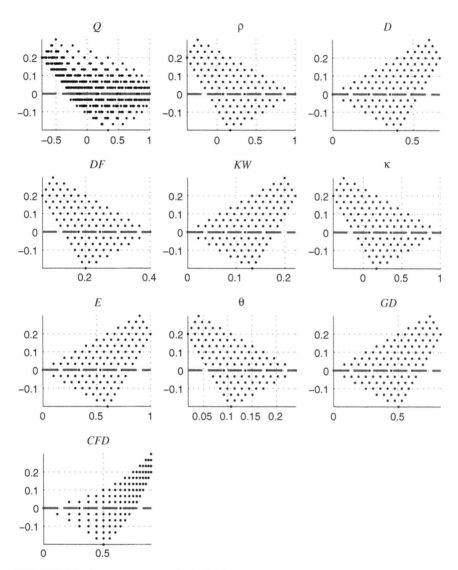

FIGURE 8.6   Improvement on the individual accuracy ($P_{maj} - p$) versus the 10 diversity measures for data size $N = 30$ and individual accuracy $p = 0.6$. The dashed line at $P_{maj} - p = 0$ separates the zones of improvement and deterioration.

If we do not enforce diversity, the ensemble is most likely to appear as a dot toward the right side of the graph. For such ensembles, the improvement on the individually best accuracy is usually negligible. The pronounced relationship in Figure 8.7 was obtained under quite artificial circumstances. When the members of the ensemble have different accuracies and different pairwise diversities, such a relationship does not exist.

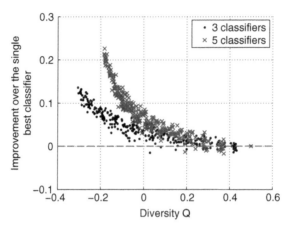

FIGURE 8.7   Plot of $(P_{maj} - P_{max})$ versus Q_{av} for $L = 3$ and $L = 5$ classifiers with accuracy $p = 0.6$ and approximately equal pairwise Q.

Recall the "pattern of success" and "pattern of failure" for the majority vote combiner, discussed earlier. The arguments and examples here reinforce the finding that independence is not necessarily the best scenario for combining classifiers [87, 240]. "Negatively dependent" classifiers may offer an improvement over independent classifiers. While this result has been primarily related to majority voting [292], the same holds for other combiners, for example, Naïve Bayes [16].

8.3.3   A Caveat: Independent Outputs $\neq$ Independent Errors

It is important to have a clear understanding of terminology. When we talk about independent *classifiers*, we often assume independence of the *classification errors* committed by the classifiers. The problem is that the classifier outputs may be independent while the errors are not [404]. In most cases in this book, it is clear from the context that we use the oracle output which takes value 1 if the object is correctly recognized, and 0 otherwise. Then "classification error," "classifier output," and simply "classifier" become synonyms for the purposes of diversity evaluation, and the concept of independence applies to all.

However, if the outputs are class labels (not the oracle), it is possible to find a counterexample where the majority vote of *independent classifiers* has higher error than any of the individual classifiers. Taking Vardeman and Morris's counterexample [404] as a guide, below we generate one of our own (the MATLAB code is provided in Appendix 8.A.3).

Consider three classifiers, D_1, D_2, and D_3 with *independent outputs* for a two-class problem. Just for fun, let us name the classes after two whimsical characters: Bart Simpson and Lisa Simpson.[2] Suppose that each classifier output is a random variable

---

[2]Cited with gratitude: The Simpsons LaTeX font by Raymond Chen (rjc@math.princeton.edu)

TABLE 8.5   Classifier Outputs (D_1, D_2, D_3) and Probability Mass Functions ($p = 0.8$)

Outputs								
D_1	D_2	D_3	$P(y_1, y_2, y_3)$		(Lisa)	(Bart)	MV	BL
L	L	L	$(1-p)^3$	0.008	0.008	0.000	L	L
L	L	B	$(1-p)^2 p$	0.032	0.001	0.031	L	B
L	B	L	$(1-p)^2 p$	0.032	0.001	0.031	L	B
L	B	B	$(1-p)p^2$	0.128	0.127	0.001	B	L
B	L	L	$(1-p)^2 p$	0.032	0.001	0.031	L	B
B	L	B	$(1-p)p^2$	0.128	0.127	0.001	B	L
B	B	L	$(1-p)p^2$	0.128	0.127	0.001	B	L
B	B	B	p^3	0.512	0.000	0.512	B	B

Notes: MV = majority vote labels; BL = Bayes (optimal) labels

(Lisa) $= P(\text{Lisa}, y_1, y_2, y_3)$ (Bart) $= P(\text{Bart}, y_1, y_2, y_3)$.

$y_i \in \{\text{Bart, Lisa}\}$, $i = 1, 2, 3$, following a Bernoulli distribution with probability p for class Bart. Taking the three variables as independent outcomes of an experiment, we create a binomial distribution with parameters $n = 3$ and probability of success p. The classifier outputs and the pmf of the joint distribution $P(y_1, y_2, y_3)$ are shown in Table 8.5.

Take for example output LBL. The probability that the three (independent) classifiers will come up with this output is $(1 - p)p(1 - p) = (1 - p)^2 p = 0.032$. It is up to us now how to split this probability into $P(\text{Lisa, LBL})$ and $P(\text{Bart, LBL})$ so that the two sum up to 0.032. The majority vote (MV) label for this combination of outputs will be Lisa. To make the matter as bad as possible for the majority vote (as we are constructing a counterexample), we can assign a tiny amount to $P(\text{Lisa, LBL})$ and the rest of the probability to $P(\text{Bart, LBL})$. In this example, the "tiny amount" was set to 0.001, but this can be any sufficiently small constant (tunable in the MATLAB code). Then the Bayes class label for the LBL combination will be Bart, as this class has a higher joint probability.

Applying the same destructive approach to all nonunanimous outputs, majority vote will be wrong for six out of the eight possible combinations, and correct only for LLL and BBB. Therefore, the majority vote error will be the sum of the maximum of $P(\text{Lisa}, .)$ and $P(\text{Bart}, .)$ for the six nonunanimous combinations

$$P_{\text{MV}} = 0.031 + 0.031 + 0.127 + 0.031 + 0.127 + 0.127 = 0.474.$$

The individual classifiers have identical error rates. For D_1, for example, the first four labels are L, therefore the respective error probabilities will be taken from the Bart column, and the remaining four (label B), from the Lisa column

$$P_{\text{ind}} = 0 + 0.031 + 0.031 + 0.001 + 0.001 + 0.127 + 0.127 + 0 = 0.318.$$

TABLE 8.6   Classifier Outputs (D_1, D_2) and Classification Errors

D_1	D_2			Probability for the oracle output			
				11	10	01	00
L	L	0.008	0.000	0.08	0	0	0
L	L	0.001	0.031	0.001	0	0	0.031
L	B	0.001	0.031	0	0.001	0.031	0
L	B	0.127	0.001	0	0.127	0.001	0
B	L	0.001	0.031	0	0.031	0.001	0
B	L	0.127	0.001	0	0.001	0.127	0
B	B	0.127	0.001	0.001	0	0	0.127
B	B	0.000	0.512	0.512	0	0	0
				0.522	0.160	0.160	0.158

Finally, the Bayes error for the example is the sum of the minima of the two columns

$$P_B = 6 \times 0.001 = 0.006.$$

We can add to this calculation the error of the largest prior classifier, that is, the classifier which assigns all objects to the most probable class. The probability of the two classes are obtained as the sum of columns Lisa and Bart, respectively, so $P(\text{Lisa}) = 0.392$ and $P(\text{Bart}) = 0.608$. The error of the largest prior classifier will be $P_{LP} = 0.392$.

Thus far, the example showed that combining *independent classifier outputs* increased the majority vote error dramatically, outreaching even the error of the largest prior classifier. Why did this happen? Consider now the classification *errors* of the individual classifiers. Since the problem is completely symmetric, it will suffice to show error dependency between D_1 and D_2. The error patterns of the two classifiers (oracle outputs) are shown in Table 8.6. Symbol 1 in the title row indicates correct classification, and 0 indicates an error.

For example, column number 6 in the table has a header 10. In this column, we list the probabilities of the events where D_1 is correct and D_2 is wrong. The top two and the bottom two entries in this column are 0s because both the classifiers give the same label. In row 3 of column "10," the probability is 0.001. This is the probability of class Lisa because D_1 gives label L (the correct one) and D_2 gives label B (the wrong one). The rest of the table is filled in the same way. The bottom row of the table is the sum of the columns, and gives the total probabilities for a 2×2 contingency table. Using Equation 8.2, the correlation between the two oracle outputs (correlation between the classification errors) is

$$\rho = \frac{0.522 \times 0.158 - 0.16 \times 0.16}{\sqrt{(0.522 + 0.16)(0.16 + 0.158)(0.522 + 0.16)(0.16 + 0.158)}}$$

$$= 0.2623. \tag{8.32}$$

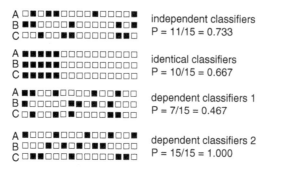

MAJORITY VOTE

3 classifiers: A, B, C
15 objects, ■ wrong vote, □ correct vote
individual accuracy = 10/15 = 0.667
P = ensemble accuracy

independent classifiers
P = 11/15 = 0.733

identical classifiers
P = 10/15 = 0.667

dependent classifiers 1
P = 7/15 = 0.467

dependent classifiers 2
P = 15/15 = 1.000

FIGURE 8.8   Four vote distributions for classifiers A, B, and C: independent classifiers, identical classifiers, and two dependency patterns.

It is clear now that, although the classifier outputs are independent, their errors are not. The example was tailored to disadvantage the majority vote thereby serving as a warning that imprecise terminology may lead to confusion and spawn ill-grounded research.

8.3.4   Independence Is Not the Best Scenario

The prevailing opinion is that we should strive to train classifiers so that they produce independent errors. In fact, independence is not the best scenario. To illustrate this, consider three classifiers and a data set of 15 objects. Each classifier labels correctly 10 of the 15 objects, hence the individual accuracy is 0.667. Figure 8.8 shows four vote distributions for the three classifiers: independent classifiers, identical classifiers, and two dependency patterns.

While the independent outputs improve on the individual accuracy, dependency may greatly improve or destroy the majority vote result. This brings the question about "good" diversity and "bad" diversity [54]. Good diversity helps to achieve correct majority with the minimum number of votes while bad diversity wastes the maximum number of correct votes without reaching majority.

Take, for example, an object and the possible distribution of votes of an ensemble of seven classifiers. Denoting a correct vote by 1 and an incorrect vote by 0, the two trivial nondiverse ensembles give votes [1111111] and [0000000]. The largest diversity with respect to this single object is obtained for the smallest possible voting margin of 1/7. The two ensemble outputs with the largest diversity are [1111000] and [1110000]. The first of these outputs leads to the correct majority vote, hence the large diversity is welcome. The latter output wastes three correct votes on a wrong decision, which makes the large diversity a drawback.

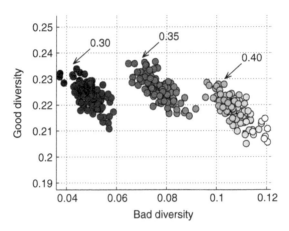

FIGURE 8.9   300 ensembles of $L = 7$ classifiers in each one. The oracle outputs were generated independently, with probability of classification error as indicated. The gray-level intensity signifies the majority vote error. Lighter color corresponds to larger error.

Brown and Kuncheva [54] propose a decomposition of the majority vote which includes terms for good and bad diversities

$$E_{\text{maj}} = (1 - \bar{p}) - \underbrace{\frac{1}{NL} \sum_{l_i \geq \frac{L+1}{2}} (L - l_i)}_{\text{good diversity}} + \underbrace{\frac{1}{NL} \sum_{l_i < \frac{L+1}{2}} l_i}_{\text{bad diversity}}, \tag{8.33}$$

where N is the number of objects in data set $\mathbf{Z}$, l_i is the number of correct votes for object $\mathbf{z}_i \in \mathbf{Z}$, L is the ensemble size, and $\bar{p}$ is the average individual accuracy. The equation shows that good diversity decreases the error whereas bad diversity adds to it. Figure 8.9 shows a scatterplot of 300 ensembles of seven classifiers in the space of good and bad diversity. Each cloud contains 100 points. The outputs were generated independently, where each output bit was set to 0 (error) with the probability displayed above the respective cloud (the average individual error, $1 - \bar{p}$). The gray level intensity signifies the majority vote error. Lighter color corresponds to larger error.

Interestingly, while there is a small variation within each cloud, the major factor for improving the ensemble error seems to be the reduction of bad diversity. The good diversity is nearly constant. The minimum and maximum ensemble errors for the three values of the individual error were

Individual	Minimum E_{maj}	Maximum E_{maj}
0.30	0.096	0.148
0.35	0.173	0.235
0.40	0.261	0.327

8.3.5   Diversity and Ensemble Margins

Several authors have spotted and explored the link between ensemble diversity and the ensemble margins [376, 379, 415]. Stapenhurst [376] offers a comprehensive and insightful review of the state of the art, and proposes a unified viewpoint for diversity measures.

Consider an ensemble of L classifiers whose diversity is estimated on a data set $\mathbf{Z} = \{\mathbf{z}_1, \ldots, \mathbf{z}_N\}$ of N objects. Let l_i be the number of classifiers in the ensemble which gives a correct label for $\mathbf{z}_i$, $i = 1, \ldots, N$. Since the total number of correct votes is $\sum_{i=1}^{N} l_i$, the average individual accuracy of the ensemble members is

$$\bar{p} = \frac{1}{LN} \sum_{i=1}^{N} l_i. \tag{8.34}$$

The *voting margin* of the ensemble for object $\mathbf{z}_i$ is

$$m_i = \frac{l_i - (L - l_i)}{L} = \frac{2l_i - L}{L}. \tag{8.35}$$

The mean margin is

$$m = \frac{1}{N} \sum_{i=1}^{N} m_i = \frac{1}{NL} \sum_{i=1}^{N} 2l_i - L = 2\bar{p} - 1. \tag{8.36}$$

Since the minimum margin is always smaller than or equal to the mean margin, for a fixed $\bar{p}$, it reaches maximum when all m_i are identical and equal to $2\bar{p} - 1$. For this to hold,

$$m_i = 2\bar{p} - 1 = \frac{2l_i - L}{L}, \tag{8.37}$$

hence the maximum of the minimal margin is achieved when

$$l_i = L\bar{p}. \tag{8.38}$$

Tang et al. [379] proved this relationship and named the case when all training objects are classified correctly by the same number of base classifiers *the uniformity condition*. The above value of l_i maximizes not only the margin but also various diversity measures, alluding to a possible link between ensemble margin and diversity. Tang et al. [379] derived expressions for six diversity measures in terms of l_i. To illustrate this derivation, here we reproduce the result for the disagreement measure (8.7).

For object $\mathbf{z}_i$, there are l_i correct classifiers and $L - l_i$ incorrect ones. Therefore, the number of pairs of classifiers that will disagree for this object is $l_i(L - l_i)$.

Averaging across all N objects and scaling for all pairs of classifiers, the disagreement measure is[3]

$$D = \frac{2}{NL(L-1)} \sum_{i=1}^{N} l_i(L - l_i) = \frac{2L}{L-1}\bar{p} - \frac{2}{NL(L-1)} \sum_{i=1}^{N} l_i^2. \qquad (8.39)$$

To maximize D with respect to a single l_i, take the derivative and set it to zero. In addition, we must enforce Equation (8.34), which is done through the Lagrange multiplier

$$\frac{\partial}{\partial l_i}\left(\frac{2L}{L-1}\bar{p} - \frac{2}{NL(L-1)} \sum_{i=1}^{N} l_i^2 + \lambda\left(\bar{p} - \frac{1}{LN}\sum_{i=1}^{N} l_i\right)\right) \qquad (8.40)$$

$$= -\frac{4l_i}{NL(L-1)} - \frac{\lambda}{LN} = 0. \qquad (8.41)$$

Solving for l_i,

$$l_i = -\frac{\lambda(L-1)}{4}. \qquad (8.42)$$

Substituting in the constraint (8.34) and solving for λ,

$$\bar{p} = \frac{1}{LN} \sum_{i=1}^{N} -\frac{\lambda(L-1)}{4} \qquad (8.43)$$

$$\lambda = -\frac{4L\bar{p}}{L-1}. \qquad (8.44)$$

Returning λ in Equation (8.42), we arrive at l_i which maximizes the disagreement diversity measure

$$l_i = L\bar{p}. \qquad (8.45)$$

The second derivative of D with the Lagrangian multiplier is negative, therefore the expression for l_i corresponds to a maximum. Tang et al. [379] prove this relationship for five more diversity measures: double fault, KW, kappa, generalized diversity, and difficulty. They further observe that, unlike maximizing the classification margins, maximizing diversity does not bring the desired improvement of the generalization

---

[3]The discrepancy with the results in reference [379] is due to different notation. To keep consistency, here we denoted by l_i the number of correct votes, while in their notation l_i is the number of wrong votes. The theoretical results are identical.

TABLE 8.7   Some Diversity Measures Expressed in Terms of the Voting Margin [376]

Measure	Expression	
Disagreement	$D = \dfrac{L}{2(L-1)}(1 - \overline{m^2})$	(8.46)
Double fault	$DF = \dfrac{1}{2}(1 - \overline{m}) - \dfrac{L}{4(L-1)}(1 - \overline{m^2})$	(8.47)
KW variance	$KW = \dfrac{1}{4}(1 - \overline{m^2})$	(8.48)
Entropy	$E = \dfrac{L}{L-1}(1 - \overline{|m|})$	(8.49)
Difficulty	$\theta = \dfrac{1}{4}(\overline{m^2} - \overline{m}^2)$	(8.50)
Nonpairwise κ	$\kappa = 1 - \dfrac{L}{L-1}\left(\dfrac{1 - \overline{m^2}}{1 - \overline{m}^2}\right)$	(8.51)
GD	$GD = \dfrac{L}{L-1}\left(\dfrac{1 - \overline{m^2}}{2(1 - \overline{m})}\right)$	(8.52)
CFD	$CFD = \dfrac{L}{L-1}\left(1 - \dfrac{1 - \overline{m}}{2\left(1 - \frac{1}{N}\sum_{i=1}^{N}\delta[m_i = 1]\right)}\right)$	(8.53)

accuracy. Their conclusions are that

- For a given $\bar{p}$, the uniformity condition is usually not achievable.
- In the general case, when the uniformity condition does not hold, the minimum margin of an ensemble is not strongly related to diversity.

Leaving the *minimum* margin aside, and taking diversity–margin correspondence further, Stapenhurst [376] shows that most diversity measures can be expressed as functions of some form of the *average* ensemble margin. Some of these measures are shown in Table 8.7. The notations are as follows:

- $m \in [-1, 1]$ average margin as in Equation (8.36).
- $\overline{|m|} \in [0, 1]$ average absolute margin.
- $\overline{m^2} \in [0, 1]$ average squared margin.

To illustrate the derivation of the expressions in Table 8.7, here we continue the example with the disagreement measure. Expressing l_i from Equation (8.37),

$$l_i = \frac{L(m_i + 1)}{2}. \tag{8.54}$$

Substituting in Equation (8.39), we have

$$D = \frac{2}{NL(L-1)} \sum_{i=1}^{N} l_i(L - l_i) \tag{8.55}$$

$$= \frac{2}{NL(L-1)} \sum_{i=1}^{N} \frac{L(m_i + 1)}{2} \left(L - \frac{L(m_i + 1)}{2} \right) \tag{8.56}$$

$$= \frac{L}{2(L-1)} (1 - \overline{m^2}), \tag{8.57}$$

which is the value in the table.

Stapenhurst [376] disputes Tang's hypothesis that high diversity is related to large minimum margin. He observes instead that for a fixed value of the training *ensemble* accuracy, diversity is more often harmful than beneficial. Looking at the expression of the diversity measures through the voting margin, it is clear that the symmetric measures include the absolute or the squared margin, whose sign does not distinguish between correct and wrong labels. The margin is involved in such a way that the smallest margin corresponds to the largest diversity. Thus, in order to have high diversity, we must have small voting margins. This is the case with the patterns of success and failure, both of which are designed to make use of very small margins (high diversity). Given that this high diversity may lead to a superb ensemble or a disastrous one, if we were to choose an ensemble by maximizing diversity, the results may be unpredictable. We may increase the good and the bad diversity in a different (unknown) proportion. This goes some way to explain the disappointing results in numerous studies where diversity was used explicitly to select the ensemble from a pool of classifiers.

Nonsymmetric measures such as double fault, *CFD*, and *GD* include the average margin with its sign, and this brings them closer to the ensemble accuracy. But the question remains whether we want a proxy for the ensemble accuracy in addition to the estimate that we can calculate from the training data anyway.

8.4   USING DIVERSITY

8.4.1   Diversity for Finding Bounds and Theoretical Relationships

Assume that classifier outputs are estimates of the *posterior probabilities*, $\hat{P}_i(\omega_s|\mathbf{x})$, $s = 1, \ldots, c$, $i = 1, \ldots, L$, so that the estimate $\hat{P}_i(\omega_s|\mathbf{x})$ satisfies

$$\hat{P}_i(\omega_s|\mathbf{x}) = P(\omega_s|\mathbf{x}) + \eta_s^i(\mathbf{x}), \tag{8.58}$$

where $\eta_s^i(\mathbf{x})$ is the error for class ω_s made by classifier D_i. The outputs for each class are combined by averaging, or by an order statistic such as minimum, maximum, or median. Tumer and Ghosh [394] derive an expression about the added classification error (the error above the Bayes error) of the ensemble under a set of assumptions

$$E_{\text{add}}^{\text{ave}} = E_{\text{add}} \left(\frac{1 + \delta(L - 1)}{L} \right), \tag{8.59}$$

where E_{add} is the added error of the individual classifiers (all have the same error), and δ is a correlation coefficient (the measure of diversity of the ensemble).[4] Breiman [50] derives an upper bound on the generalization error of random forests using the averaged pairwise correlation, which also demonstrates that lower correlation leads to better ensembles.

8.4.2   Kappa-error Diagrams and Ensemble Maps

8.4.2.1   Kappa-Error Diagrams   Margineantu and Dietterich suggest the kappa-error plots [271]. Every pair of classifiers is plotted as a dot in a two-dimensional space. The pairwise measure kappa (8.4) is used as the *x*-coordinate of the point and the average of the individual training errors of the two classifiers is used as the *y*-coordinate. Thus, for an ensemble of L classifiers, there are $L(L - 1)/2$ points in the scatterplot. The best pairs are situated in the left bottom part of the plot: they have low error and low kappa (low agreement = high diversity).

⬛ **Example 8.2   Kappa-error plot of four ensemble methods**
Figure 8.10 shows the kappa-error diagram for four ensemble methods: AdaBoost, bagging, random forest, and rotation forest applied to the UCI letters data set. A random half of the data set was used for training and the other half, for testing. The base classifier was a decision tree and the ensemble size was $L = 25$, generating 300 points of classifier pairs in each ensemble cloud. The testing ensemble errors are shown in the caption of the figure.

AdaBoost appears to be the best ensemble in this example. Its cloud of points shows that diversity pays off. The two ensembles with more accurate individual classifiers but higher kappa (less diversity)—bagging and rotation forest—have larger testing error. The rotation forest cloud of points lies slightly lower and to the left of the bagging cloud, which indicates lower individual error and marginally higher diversity. This combination results in a smaller testing error compared to that of bagging.

Desirable as it may be, the exact left bottom corner at $(-1, 0)$ is not achievable. Classifiers that are ideally accurate will be identical, therefore $\kappa = 1$. For each ensemble, a compromise between diversity and individual accuracy must be negotiated. The clouds corresponding to different ensembles, plotted on the same diagram, usually form a "belly" whereby ensembles with higher diversity have members with higher individual errors and vice versa. It is curious to find out why this belly-shaped pattern

---

[4] Averaged pairwise correlations between $P_i(\omega_s|\mathbf{x})$ and $P_j(\omega_s|\mathbf{x})$, $i, j = 1, \ldots, L$ are calculated for every s, then weighted by the prior probabilities $\hat{P}(\omega_s)$ and summed.

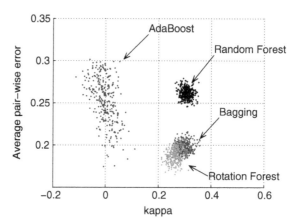

FIGURE 8.10   A kappa-error diagram of four ensemble methods for the letter data set. Testing errors: AdaBoost 5.76%, bagging 9.46%, random forest 9.26%, and rotation forest 8.47%.

exists, and how close a pair of classifiers can be to the bottom left corner of the diagram. The following lower bound relating κ (Equation 8.6) and the average pairwise error e can be proved [243] (the derivation is shown in Appendix 8.A.4)

$$\kappa_{\min} = \begin{cases} 1 - \frac{1}{1-e}, & \text{if } 0 < e \leq 0.5 \\ 1 - \frac{1}{e}, & \text{if } 0.5 < e < 1. \end{cases} \tag{8.60}$$

The bound is tight. It is achievable for a pair of classifiers if they have the same individual error rate $e < 0.5$ and make no simultaneous errors. The bound is plotted in Figure 8.11a. The upper branch ($e > 0.5$), plotted with a dashed line, is of less interest because it corresponds to individual error for the pair of classifiers $e > 0.5$. The lower branch ($e \leq 0.5$) is the "target" part of the bound, where better ensembles are expected to be found. Figure 8.11a shows 20,000 simulated classifier pairs. The number of data points was fixed at $N = 200$ for each contingency table. The N points were randomly split to fill in the a, b, c, and d values in the contingency table. Each classifier pair is a point on the plot, where the coordinates κ and e are calculated as in Equations 8.A.32 and 8.A.33.

The bound itself is not directly related to the ensemble performance. It is expected that ensembles that have classifier pairs closer to the bound will fare better than ensembles that are far away.

Next we generated randomly 1000 ensembles of $L = 3$ classifiers. Each ensemble was a three-way contingency table with eight entries: $N_{000}, N_{001}, \ldots, N_{111}$. The value N_{xxx} is the number of data points that have been classified correctly ($x = 1$) or wrongly ($x = 0$) by classifiers 1, 2, and 3, respectively. For example, N_{011} is the number of points classified correctly by classifiers 2 and 3 and misclassified by classifier 1. The integers N_{xxx} were generated randomly so that $\sum_{xxx} N_{xxx} = N$. The majority vote

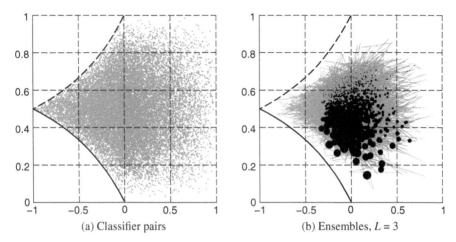

(a) Classifier pairs

(b) Ensembles, $L = 3$

FIGURE 8.11 Illustration of the bound of the kappa-error diagram. (a) 20,000 pairs of classifiers; (b) ensembles of three classifiers, where larger marker indicates higher accuracy.

accuracy can be calculated from the three-way contingency table as

$$P_{maj} = \frac{1}{N}(N_{110} + N_{101} + N_{011} + N_{111}). \tag{8.61}$$

Figure 8.11b illustrates the random ensembles. Each ensemble is depicted as a triangle where the three classifier pairs in the ensemble (points) are connected with lines. In the geometric center of each triangle, a black dot is plotted to indicate the center of the ensemble "cloud." The size of the dot is a gauge of the ensemble accuracy. Ensembles with higher majority vote accuracy are shown with larger dots. A tendency can be observed: ensembles that have more accurate individual classifiers (the triangle is lower down on the y-axis) are better. This tendency is mirrored in the experiments with real data and with ensembles of size $L = 1000$, shown later. Interestingly, diversity does not play as big a role as might be expected. The size of the points increases slightly to the left (toward smaller κ, hence large diversity) but the error-related tendency is much more pronounced. This suggests that in order to create small ensembles with high majority vote accuracy, we should strive to obtain accurate individual classifiers and be less concerned about their diversity. We note that, while the bound on the diagram is valid for any ensemble method, Figure 8.11 gives insights only about the majority vote of ensembles of three classifiers.

Looking for a general pattern across ensemble methods and data sets (however inappropriate such an approach might be), we put together an experiment with 31 UCI data sets and 5 ensemble methods: bagging, AdaBoost, random subspace, random forest, and rotation forest. Each ensemble consisted of 1000 linear classifiers, thereby generating a cloud of 499,500 points on the kappa-error diagram. The number of such clouds is 31 (data sets) × 5 (ensemble methods) = 155. Figure 8.12 shows the kappa-error plot and the derived bound [243]. The ensemble accuracy is indicated by color. Lighter color signifies lower accuracy.

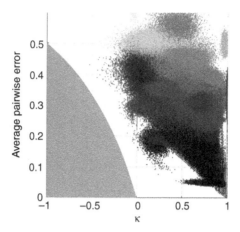

FIGURE 8.12 Kappa-error diagram for 31 data sets and 5 ensemble methods ($L = 1000$). Lighter color signifies lower accuracy.

The plot demonstrates several general tendencies:

- Ensemble accuracy is higher (darker color) for clouds closer to the bound.
- The darker color toward the bottom right corner confirms the result observed in the simulations: the individual accuracy is the dominant factor for better ensemble accuracy.
- There is feasible unoccupied space in the diagram, closer to the boundary, where ensembles of higher accuracy may be engineered.

The bound helps by giving additional insight about the extent of theoretically possible improvement of the ensemble members. It does not however prescribe the way of creating these classifiers.

8.4.2.2 Ensemble Maps Diversity measures have been used to find out what is happening within the ensemble. Pękalska and coauthors [306] look at a two-dimensional plot derived from the matrix of pairwise diversity. Each classifier is plotted as a dot in the two-dimensional space found by Sammon mapping which preserves the distances between the objects. Each point in the plot represents a classifier and the distances correspond to pairwise diversities. The ensemble is a classifier itself and can also be plotted. Any method of combination of the individual outputs can also be mapped. Even more, the true-label "classifier" can be plotted as a point to complete the picture.

■ **Example 8.3   Ensemble map**
The two-dimensional fish data was used again so that we can match visually the classifier decision boundaries to the classifiers' representations in the ensemble map. Figure 8.13a shows the classification boundaries of seven classifiers forming the ensemble. Each classifier was trained on 10 points sampled at random from the

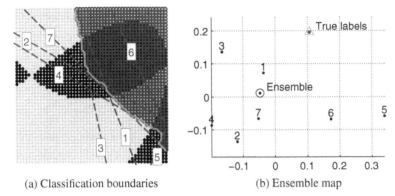

(a) Classification boundaries   (b) Ensemble map

FIGURE 8.13   Classification boundaries and an ensemble map for seven linear classifiers.

2500 labeled grid points. The classifiers are numbered from 1 to 7. The ensemble classification region for class "fish" (black dots) is shaded.

Figure 8.13b shows the ensemble map calculated from the disagreement measure D, using nonmetric multidimensional scaling.[5] Each individual classifier is shown with its number. We used the pairwise disagreement matrix as the entry to the scaling function. Before we calculated the disagreement matrix, we added two more columns to the matrix with the oracle outputs of the individual classifiers: the ensemble oracle output, and a column with 1s representing the ideal classifier which recognizes all objects correctly.

Although a truthful picture of the classifier dependencies is not guaranteed by any means, the scatterplot in Figure 8.13b matches to some degree the appearances in Figure 8.13a. Classifiers 1 and 3 have close boundaries and appear fairly close in Figure 8.13b. The same holds for classifiers 2, 4, and 7. Closest to the ensemble are classifiers 1 and 7, whose boundaries are not far from the ensemble's one. The seven individual errors (in %) were: (1) 30.12, (2) 35.96, (3) 33.68, (4) 36.12, (5) 34.16, (6) 33.36, (7) 33.16. The ensemble error was 28.40%. Of the individual classifiers, classifier 1 is closest to the "true labels" point (the ideal classifier). The proximity to the ideal classifier on the plot cannot be associated directly with the classification error but some weaker association exists. Classifiers 2 and 4 have the largest individual error, and appear farthest from "true labels". On the other hand, 1 and ensemble have the lowest error and are closest to true labels. More interestingly, the ensemble point is amidst points 1–7, indicating its relationship with all ensemble members.

8.4.3   Overproduce and Select

Bigger ensembles are not necessarily better ensembles. It has been reported that small ensembles may be just as good [64,181]. Then it stands to reason to select an ensemble from a pool of classifiers trained with the chosen diversifying heuristics. Originally,

---

[5]MATLAB function `mdscale` was used.

diversity was considered a highly desirable criterion, and was incorporated explicitly within such selection methods [6, 25, 94, 154, 155, 271, 336, 361, 431]. It transpired, however, that diversity was not as useful as hoped. Regardless of the doubts and the limited initial success, the overproduce-and-select approach, also called "ensemble pruning," is flourishing [270, 273, 279, 300, 301, 333, 436]. This subject is discussed in detail in the recent monographs by Zhou [439] and Rokach [335] and will be only briefly illustrated here.

Selection of ensemble members can be thought of as feature selection in the intermediate feature space, where the classifier outputs are the new features. Therefore the same old questions apply:

1. How do we choose the criterion for evaluating an ensemble?
2. How do we traverse the possible candidate subsets?

Several criteria have been explored, some of which explicitly include diversity. But the real wealth of ingenious ideas comes from answering the second question. The methods there range from ranking and cutting the list, forward and backward sequential search, to clustering and electing prototypes, genetic algorithms and analytical optimization. Whatever the answers to the two questions are, one of the most important factors for the success of ensemble pruning is the availability of validation data.

Without a particular reason, apart from simplicity, we chose to illustrate the following ensemble pruning methods.

- *Random order.* Evaluate ensembles of increasing size from 1 to L, adding one classifier at a time, randomly chosen from the pool of trained classifiers. Return the ensemble with the minimum error on the validation set.
- *Best first.* Sort the individual classifiers based on their training error, starting with the best classifier. Evaluate ensembles of increasing size from 1 to L, adding one classifier from the sorted list at a time. Return the ensemble with the minimum error on the validation set.
- *Sequential forward selection (SFS).* Start with the best classifier and add one classifier at a time until all classifiers are included. The classifier to add is chosen from among the remaining classifiers by checking separately each one as an addition to the current ensemble. The classifier which leads to the smallest ensemble error on the validation set is chosen to augment the ensemble. Return the ensemble with the minimum error on the validation set.
- *Kappa-error convex hull pruning* [94, 271]. In kappa-error diagrams, the most desirable pairs of classifiers are situated toward the lower left corner of the plot. Then we can use the convex hull of the ensemble cloud [271] and select only the classifiers within.
- *Pareto pruning.* The ensemble consists of all classifiers which define the Pareto frontier. Figure 8.14 shows the Pareto frontier and the convex hull for an ensemble cloud in a kappa-error diagram. The calculation of the Pareto frontier is explained in Appendix 8.A.5. MATLAB code is also given.

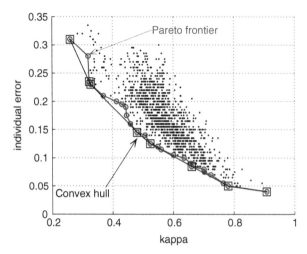

FIGURE 8.14   An example of the convex hull and Pareto frontier of an ensemble cloud in a kappa-error diagram.

We used the fish data set and the rotated checker board data. Fifty Naïve Bayes classifiers were trained on small random subsamples of the training set. The combination method was BKS, as explained in Chapter 4. The procedure was repeated 50 times, and the training, validation, and testing errors were averaged across the repetitions. Figure 8.15a shows the testing errors for the fish data set, and Figure 8.15b, for the checker board data set.

Each of the 50 runs returns a value for the ensemble size and the corresponding testing error for all methods. Figures 8.16 and 8.17 show the average ensemble size and testing error for the five overproduce-and-select methods. Ellipses at one standard deviation are also plotted.

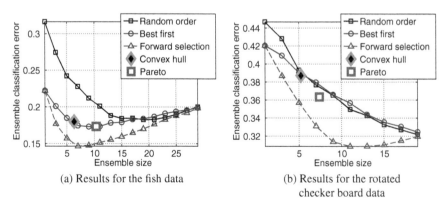

(a) Results for the fish data

(b) Results for the rotated checker board data

FIGURE 8.15   Averaged testing errors from 50 runs of the five overproduce-and-select methods.

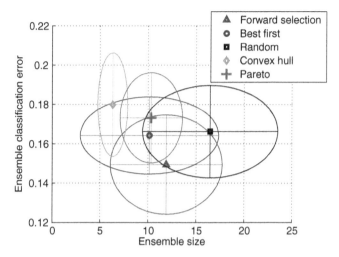

FIGURE 8.16 Results with the fish data. Ensemble size and testing error for the five overproduce-and-select methods, averaged across 50 runs. The ellipses mark one standard deviation of the points from the mean.

It can be seen that, for this combination of data and experimental protocol, the best selection method is the forward search, whose criterion is the validation ensemble error (the lowest positioned ellipse). However, the improvement over the random order ensemble is disappointingly small. The two methods based on diversity and individual accuracy, Pareto selection and convex hull selection, identify smaller ensembles but

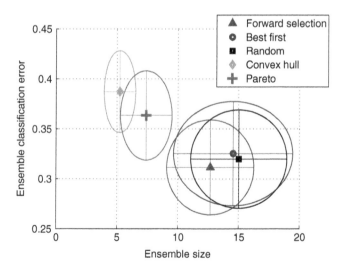

FIGURE 8.17 Results with the rotated checker board data. Ensemble size and testing error for the five overproduce-and-select methods, averaged across 50 runs. The ellipses mark one standard deviation of the points from the mean.

with a larger error. This observation is in tune with the general consensus that the selection criterion should be predominantly based on some estimate of the ensemble error with less contribution from diversity, if any.

8.5 CONCLUSIONS: DIVERSITY OF DIVERSITY

Diversity is an intuitive rather than a precise concept, which is why there are so many measures and definitions. It must be clear what we intend to measure. Consider the following three views.

1. *Diversity as a characteristic of the set of classifiers.* We have a set of classifiers and we have not decided yet which combiner to use. Also, we do not involve information about whether or not the classifier votes are correct. This view seems to be the "cleanest." It would provide information in addition to the individual error rates and the ensemble error rate. In a way, we measure diversity to *discover* whether it contributes to the success of the ensemble.

2. *Diversity as a characteristic of the set of classifiers and the combiner.* In this case the ensemble output is also available. Thus, we can find out which classifier deviates the most, and which deviates the least from the ensemble output. "Individual" diversity can be gauged on this basis. Different combiners might lead to different diversity values for the same set of classifiers.

3. *Diversity as a characteristic of the set of classifiers, the combiner, and the errors.* Here we can also use the oracle information available.

The latter is the most useful perspective because it can be related to the ensemble error, and is the basis of the diversity measures introduced and discussed in this chapter.

Diversity has been intensively studied, and several general frameworks and unifying viewpoints have been proposed [56]. Among these are the information-theoretic framework [52, 440] and the voting margin framework [376, 380]. Compared to 10 years ago, we now have a deeper understanding of ensemble diversity. Below we list five statements which have acquired somewhat axiomatic importance.

1. Diversity is important.

2. There are many (too many?) ways to measure diversity, and no general consensus on a single measure. Some studies on diversity attempt to create measures which relate diversity to the ensemble accuracy. Lucrative as this pursuit is, the more we involve the ensemble performance into defining diversity, the more we are running onto the risk of replacing a simple calculation of the ensemble error by a clumsy proxy which we call diversity. Interpretability of the measure as *diversity* might be lost on the way to trying to tie it up with the ensemble error.

3. Diversity and independence are not synonyms.

4. Diversity may be beneficial or detrimental.

5. So far, explicit inclusion of diversity in ensemble creation/selection has had limited success but ensemble methods which intuitively or heuristically increase diversity fare very well.

APPENDIX

8.A.1   DERIVATION OF DIVERSITY MEASURES FOR ORACLE OUTPUTS

8.A.1.1   Correlation ρ

Table 8.1 containing the relationship between the oracle outputs of two classifiers is reproduced below for easier reference:

	D_j correct (1)	D_j wrong (0)
D_i correct (1)	a	b
D_i wrong (0)	c	d

Total, $a + b + c + d = 1$

Consider the two classifier outputs to be binary random variables X for D_i and Y for D_j. In terms of covariance (Cov) and variance (Var), the correlation between X and Y is

$$\rho = \frac{Cov(X, Y)}{\sqrt{Var(X)}\sqrt{Var(Y)}}.$$

Then

$$\rho = \frac{\sum_{x,y}(x - \bar{x})(y - \bar{y})P(x, y)}{\sqrt{\sum_x(x - \bar{x})^2 P_x(x)}\ \sqrt{\sum_y(y - \bar{y})^2 P_y(y)}}, \tag{8.A.1}$$

where

- $x, y \in \{0, 1\}$,
- $P(x, y)$ is the pmf for the pair of values (x, y),
- $\bar{x}, \bar{y}$ are the respective expectations, and
- P_x and P_y are the marginal distributions for X and Y, respectively.

The space jointly spanned by X and Y is $\{0, 1\} \times \{0, 1\}$, and the joint probabilities are the table entries. Then

$$\bar{x} = 1 \times (a + b) + 0 \times (c + d) = a + b \tag{8.A.2}$$

$$\bar{y} = 1 \times (a + c) + 0 \times (b + d) = a + c. \tag{8.A.3}$$

The numerator of ρ is

$$
\begin{aligned}
Cov(X, Y) &= (1 - (a + b))(1 - (a + c))a + (1 - (a + b))(0 - (a + c))b \\
&\quad + (0 - (a + b))(1 - (a + c))c + (0 - (a + b))(0 - (a + c))d \\
&= (c + d)(b + d)a - (c + d)(a + c)b \\
&\quad - (a + b)(b + d)c + (a + b)(a + c)d \\
&= (c + d)(ab + ad - ab - bc) + (a + b)(ad + cd - bc - cd) \\
&= (c + d)(ad - bc) + (a + b)(ad - bc) \\
&= \underbrace{(a + b + c + d)}_{1}(ad - bc) = ad - bc.
\end{aligned}
$$

Further on,

$$
\begin{aligned}
Var(X) &= (1 - (a + b))^2(a + b) + (0 - (a + b))^2(c + d) \\
&= (c + d)^2(a + b) + (a + b)^2(c + d) \\
&= (c + d)(a + b)(a + b + c + d) = (a + b)(c + d),
\end{aligned}
$$

and

$$
\begin{aligned}
Var(Y) &= (1 - (a + c))^2(a + c) + (0 - (a + c))^2(b + d) \\
&= (b + d)^2(a + c) + (a + c)^2(b + d) \\
&= (b + d)(a + c)(a + b + c + d) = (a + c)(b + d).
\end{aligned}
$$

Assembling the numerator and the denominator, we arrive at

$$
\rho = \frac{ad - bc}{\sqrt{(a + b)(c + d)(a + c)(b + d)}}. \tag{8.A.4}
$$

8.A.1.2   Interrater Agreement κ

Again denote the binary random variables associated with the two classifier outputs by X and Y, respectively. The calculation of the pairwise κ follows the equation

$$
\kappa = \frac{\text{Observed agreement} - \text{ABC}}{1 - \text{ABC}},
$$

where ABC stands for agreement by chance. The observed agreement is the probability that both classifiers produce identical outputs (both correct or both incorrect). Using again Table 8.1, the observed agreement is

$$
\text{Observed agreement} = a + d. \tag{8.A.5}
$$

For ABC, we must calculate the probability of identical outputs assuming that X and Y are independent. The probability for $X = 1$ and $Y = 1$ is calculated from the marginal pmfs as

$$P_x(X = 1)P_y(Y = 1) = (a + b)(a + c),$$

and the probability for $X = 0$ and $Y = 0$ is

$$P_x(X = 0)P_y(Y = 0) = (c + d)(b + d).$$

Then

$$ABC = (a + b)(a + c) + (c + d)(b + d).$$

The numerator of κ becomes

$$Nu = a + d - (a + b)(a + c) - (c + d)(b + d) \tag{8.A.6}$$

$$= a + d - a^2 - ab - ac - ad + ad \tag{8.A.7}$$

$$+ ad - ad - bd - cd - d^2 - 2bc \tag{8.A.8}$$

$$= a + d - a(a + b + c + d) + ad \tag{8.A.9}$$

$$+ ad - d(a + b + c + d) - 2bc \tag{8.A.10}$$

$$= 2(ad - bc). \tag{8.A.11}$$

The denominator becomes

$$De = 1 - (a + b)(a + c) - (c + d)(b + d) \tag{8.A.12}$$

$$= (a + b) + (c + d) - (a + b)(a + c) - (c + d)(b + d) \tag{8.A.13}$$

$$= (a + b)(1 - (a + c)) + (c + d)(1 - (b + d)) \tag{8.A.14}$$

$$= (a + b)(b + d) + (c + d)(a + c). \tag{8.A.15}$$

$$\tag{8.A.16}$$

Putting the two together, we obtain

$$\kappa = \frac{2(ad - bc)}{(a + b)(b + d) + (a + c)(c + d)}. \tag{8.A.17}$$

8.A.2   DIVERSITY MEASURE EQUIVALENCE

Here we prove the equivalence between the averaged disagreement measure D_{av} and Kohavi–Wolpert variance KW.

Recall that $l(\mathbf{z}_j)$ is the number of correct votes (1s) for object $\mathbf{z}_j$. The Kohavi–Wolpert variance [215], in the case of two alternatives, 0 and 1, is

$$KW = \frac{1}{NL^2} \sum_{j=1}^{N} l(\mathbf{z}_j)(L - l(\mathbf{z}_j)) \tag{8.A.18}$$

$$= \frac{1}{NL^2} \sum_{j=1}^{N} \left(\sum_{i=1}^{L} y_{j,i} \right) \left(L - \sum_{i=1}^{L} y_{j,i} \right) = \frac{1}{NL^2} \sum_{j=1}^{N} A_j, \tag{8.A.19}$$

where

$$A_j = \left(\sum_{i=1}^{L} y_{j,i} \right) \left(L - \sum_{i=1}^{L} y_{j,i} \right). \tag{8.A.20}$$

The disagreement measure between D_i and D_k used in [368] can be written as

$$D_{i,k} = \frac{1}{N} \sum_{j=1}^{N} (y_{j,i} - y_{j,k})^2. \tag{8.A.21}$$

Averaging over all pairs of classifiers i, k,

$$D_{\text{av}} = \frac{1}{L(L-1)} \sum_{i=1}^{L} \sum_{k=1_{i \neq k}}^{L} \frac{1}{N} \sum_{j=1}^{N} (y_{j,i} - y_{j,k})^2 \tag{8.A.22}$$

$$= \frac{1}{NL(L-1)} \sum_{j=1}^{N} \sum_{i=1}^{L} \sum_{k=1_{i \neq k}}^{L} (y_{j,i} - y_{j,k})^2$$

$$= \frac{1}{NL(L-1)} \sum_{j=1}^{N} B_j, \tag{8.A.23}$$

where

$$B_j = \sum_{i=1}^{L} \sum_{k=1_{i \neq k}}^{L} (y_{j,i} - y_{j,k})^2. \tag{8.A.24}$$

Dropping the index j for convenience and noticing that $y_i^2 = y_i$,

$$A = L \left(\sum_{i=1}^{L} y_i \right) - \left(\sum_{i=1}^{L} y_i \right)^2 \tag{8.A.25}$$

$$= L \left(\sum_{i=1}^{L} y_i \right) - \left(\sum_{i=1}^{L} y_i^2 \right) - \left(\sum_{i=1}^{L} \sum_{k=1 \, i \neq k}^{L} y_i y_k \right) \tag{8.A.26}$$

$$= (L-1) \left(\sum_{i=1}^{L} y_i \right) - \left(\sum_{i=1}^{L} \sum_{k=1 \, i \neq k}^{L} y_i y_k \right). \tag{8.A.27}$$

On the other hand,

$$B = \sum_{i=1}^{L} \sum_{k=1 \, i \neq k}^{L} (y_i^2 - 2 y_i y_k + y_k^2) \tag{8.A.28}$$

$$= 2(L-1) \left(\sum_{i=1}^{L} y_i \right) - 2 \left(\sum_{i=1}^{L} \sum_{k=1 \, i \neq k}^{L} y_i y_k \right) \tag{8.A.29}$$

$$= 2\mathcal{A}. \tag{8.A.30}$$

Therefore,

$$KW = \frac{L-1}{2L} D_{\text{av}}. \tag{8.A.31}$$

Since the two diversity measures differ by a coefficient, their correlation with $P_{\text{maj}} - P_{\text{mean}}$ will be the same.

8.A.3   INDEPENDENT OUTPUTS $\neq$ INDEPENDENT ERRORS

The code below calculates the example in Section 8.3.3. In addition to the values needed for Table 8.5, the program prints the individual errors, the majority vote error, the error of the largest prior classifier, and the Bayes error.

It also verifies the results by a numerical simulation. One thousand classifier outputs are generated simulating the desired distribution. Any probability estimates from the data are identical to those in Table 8.5. The correlation coefficients are calculated and displayed for the outputs and for the errors, showing that uncorrelated outputs (in this case they are necessarily independent) may have correlated errors rendering majority vote useless.

```
1  %----------------------------------------------------------------%
2
3  p = 0.8; % Bernoulli probability for class 1
4  w = 0.001; % fixed error for the wrong class
5  t = dec2bin(0:7,3);
6  t = reshape(str2num(t(:)),8,3); % labels (0 Lisa,1 Bart)
7  pmf = prod(p.^t.*(1-p).^(1-t),2); % P(y1,y2,y3)
8  z = sum(t,2); % majority vote score for class 1
```

```
 9
10  % Y1 = P(Bart,y1,y2,y3), Y2 = P(Lisa,y1,y2,y3)
11  % .... only unanimity is correct
12  Y1(z == 3) = pmf(z == 3); Y1(z == 0) = 0;
13  Y2(z == 0) = pmf(z == 0); Y2(z == 3) = 0;
14  % .... make majority vote wrong
15  Y1(z == 2) = w; Y2(z == 2) = pmf(z == 2) - w;
16  Y1(z == 1) = pmf(z == 1) - w; Y2(z == 1) = w;
17  Y = [Y1' Y2']; [Ymax,tl] = max(Y');
18
19  % individual errors
20  eri(1) = sum(Y2(find(t(:,1)))+Y1(find(~t(:,1))));
21  eri(2) = sum(Y2(find(t(:,2)))+Y1(find(~t(:,2))));
22  eri(3) = sum(Y2(find(t(:,3)))+Y1(find(~t(:,3))));
23  erm = sum(Y2(z>1) + Y1(z<=1)); % majority vote error
24  erb = 1 - sum(Ymax); % Bayes error
25
25  fprintf('Individual errors %.3f %.3f %.3f\n',eri)
27  fprintf('Majority vote error %.3f\n',erm)
28  fprintf('Largest-prior classifier error %.3f\n',min(sum(Y)))
29  fprintf('Bayes error %.3f\n\n',erb)
30
31  %% Experimental validation
32
33  [data1,data2] = deal([]);
34  for i = 1:8
35      data1 = [data1;repmat(t(i,:),round(Y1(i)*1000),1)];
36      data2 = [data2;repmat(t(i,:),round(Y2(i)*1000),1)];
37  end
38  data = [data1;data2];
39  labels = [ones(size(data1,1),1);zeros(size(data2,1),1)];
40  % experimental individual errors
41  eri_exp = data~=repmat(labels,1,3);
42  fprintf('---- Calculated from data ----\n')
43  fprintf('Individual errors %.3f %.3f %.3f\n',...
44      mean(eri_exp))
45  fprintf('Majority vote error %.3f\n\n',...
46      mean((sum(data,2)>1)~=labels))
47
48  fprintf('Correlation between outputs ----\n')
49  disp(corrcoef(data))
50
51  fprintf('Correlation between errors ----\n')
52  disp(corrcoef(eri_exp))
53  %------------------------------------------------------------------%
```

8.A.4   A BOUND ON THE KAPPA-ERROR DIAGRAM

Consider N data points and the contingency table of two classifiers, C_1 and C_2:

	C_2 correct	C_2 wrong
C_1 correct	a	b
C_1 wrong	c	d

where the table entries are the number of points jointly classified as indicated, and $a + b + c + d = N$. The averaged individual error for the pair of classifiers is

$$e = \frac{1}{2}\left(\frac{c+d}{N} + \frac{b+d}{N}\right) = \frac{b+c+2d}{2N}. \tag{8.A.32}$$

Recall the expression for κ

$$\kappa = \frac{2(ad - bc)}{(a+b)(b+d) + (a+c)(c+d)}. \tag{8.A.33}$$

To facilitate further analyses, it will be convenient to express κ in terms of e and N. We can express a and d as functions of b, c, e, and N and substitute in Equation 8.A.33, which leads to

$$\kappa = 1 - \frac{2N(b + c)}{4N^2 e(1 - e) + (b - c)^2}. \tag{8.A.34}$$

The only restrictions on the values of a, b, c and d so far are that each is nonnegative and they sum up to N. For a fixed e and $(b + c)$, if $b \neq c$, there will be a positive term $(b - c)^2$ in the denominator, which will decrease the fraction, and therefore increase κ. By requiring that $b = c$, and hence dropping the respective term from the denominator, a smaller κ is obtained

$$\kappa' = 1 - \frac{2b}{2Ne(1 - e)} = 1 - \frac{b}{Ne(1 - e)} \leq \kappa. \tag{8.A.35}$$

The minimum value of kappa will be obtained for the largest possible b for the fixed e. To find this value, consider the following system of equations and inequalities:

$$e = \frac{b+c+2d}{2N} = \frac{b+d}{N} \qquad \text{error} \tag{8.A.36}$$

$$2b + d \leq N \qquad \text{total count} \tag{8.A.37}$$

$$d \geq 0 \qquad \text{nonnegativity} \tag{8.A.38}$$

Expressing d from Equation 8.A.36, $d = Ne - b$, and substituting in Equation 8.A.37, we obtain

$$b \leq N(1 - e).$$

On the other hand, substituting in Equation 8.A.38,

$$b \leq Ne.$$

Since both must be satisfied,

$$b_{\max} = \min\{N(1 - e), Ne\}.$$

If $e \leq 0.5$, $b_{\max} = Ne$ and for $e > 0.5$, $b_{\max} = N(1 - e)$. Then, the minimum κ is given by

$$\kappa_{\min} = \begin{cases} 1 - \frac{1}{1-e}, & \text{if } 0 < e \leq 0.5 \\ 1 - \frac{1}{e}, & \text{if } 0.5 < e < 1. \end{cases} \tag{8.A.39}$$

Note that the bound is tight. It is achievable for $b = c$ and $d = \max\{0, (e - 0.5)N\}$.

8.A.5   CALCULATION OF THE PARETO FRONTIER

Let $A = \{a_1, \ldots, a_m\}$ be a set of alternatives (pairs in our case) characterized by a set of criteria $C = \{C_1, \ldots, C_M\}$ (low kappa and low error in our case). Let $C_k(a_i)$ be the value of criterion C_k for alternative a_i. Without loss of generality, assume that lower values are preferable. The Pareto optimal set $S^* \subseteq S$ contains all nondominated alternatives. An alternative a_i is nondominated if there is no other alternative $a_j \in S$, $j \neq i$, such that

$$C_k(a_j) \leq C_k(a_i),$$

where at least one of these inequalities is strict. The concept is illustrated in Figure 8.A.1a. Suppose that the figure is a zoom in the kappa-error diagram, and points A and B are in the convex hull. The x-axis is kappa, and the y-axis is the average individual error of the classifier pairs. Point C is not in the convex hull because it is "behind" the segment AB. However, C is better than A on the error criterion and better than B on the kappa criterion. Therefore, C is nondominated, so it is in the Pareto optimal set.

The MATLAB function `pareto_n`, given below, calculates the indices of the points in the Pareto frontier for n criteria. Use the script below to check out how the function operates. The code generates a sphere of points, and then approximates a

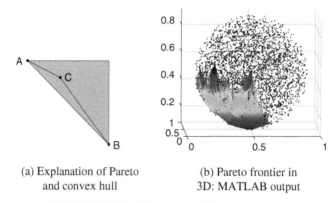

(a) Explanation of Pareto
and convex hull

(b) Pareto frontier in
3D: MATLAB output

FIGURE 8.A.1   Illustration of Pareto optimality.

surface on the Pareto optimal set of points assuming that small values are preferable.
Figure 8.A.1b shows the MATLAB output.

```
1  %--------------------------------------------------------------------%
2  % Check Pareto
3  a = rand(10000,3); sp = sum((a-0.5).^2,2) < 0.18;
4  a = a(sp,:); P = pareto_n(-a);
5  plot3(a(:,1),a(:,2),a(:,3),'k.'), hold on
6  [x,y] = meshgrid(0:0.01:1,0:0.01:1);
7  z = griddata(a(P,1),a(P,2),a(P,3),x,y);
8  surf(x,y,z), shading flat, grid on, axis equal
9  rotate3d
10 %--------------------------------------------------------------------%
```

```
1  %--------------------------------------------------------------------%
2  function P = pareto_n(a)
3  % --- Pareto-optimal set of alternatives
4  % Input: -------------------------------------
5  %        a:  criteria values, the larger the better
6  %            = matrix N (alternatives) by n
7  %            (criterion values)
8  % Output:  -------------------------------------
9  %        P:  indices of the alternatives in the
10 %            Pareto set
11
12 [N,n] = size(a); Mask = zeros(1,N); Mask(1) = 1;
13 for i = 2:N % check each alternative for non-dominance
14     flag = 0; % alternative i is not dominated
15     SM = sum(Mask); P = find(Mask);
```

```
16    for j = 1:SM
17        if sum(a(i,:) <= a(P(j),:)) == n
18            flag = 1; % i is dominated
19        end
20    end
21    if flag == 0 % still not dominated!
22        % eliminate members of P which i dominates
23        for j = 1:SM
24            if sum(a(P(j),:) <= a(i,:)) == n
25                Mask(P(j)) = 0;
26            end
27        end
28        Mask(i) = 1; % add alternative i
29    end
30 end
31 P = find(Mask);
32 %------------------------------------------------------------------%
```

9

ENSEMBLE FEATURE SELECTION

9.1 PRELIMINARIES

9.1.1 Right and Wrong Protocols

It is important that feature selection experiments are "clean," an issue that has been often overlooked [370]. In this context, "clean" means that the testing data which evaluates the quality of a classifier *and* a feature subset *must not* have been seen at any point during the training. This concern is valid for any feature selection protocol, ensemble based and nonensemble based alike. A typical example of a wrong protocol is shown in Figure 9.1. Suppose that we are interested in evaluating a ranking method R using a labeled data set $\mathbf{Z}$, and the parameter that needs tuning is the number of selected features d out of n. The steps in the protocol shown in the figure are as follows:

A. Carry out a cross-validation on $\mathbf{Z}$. Train a ranker on the training part of the ith fold, and estimate the number of features d_i on the testing part of the fold. Average the results to find one final recommended value of d.

B. By this point, the parameter d is tuned, and can be applied to the whole data set to find an optimal feature subset S. The ranker is applied to $\mathbf{Z}$, and the top d features are retained as S.

C. A new cross-validation on $\mathbf{Z}$ is carried out to evaluate the testing error of classifier C using only subset S.

Combining Pattern Classifiers: Methods and Algorithms, Second Edition. Ludmila I. Kuncheva.
© 2014 John Wiley & Sons, Inc. Published 2014 by John Wiley & Sons, Inc.

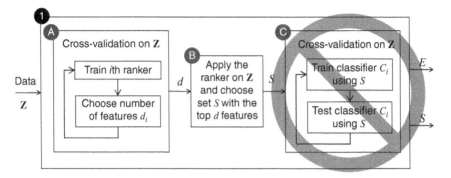

FIGURE 9.1   Incorrect but often used protocol for feature selection and classification.

The misconception here is that the cross-validation protocol, applied for finding both d and C will give us a faithful estimate of the error of C using S. This error estimate is denoted by E in Figure 9.1.

Unfortunately, this is not the case. The protocol is sound up to finding d and S, but is *contaminated* when estimating the error of C. The problem is that $\mathbf{Z}$ has already been used to select S. This practice has been flagged as a problem quite a few times, for example, in fMRI studies, where it is called "peeking" [308]. Sometimes the possible optimistic bias of the error is acknowledged [149] but quite often no such awareness is demonstrated. This casts a doubt on feature selection studies, especially in application areas where pattern recognition and machine learning are still a luxury tool rather than the norm.

In defence of the peeking practice, the optimistic bias may not be very large, and the results may still be valid. Nonetheless, credible claims can only be made on a clean experiment.

Figure 9.2 shows an example of a noncontaminated protocol. To start with, note in both Figures 9.1 and 9.2 black markers with numbers inside. Number 1 indicates the overall training protocol and number 2, the calculation of the output returned to the user. The top part in Figure 9.2 shows a standard classifier training protocol with steps 1 and 2. The purpose of step 1 is to evaluate the classifier error E. Once this is completed, the classifier to be returned to the user is trained on the *whole* of $\mathbf{Z}$. The "guarantee" that goes out with this classifier is the error rate E. The vital presumption here is that a classifier trained on a larger set than those used in step 1 will have a generalization error no worse than E.

A possible step 2 for the protocol in Figure 9.1 is to use S found in step 1 and train a classifier on the whole of $\mathbf{Z}$. However, the returned E may not be a truthful estimate of the generalization error.

The bottom part of Figure 9.2 shows an example of a clean protocol for feature selection and estimation of the classification error, applicable for both ensemble-based and nonensemble-based feature selection. This time the sub-steps inside step 1

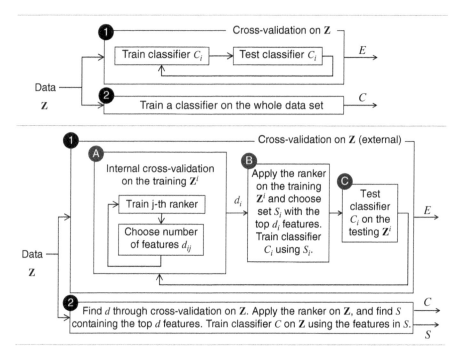

FIGURE 9.2   An example of a noncontaminated experimental protocol for feature selection and classification.

are as follows:

- Carry out a K-fold cross-validation using $\mathbf{Z}$. Denote the ith training fold $\mathbf{Z}^{i+}$, and the ith testing fold $\mathbf{Z}^{i-}$ ($\mathbf{Z}^{i+} \cup \mathbf{Z}^{i-} = \mathbf{Z}$). For each i, $i = 1 \dots, K$:

    A. Carry out an *internal* cross-validation on $\mathbf{Z}^{i+}$. Train a ranker on the training part of the jth internal fold, and estimate the number of features d_{ij} on the testing part of the fold. Average the results to find one final recommended value of d_i for the outer cross-validation fold.
    B. The ranker is applied to $\mathbf{Z}^{i+}$, and the top d_i features are retained as S_i. Train a classifier C_i on $\mathbf{Z}^{i+}$ using only subset S_i.
    C. Test C_i on $\mathbf{Z}^{i-}$.

- Calculate the error E as the average of the testing errors of the K folds.

Note that the testing set of each fold, $\mathbf{Z}^{i-}$, is reserved for estimating the accuracy of the whole sequence of estimation of d_i and training of C_i. The testing set is not seen through any of these steps. Therefore, it estimates the error of the whole package ⟨feature selection+classifier training⟩.

The correct protocol allows for a straightforward step 2 to be executed. Cross-validation of the whole of **Z** is carried out to determine d as in step A in Figure 9.1. The feature set S is determined by running the ranker through **Z** and retaining the top d features, as in step B of the same figure. Finally, a classifier is trained on **Z** using S. The selected feature set S and the final classifier C are exactly the same as the ones in step 2 with the wrong protocol. The difference is in the estimate E. Although this seems to be a minor glitch, the results returned to the user may be misleading, and may have unpleasant implications if taken at face value and applied to real-life problems. The magnitude of the bias will likely depend on the types of the feature selector and classifier, as well as on some properties of the data.

▣ Example 9.1   Optimistic bias of the wrong protocol

We ran an experiment on the "spam" data set from the UCI collection [22]. The data set has the following characteristics

Number of objects N	4601
Number of features n	57
Number of classes c	2
Percentage of data in the larger class	60.6%

The problem is to distinguish between spam and legitimate e-mail. The first 54 features are the frequencies of each of the 48 words and each of the 6 characters. The remaining three features are the average length of uninterrupted sequences of capital letters, the length of the longest uninterrupted sequence of capital letters, and total number of capital letters in the e-mail. All features are continuous-valued.

Tenfold cross-validations were run with the protocol in Figure 9.1, and as the internal and the external loops in Figure 9.2. One hundred runs were carried out where 200 data points were randomly chosen to be the data set **Z**, and the remaining 4401 data points were left for testing. The classification error E was calculated once through the wrong protocol (E_w) and once through the correct one (E_c). For each 200/4401 split, the corresponding error E was calculated from the testing set left aside. A two-sided signed rank test was carried out, once for E_w and E, and once for E_c and E. The null hypothesis is that the difference between the two variables comes from a distribution with median zero, or, in other words, there is no bias. The p-value for the test (E_w, E) was $p = 0.0357 < 0.05$, rejecting the null hypothesis and suggesting that there is bias. The p-value for the test (E_c, E) was $p = 0.0891 > 0.05$, so the null hypothesis cannot be rejected. The mean values for the wrong protocol were $\bar{E}_w = 30.47\%$ and $\bar{E} = 31.37\%$, and for the correct protocol, $\bar{E}_c = 32.27\%$ and $\bar{E} = 32.84\%$.

The figures and the examples are about feature rankers and tuning the number of features d, but the concerns hold for any feature selection method or parameter tuning. The internal and external training–testing protocol does not have to be

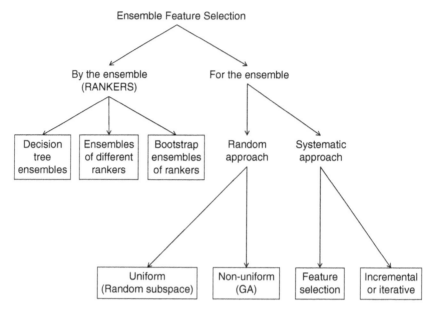

FIGURE 9.3   A summary diagram of ensemble feature selection approaches.

cross-validation. The message here is that the testing data must not be seen at *any* part of the training, including feature selection and parameter tuning.

9.1.2   Ensemble Feature Selection Approaches

Continuing from 1.6.2, here we introduce approaches and methods for ensemble feature selection as summarized in Figure 9.3. The first question is about the goal of the feature selection.

Sometimes the features are interpretable, and the end user is interested in finding a highly informative subset. We call this "feature selection *by the ensemble*." For example, in functional magnetic resonance imaging (fMRI), each feature is a voxel in the brain image. A subset of informative features may reveal parts of the brain whose activity is related to a certain cognitive task. The classification accuracy in such problems is of secondary interest, and serves to ensure the quality of the selected feature subset. The output of this approach is the feature set, which can be used subsequently with classifiers or ensembles of choice.

The alternative approach, called here "feature selection *for the ensemble*" aims at selecting features which improve the performance of the ensemble. The output is the ensemble itself, and the feature selection is rather a means to an end.

9.1.3   Natural Grouping

In some problems the features are naturally grouped. For example, in text-independent speaker identification, two groups of features are related to the pitch of the signal, and the speech spectrum. The speech spectrum can be further characterized by the linear predictive coefficients, the cepstrum, and so on [72]. In handwritten digit recognition, an image can be viewed from different perspectives such as pixels, morphological features, Fourier coefficients of the character shapes, and so on [109, 399]. Emotion recognition is typically a multi-modal classification problem where the input comes from various modalities: physiological measurements of the peripheral nervous system such as electrodermal skin response and heart rate, electroencephalography (EEG), behavioral cues, voice modulation, facial expressions, and so on [63]. Sometimes the groups of features are measured at different geographical locations, for example, radar images of a flying object. Instead of transmitting all the features and making a decision centrally, individual classifiers can be built and only their decisions will have to be transmitted.

In terms of the methods' diagram in Figure 9.3, having naturally grouped features will amount to a ready-made feature selection *for the ensemble*. Each ensemble member is trained on its bespoke feature set [208, 398].

9.2   RANKING BY DECISION TREE ENSEMBLES

The decision tree classifier offers several ways of measuring the importance of individual features [50, 149, 267, 378].

9.2.1   Simple Count and Split Criterion

Consider a bagging ensemble of decision trees. The features in the problem can be ranked by the number of times they have been used to split a node.

To refine and stabilize the measure of feature importance, we can include the value of the criterion at the split. This value should be weighted by the number of training points that arrived at the node. For example, suppose that the reduction of the Gini impurity at node t is $\Delta(t) = 0.1$, and the number of points at the node is $N_t = 60$. Then a value of $N\Delta(t) = 6$ will be added to the score for the feature responsible for the split at t. The total score for a feature is calculated by summing up all such values across all trees in the ensemble.

■ Example 9.2   Simple count and Gini-sum feature ranking
To illustrate the two ranking methods we used the "mfeat" UCI data set. The set consists of features extracted from handwritten digits from 0 to 9. The "'pix" version of data set was used because the feature importance can be easily visualized. Each object is described by 240 features which are the gray level intensities of the pixels in a 16-by-15 matrix containing the image. This allows us to use a matrix of the same size where the intensity of each pixel will correspond to that feature's importance.

We chose only three classes: 3, 6, and 8, examples of which are shown in Figure 9.4. The reason for this choice is that we know where the differences should be and can judge visually whether the feature importance is adequate.

FIGURE 9.4 Examples of handwritten digits in the three classes.

The data was split into training and testing halves. Twenty-five decision trees were trained on the training part, and the features were ranked according to their importance by the simple count (ranking R_1) and the reduction of the Gini impurity (ranking R_2). Figures 9.5a and 9.5b show the feature importance as gray level intensities of the corresponding pixels. The results were obtained as the average of 25 runs. Both plots place the most important pixels in two locations; row 6 column 12, which distinguishes between 6 and (3,8) and row 11 column 3, which distinguishes between 3 and (6,8). Arguably, the Gini-impurity ranking is less noisy than the simple count ranking.

Figure 9.5c shows the testing error (average of 25 runs) of the two rankings and of a random permutation of the features. The linear discriminant classifier (LDC)

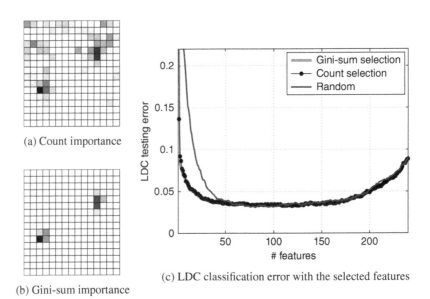

(a) Count importance

(b) Gini-sum importance

(c) LDC classification error with the selected features

FIGURE 9.5 Results with the two feature ranking methods. Darker color in (a) and (b) signify higher importance.

was chosen for the task. A point on the curve at number of features d is the testing error of the LDC using a feature set with the top d features of the respective ranking. The random permutation curve is above the other two for a small number of features. However, our experiment did not reveal any difference between the two ranking methods. This is not to say that such difference may never be found. The result merely brings up again the pattern recognition mantra that simple methods usually work very well for the majority of problems.

9.2.2 Permuted Features or the "Noised-up" Method

The hypothesis for this approach is that if a feature is important, then shuffling its values in the testing data will increase the error dramatically. Conversely, if the feature is irrelevant, shuffling its values will not have a great effect on the error. The approach was first proposed for the random forest ensemble [50] but can be applied to any ensemble that provides out-of-bag (OOB) data. Feature selection is done together with ensemble training, without seeing testing or validation data. The algorithm is illustrated in Figure 9.6.

Feature evaluation happens along the testing of each classifier on the respective OOB. The values of each feature are permuted within the OOB and the classifier is tested again. The labels are stored for further use. At the end of the training, the ensemble error e is evaluated on the OOB. Next, n more ensemble errors $e_1, \dots, e_n$ are calculated, one for each noised-up feature. The merit of the feature is $F_j = e_j - e$.

NOISED-UP FEATURE RANKING

Given is a labeled data set $\mathbf{Z} = \{\mathbf{z}_1, \dots, \mathbf{z}_N\}$ described by n features in the set $X = \{X_1, \dots, X_n\}$.

1. Choose the ensemble size L.
2. For each $i = 1, \dots, L$.
   (a) Take a bootstrap sample S_i from $\mathbf{Z}$ and train classifier D_i on it. Denote by O_i the out-of-bag (OOB) set. Test D_i on O_i and store the assigned labels.
   (b) For every feature $X_j, j = 1, \dots, n$,
       i. Permute the values of X_j in O_i. Denote the new set O_i'.
       ii. Test D_i on O_i', and store the assigned labels.
3. For each $j = 1, \dots, N$, derive the ensemble label for $\mathbf{z}_i$ using the labels assigned by all classifiers which contained $\mathbf{z}_i$ in their OOB sets. Calculate the ensemble error e.
4. For each feature $X_j, j = 1, \dots, n$, calculate the OOB ensemble error e_j with the noised-up X_j. Calculate the feature importance as the error difference

$$F_j = e_j - e.$$

Return the feature scores $F_1, \dots, F_n$.

FIGURE 9.6   Noised-up feature selection from a bagging/random forest ensemble [50].

Breiman illustrated the noised-up ranking method on two data sets with fairly small n (8 and 16) while building a sizeable ensemble with $L = 1000$ classifiers. One problem with larger n is the potentially large number of ties, especially for "wide" data sets where the number of features exceeds by orders of magnitude the number of objects. Svetnik et al. [378] propose to use the ensemble margins instead of the ensemble error. If the average ensemble margin is m, and the average margin with the noised-up feature X_j is m_j, the score for X_j is $F_j = m - m_j$.

🖳 Example 9.3   The noised-up feature ranking method

The UCI spam data set was used for this example. Twenty-five decision trees were trained on a random sample of 300 objects. The noised-up ranking was obtained from the ensemble. The count ranking and the Gini-sum ranking were also computed. Finally a random permutation of the features was also created as a benchmark. Figure 9.7 shows the LDC classification error with the ranked features. The curves are the average of 10 runs of the same experiment with different training–testing splits.

The results with only the top 30 features are shown because further on the covariance matrix of the LDC becomes ill-defined, and a regularized version of the classifier has to be used. A rapid jump of the error is observed, which obscures the point we want to make with this illustration.

This time the random ranking has pronouncedly larger error than the other three ranking methods. Besides, the count selection method seems to have a consistently higher error than the other two methods. To explore this further, we ran sign rank test between the Gini ranker and the count ranker, once for each value of the number of features d from 1 to 30. Figure 9.8a shows the results. The same two curves for the respective rankers are plotted, and the points where significant differences were found by the test ($p < 0.05$) are marked with circles.

Figure 9.8b shows the results from the sign rank test comparing the count ranker and the "noised-up" ranker. The points where significant difference was found ($p < 0.05$) are marked with squares.

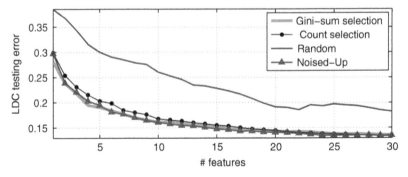

FIGURE 9.7   Feature selection by tree ensembles on the spam data [22] (average of 10 runs). Training: $N_{tr} = 300$ objects; testing with linear discriminant classifier (LDC): $N_{tr} = 4301$ objects; ensemble size $L = 25$ classifiers.

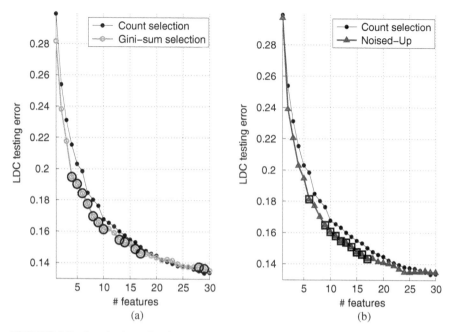

FIGURE 9.8   Results from the sign rank test between the count ranker and: (a) the Gini-sum ranker; and (b) the noised-up ranker. The points where significant difference was found are marked with circles (a) and squares (b).

Why are we not using the protocol advocated in Section 9.1.1? The reason is that we do not need the step to tune the number of features d or recommend a feature subset. The purpose of this example was to explore the behavior of the classification error, which did not require an internal loop.

9.3   ENSEMBLES OF RANKERS

9.3.1   The Approach

The rationale for using ensembles instead of simple feature selection procedures is that the ensemble solution has been found to be much more stable and of the same or better accuracy [1, 197, 401, 408]. The generic diagram of an ensemble of feature rankers is shown in Figure 9.9.

As with classifier ensembles, any diversifying heuristic can be employed to create the rankers. For example, bootstrap samples can be taken as in bagging, and a ranker can be trained on each replicate [1]. Alternatively, different ranking methods can be applied to the unaltered data set [374].

Once the rankers have been trained, the "aggregator" fuses them into a final ranking. The simplest way to do this is to average the L ranks obtained by each feature.

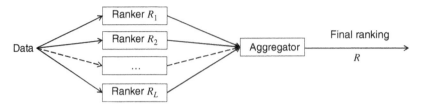

FIGURE 9.9 The generic diagram of an ensemble of feature rankers.

▣ Example 9.4 A numerical example of a feature ranking ensemble

Consider a problem with $n = 10$ features and an ensemble of $L = 4$ rankers. Suppose that each ranker arranges the features by importance, starting with the most important one. Let the four ranking lists be

$$
\begin{array}{lllllllllll}
R_1: & 4 & 7 & 9 & 6 & 8 & 2 & 1 & 3 & 5 & 10 \\
R_2: & 4 & 2 & 1 & 5 & 7 & 9 & 8 & 3 & 10 & 6 \\
R_3: & 9 & 6 & 4 & 2 & 7 & 1 & 8 & 3 & 5 & 10 \\
R_4: & 9 & 5 & 2 & 4 & 3 & 1 & 8 & 7 & 6 & 10
\end{array}
$$

The ranks of the features are calculated from this arrangement as follows. The top ranked feature receives rank 1, the next one rank 2, and so on. The last feature receives rank n, where n is the total number of features. The ranks of the 10 features in the example are shown below:

Feature No. :	1	2	3	4	5	6	7	8	9	10
From R_1 :	7	6	8	1	9	4	2	5	3	10
From R_2 :	3	2	8	1	4	10	5	7	6	9
From R_3 :	6	4	8	3	9	2	5	7	1	10
From R_4 :	6	3	5	4	2	9	8	7	1	10
Average rank	5.50	3.75	7.25	2.25	6.00	6.25	5.00	6.50	2.75	9.75

For instance, feature number 1 appears at the seventhth place in ranking R_1, therefore its rank is 7. The bottom row shows the average ranks. According to these scores, the ensemble ranking is

$$\text{Final ranking } R: \quad 4 \quad 9 \quad 2 \quad 7 \quad 1 \quad 5 \quad 6 \quad 8 \quad 3 \quad 10$$

9.3.2 Ranking Methods (Criteria)

Many pattern recognition problems consider only two classes, for example, malignant or benign tissue, fraudulent or legitimate transaction, network attack or normal service, face or nonface sub-image, and many more. The class of interest is usually called the "positive" class, and the other, the "negative" class. Here we consider feature ranking criteria for two classes. Extensions to multi-class problems are possible, for example, by using a one-versus-all approach, and average the values of the criterion.

If the feature of interest x is continuous-valued, the distributions for the two classes can be estimated and compared. The larger the difference, the better the feature. Given the relatively small sample sizes in wide data, the typical choices are the t-test and the Mann–Whitney U test. They both provide an instant feature ranking based on the respective test statistic.

A continuous-valued x can be thresholded to give a positive or a negative label to the object. This allows for using the area under the receiver operating characteristic (ROC) curve as another measure of quality of the features. If x is binary, many more ranking criteria can be added. A wealth of binary ranking criteria have been studied for text classification [129]. Using binary values will reduce sensitivity of the feature but it will also reduce the noise, which, on balance, may prove to be beneficial.

The optimal threshold for splitting a feature can be derived from the training data [15]. In a training data set of N objects, assuming that all values of x are different, there are N possible split points. Ideally, all N values should be checked, and the most favorable threshold should be taken forward to define the binarization.

A split of x at a given threshold T leads to a binary feature x' and the following contingency table with *proportions* calculated from the training data.

	$x' = 1$	$x' = 0$	
positive	a (true positive)	b (false negative)	
negative	c (false positive)	d (true negative)	(9.1)

$$a + b + c + d = 1.$$

Here we do not assume that larger values of x correspond to the positive class. Therefore the labels 1 and 0 can be assigned to x' in reverse order, giving a mirror table to (9.1)

	$x' = 1$	$x' = 0$	
positive	b (true positive)	a (false negative)	
negative	d (false positive)	c (true negative)	(9.2)

$$a + b + c + d = 1.$$

When calculating the value of a feature, the better of the two assignments should be taken forward.

The literature abounds with feature ranking criteria [15, 86, 129, 229, 408], examples of which are listed below

1. Accuracy. $Acc = a + d.$

2. Probability ratio. $PR = \frac{a(c+d)}{c(a+b)}$.

3. Odds ratio. $Odds = \frac{ad}{bc}$.

4. Power. $Pow = \left(\frac{d}{c+d}\right)^k - \left(\frac{b}{a+b}\right)^k$, where k is a parameter; recommended value $k = 5$.

5. GM measure. This is the geometric mean of sensitivity and specificity

$$GM = \sqrt{sensitivity \times specificity}$$
$$= \sqrt{\frac{ad}{(a+b)(c+d)}}.$$

This measure bypasses the possible imbalance of the class prevalence.

6. F1 measure. This measure is the harmonic mean of recall and precision, often used in document retrieval

$$F1 = \frac{2 \times recall \times precision}{recall + precision} = \frac{a}{2a+b+c}.$$

7. Gini index. To calculate this index we start with Gini impurity

$$1 - (a+b)^2 - (a+c)^2 = 2(a+b)(c+d).$$

The impurity of the set for which $x' = 1$ is $\frac{2ac}{(a+c)^2}$. It must be weighted by $(a+c)$, which is the probability of $x' = 1$. Including the impurity for $x' = 0$, the overall index is defined as the reduction of impurity

$$Gini = 2(a+b)(c+d) - \frac{2ac}{(a+c)} - \frac{2bd}{(b+d)}.$$

8. Mutual information (MI). MI is one of the most advocated feature selection criteria. Brown et al. [55] offer a valuable review and propose a general framework, which accommodates most of the past MI criteria. For the 2-by-2 contingency table 9.1,

$MI(x; class)$

$$= a \log \frac{a}{(a+b)(a+c)} + b \log \frac{b}{(a+b)(b+d)}$$
$$+ c \log \frac{c}{(a+c)(c+d)} + d \log \frac{d}{(b+d)(c+d)}.$$

9. Chi square. Calculate first the table with the expected proportions assuming independence between the class label and feature x'

$$e = \{e(i,j)\} \tag{9.3}$$

$$= \frac{\begin{array}{c|cc} & x' = 1 & x' = 0 \\ \hline \text{positive} & (a+b)(a+c) & (a+b)(b+d) \\ \text{negative} & (a+c)(c+d) & (b+d)(c+d) \end{array}}{}.$$

The χ^2 statistic is calculated as

$$\chi^2 = \sum_{i=1}^{2} \sum_{j=1}^{2} \frac{(t(i,j) - e(i,j))^2}{e(i,j)},$$

where $t(i,j)$ is the respective entry in Table 9.1 ($t(1,1) = a, t(1,2) = b, t(2,1) = c$ and $t(2,2) = d$). Larger values of χ^2 indicate a better feature.

10. Binormal separation.

$$BNS = \left| \Phi^{-1} \left(\frac{a}{a+b} \right) - \Phi^{-1} \left(\frac{c}{c+d} \right) \right|,$$

where Φ^{-1} is the inverse of the cumulative probability function of the normal distribution. This ranking criterion was highly recommended by Forman [129].

11. Kolmogorov–Smirnov.

$$KS = \left| \frac{a}{a+b} - \frac{c}{c+d} \right|.$$

Criteria 1–11 require an exhaustive run through the K splits of x to find the optimal threshold T. Formally, if the criterion is denoted by $C(T)$ for threshold T, the value which is used to rank the features is

$$C^* = \max_T \{C(T)\}.$$

The next three measures treat x as a continuous-valued variable and estimate its worth in a single calculation.

12. t-test. The Student t-test statistic has been used extensively in fMRI data analysis for ranking the voxels and determining statistically significant relationships. We will use the statistic as a ranking criterion (without the test), assuming unequal variances of the two classes. Denote by $m_{(+)}$ and $m_{(-)}$ the means of x for the positive and the negative class, respectively, calculated from the training data. Denote by $s_{(+)}$ and $s_{(-)}$ the respective unbiased estimates of the standard

deviations. Since we are interested in the magnitude of the difference and not its sign,

$$t = \left| \frac{m_{(+)} - m_{(-)}}{\sqrt{\frac{s_{(+)}}{n_{(+)}} + \frac{s_{(-)}}{n_{(-)}}}} \right|, \tag{9.4}$$

where $n_{(+)}$ and $n_{(-)}$ are the class counts. Larger values of t mean a better feature.

13. Mann–Whitney U test or Wilcoxon rank-sum test is less affected by outliers compared to the t-test. To calculate the U statistics, arrange x in ascending order and calculate the ranks. Sum the ranks for the two classes separately to get $R_{(+)}$ and $R_{(-)}$. Then

$$U = \min \left\{ R_{(+)} - \frac{n_{(+)}(n_{(+)} + 1)}{2}, \quad R_{(-)} - \frac{n_{(-)}(n_{(-)} + 1)}{2} \right\}. \tag{9.5}$$

Larger values of U mean a better feature.

14. Area under the ROC curve (AUC). AUC has been one of the preferred criteria for evaluating the quality of classification algorithms. We will add it as a feature ranking criterion, acknowledging the recently published study which warns against over-trusting AUC for small sample sizes [171]. The feature values are arranged in increasing order, and a threshold is set between every pair of values. A classifier is associated with each threshold, whose sensitivity and specificity determine a point on the ROC curve. The AUC is approximated from these discrete points.

15. SVM is particularly well suited to wide data because it scales linearly along the feature dimension while tolerating the small sample size by ensuring large classification margins. The linear-kernel SVM can be used as a feature ranking algorithm. A feature's relevance is measured by the absolute value of the weight for this feature in the linear discriminant function of the trained SVM. This feature ranking method can be thought of as pseudo-multivariate and falls into the category of *embedded* methods [346]. The features are ranked by their worth if the whole feature set is used with the SVM but this does not mean that if the top k features were cut off, they will make a good subset. The recursive feature elimination (RFE) algorithm is an addition to the SVM feature selector, which brings it a step closer to a true multivariate procedure [165]. Starting with an SVM on the entire feature set, a fraction of the features with the lowest weights is dropped. A new SVM is trained with the remaining features, and subsequently reduced in the same way. The procedure stops when the set of the desired cardinality is reached. While SVM-RFE has been found to be extremely useful for wide data such as fMRI data [86], it was discovered that the RFE step is not always needed [1, 149, 401]. This may happen when the features are loosely related, and a single SVM captures adequately their relationship. In such problems, it can be expected that other single-pass ranking algorithms will fare well too.

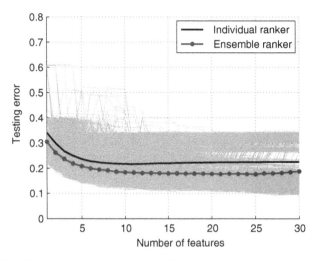

FIGURE 9.10   Testing error for the individual ranker and the ensemble ranker averaged across the 100 runs. The gray lines are plots of the errors of all individual rankers.

Example 9.5   Ensemble ranker for the spam data

A random set of 200 points was taken as training and the remaining 4401 points were used for testing. With each split, 25 t-test rankers (number 12 in the above list) were trained on bootstrap samples from the training data. Their individual testing errors were calculated and stored. The LDC was used for the testing. The ensemble ranking was found and also evaluated on the testing set. The experiment was repeated 100 times. Figure 9.10 shows the testing error for the individual ranker and the ensemble ranker averaged across the 100 runs. The gray lines are plots of the errors of all $(100 \times 25 = 2500)$ individual rankers.

The results support the expectation that ensemble of rankers improve on the testing error of the individual ranker.

9.4   RANDOM FEATURE SELECTION FOR THE ENSEMBLE

9.4.1   Random Subspace Revisited

Here we view the random subspace ensemble introduced in Chapter 6 in the light of feature selection. Each classifier in the ensemble is built upon a randomly chosen subset of features of predefined size d. Ho [182] suggests that good results are obtained for tree classifiers built upon $d \approx \frac{n}{2}$ features, where n is the total number of features. The random subspace method has been found to work well when there is redundant information but it is "dispersed" across all the features rather than concentrated in a subset of them [182, 369].

Pękalska et al. [307] consider a *dissimilarity representation* of the objects. Each of the N objects is described by N features which are the distances to all the objects (including itself). Such representation is useful when the data originally contains a prohibitive number of features or when a similarity measure between objects is easily available. Pękalska et al. proposed using LDC on randomly selected feature subsets and recommend between 4% and 30% of the (similarity) features as the cardinality of the selected set.

Latinne et al. [254] propose combining bagging with random selection. B bootstrap replicates are sampled from the training set $\mathbf{Z}$ and, for each replicate, R subsets of features are chosen. The proportion of features, K, is the third parameter of the algorithm. The ensemble consists of $L = B \times R$ classifiers. The combination between bagging and feature selection aims at making the ensemble more diverse than when using either of the methods alone.

The size of the feature subset and the ensemble size are the two parameters of the random subspace method. When the total number of features is tens of thousands, pre-selection or post-selection have been proposed. Bertoni et al. [37] eliminate redundant and irrelevant features by simple ranking, and subsequently select a feature subset on which they build the random subset ensemble. A different approach was proposed by Lai et al. [246], which can be viewed as post-selection. First the feature subsets for the ensemble classifiers are drawn from the whole feature set and then feature selection takes place separately on each feature sample.

Note that the RS ensemble does not offer an explicit final feature subset. The purpose of the feature selection is to achieve the highest possible classification accuracy. The reduction of the original feature space comes as a by-product.

9.4.2   Usability, Coverage, and Feature Diversity

The parameters of the random subspace ensemble are the ensemble size L and the cardinality of the feature subset M. Here we reproduce an argument about choosing these values based on three related concepts: usability, coverage, and feature diversity [228].

Consider a data set with a very large number of features such as data coming from fMRI. In such data sets, a small number of features often contain most of the information, while the remaining features will contribute only noise to the classifier.

Let X be the feature set of cardinality n. Assume that there are Q "important" features, set $\mathcal{I} = \{q_1, \dots, q_Q\}$, $\mathcal{I} \subset X$, where $|\mathcal{I}| = Q \ll n$, and the remaining $n - Q$ features are random noise. We also assume that the cardinality of the subspaces, M, is much smaller than n. The question is whether we can select "optimal" L and M, given n and hypothesizing Q.

We start from the postulate that accurate and diverse individual classifiers make the best ensembles. The subset of features, on which the individual classifiers are built, can serve as indirect indication for the accuracy and diversity of these classifiers. If a classifier uses only "noise" features, its accuracy will be no better than random guessing. Also, classifiers that use the same "important" features will be similar or identical, therefore redundant in the ensemble. Finally, we would like the whole of $\mathcal{I}$ to be covered, so that important information is not lost. In other words, we would like each $q \in \mathcal{I}$ to be selected at least once in the L samples of M features.

> *Definition 1.* A classifier is called *usable* if its feature subset contains at least one "important" feature $q \in \mathcal{I}$.
>
> *Definition 2.* The *usability of the ensemble*, U_e, is measured as the proportion of usable classifiers out of L. An ensemble is called *completely usable* if it contains only usable classifiers ($U_e = 1$).
>
> *Definition 3.* FSD between $S_1, S_2 \subset X$, is measured by the cardinality of the set of nonshared features $q \in \mathcal{I}$ contained within $S_1 \cup S_2$. Two classifiers are *nonidentical* if their feature subsets differ by at least one "important" feature.

We address the following three questions. Given M, L, n and Q,

1. *Usability.* What is the probability that the selected ensemble is completely usable?
2. *Coverage.* What is the probability that the whole of $\mathcal{I}$ will be covered (complete coverage)?
3. *Diversity.* What is the probability that the usable classifiers in the ensemble will be nonidentical (feature set diversity(FSD))?

9.4.2.1 Usability Denote by Y the number of important features within a single sample (without replacement) of size M from X. Y is a random variable with hypergeometric distribution. (To help with the terminology, consider that the sample is taken from an urn with a *total* of n marbles, of which Q are *black*, and the remaining $n - Q$ are white. The number of *selected* marbles in one sample is M. Then Y is the number of black marbles within the sample.) The probability mass function of Y is

$$P(Y = i) = \frac{\binom{Q}{i}\binom{n-Q}{M-i}}{\binom{n}{M}}, \quad i = 0, 1, \ldots, Q.$$

Then the probability of having a usable classifier is

$$P(\text{usable classifier}) = 1 - P(Y = 0) = 1 - \frac{\binom{n-Q}{M}}{\binom{n}{M}}.$$

Therefore, since the subsets are sampled independently, the probability of having a completely usable ensemble is

$$P(U_e = 1) = P(\text{usable classifier})^L = \left(1 - \frac{\binom{n-Q}{M}}{\binom{n}{M}}\right)^L. \tag{9.6}$$

The ratio of the two binomial coefficients can be simplified for computational purposes to give

$$P(U_e = 1) = \left(1 - \prod_{i=0}^{M-1}\left(1 - \frac{Q}{n-i}\right)\right)^L.$$ (9.7)

Since we assumed $M \ll n$, the equation can be simplified further to

$$P(U_e = 1) \approx \left(1 - \left(1 - \frac{Q}{n}\right)^M\right)^L.$$ (9.8)

This approximation is equivalent to approximating the hypergeometric distribution with a binomial distribution. The intuition is that the population from which the sample is taken is so vast that sampling *with* replacement will be approximately equivalent to sampling without replacement. If sampling is done with replacement, Y will have a binomial distribution with parameters M and $p = \frac{Q}{n}$, and the probability of a usable classifier will be $1 - \left(1 - \frac{Q}{n}\right)^M$. Then the probability of a completely usable ensemble will be as in Equation (9.8).

To calculate the expected value of the degree of usability of the ensemble, let Z be a random variable expressing the number of usable classifiers in the ensemble. Z has a hypergeometric distribution with the following parameters. The *total* is the number of all possible samples (without replacement) of size M from X, that is, $\binom{n}{M}$. The number of *usable* classifiers is calculated by taking the number of non-usable classifiers, $\binom{n-Q}{M}$, from the total. The number of *selected* classifiers at a time is L. The expected value of Z is $\frac{\text{Selected} \times \text{Usable}}{\text{Total}}$, therefore the expected usability of the ensemble is $E(U) = \frac{1}{L}E[Z]$.

$$E(U) = \frac{1}{L} \times L \times \left(1 - \frac{\binom{n-Q}{M}}{\binom{n}{M}}\right) = 1 - \frac{\binom{n-Q}{M}}{\binom{n}{M}}.$$ (9.9)

The expected usability of the ensemble is equivalent to the probability of selecting a usable classifier, and does not depend on the ensemble size L. Our hypothesis is that higher usability will lead to accurate ensembles.

9.4.2.2 *Coverage*   For calculating the probability that the whole of $\mathcal{I}$ will be covered, we will again use the binomial approximation to the hypergeometric distribution. This approximation implies that the features within the selected subset of size M are sampled independently. Consider an important feature $q \in \mathcal{I}$. The probability that a particular feature q in X is hit in M trials is $\frac{M}{n}$. Therefore, the probability of not selecting q in any of the L classifiers of the ensemble is $P(\bar{q}) = \left(1 - \frac{M}{n}\right)^L$. The probability of q being in one or more of the L selections is $1 - P(\bar{q})$, and the probability of all features being covered is

$$P(\text{Complete coverage}) = \left(1 - \left(1 - \frac{M}{n}\right)^L\right)^Q. \tag{9.10}$$

Denote by Z the number of covered features out of Q. Z has binomial distribution with parameters Q and $p = 1 - \left(1 - \frac{M}{n}\right)^L$. The expected coverage is

$$E(C) = \frac{1}{Q}\left(1 - \left(1 - \frac{M}{n}\right)^L\right)Q = 1 - \left(1 - \frac{M}{n}\right)^L. \tag{9.11}$$

The expected coverage depends on the ensemble size L and the subset size M but not on Q. Again, the hypothesis is that higher degree of coverage is a prerequisite for a good ensemble.

9.4.2.3 Feature Set Diversity As argued above, we approximate the hypergeometric distribution that underpins the selection without replacement with a binomial distribution, where the selection is done with replacement.

Let S_1 and S_2 be subsets of X, both of cardinality M. Denote by $I_1 \subseteq \mathcal{I}$ and $I_2 \subseteq \mathcal{I}$ the respective subsets of "important" features within S_1 and S_2. Define FSD by

$$\text{FSD}(S_1, S_2) = |I_1 \cup I_2| - |I_1 \cap I_2|.$$

Each feature $q \in \mathcal{I}$ may or may not contribute to FSD. A value of 1 will be added if q is in either set but not in both. Then the expected diversity for any pair of subsets S_1 and S_2 is

$$E(\text{FSD}) = \sum_{i=1}^{Q} P(q_i \in I_1)P(q_i \notin I_2) + P(q_i \notin I_1)P(q_i \in I_2). \tag{9.12}$$

Since all features in $\mathcal{I}$ have identical chance of $\frac{M}{n}$ to be selected in a subset of size M, and the subsets are drawn independently,

$$E(\text{FSD}) = 2Q\frac{M}{n}\left(1 - \frac{M}{n}\right). \tag{9.13}$$

The probability of selecting randomly two identical classifiers ($I_1 = I_2$, regardless of the nonimportant features) is

$$P(2\text{id}) = \sum_{j=1}^{\min\{Q,M\}} P(\text{Choose two sets with } j \text{ important})$$

$$\times P(\text{match})$$

$$= \sum_{j=1}^{\min\{Q,M\}} \frac{\binom{Q}{j}^2 \binom{n-Q}{M-j}^2}{\binom{n}{M}^2} \times \frac{1}{\binom{Q}{j}}$$

$$= \sum_{j=1}^{\min\{Q,M\}} \frac{\binom{Q}{j} \binom{n-Q}{M-j}^2}{\binom{n}{M}^2}$$

Finally, the probability of having an ensemble where every pair of classifiers is nonidentical is

$$P(\text{All pairs non-id}) = \left(1 - \sum_{j=1}^{\min\{Q,M\}} \frac{\binom{Q}{j} \binom{n-Q}{M-j}^2}{\binom{n}{M}^2} \right)^{\frac{L(L-1)}{2}} \tag{9.14}$$

This calculation disregards nonusable classifiers. So an ensemble can be diverse even if it contains nonusable classifiers for which $I_1 = I_2 = \emptyset$.

9.4.2.4 *Simulation Results* Figure 9.11 shows the theoretical and simulated curves for $E(U)$ (9.9), $E(C)$ (9.11), and $E(\text{FSD})$ (9.12) for $n = 1000$, $Q = 100$, and

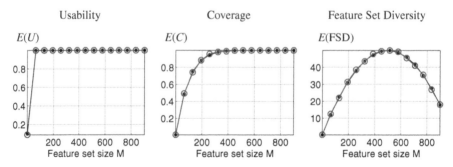

FIGURE 9.11 Theoretical and simulation curves (coinciding) for the expected values of U, C, and FSD for $n = 1000$, $Q = 100$, and $L = 10$. The empirical curve is calculated as an average of 10 ensembles with randomly sampled $L = 10$ sets of M features.

$L = 10$. Changing the value of L to 50 and 100, and Q to 10 and 50 did not lead to large differences in the shapes and positions of the curves. The results suggest that values of M close to $\frac{n}{2}$ are optimal as all three criteria reach their maxima, also observed across different ensemble sizes.

It is interesting to find out which of the three criteria has the largest impact on the ensemble error. For now, without knowing the value of Q, we take forward the wisdom that for problems with very large n, where only a small number of features may be relevant, the RS ensembles benefit from relatively small L, say $L \leq 100$, and $M \approx \frac{n}{2}$.

■ Example 9.6   A random subspace ensemble on an fMRI data set

Here we reproduce an example from a previous work [225]. We used an fMRI data set collected at the School of Psychology, University of Bangor, UK. The data consisted of the single-subject's fMRI responses to two types of stimuli: faces and places. Each presentation of a stimulus defined a point in the data set. The total number of features was 106,720 and the number of objects was 24, 12 in each class. The classification task was to predict which type of stimuli the subject is looking at, judging by the fMRI response. We trained RS ensembles with SVM as the base classifier. First, $n = 1000$ features were pre-selected by the SVM method [86]. A three-fold cross-validation was applied to test the RS ensemble for a 10×10 grid of values for M and L. M was varied from 1 to n at equal intervals, and L was varied from 1 to $n/5$. Figure 9.12 plots the surface of the ensemble error over the (L, M) grid. The best values of $M = 500$ and $L = 100$ are marked as lines on the 3D plot. The lines intersect near the minimum of the error surface, which confirms empirically the recommendation for L and M.

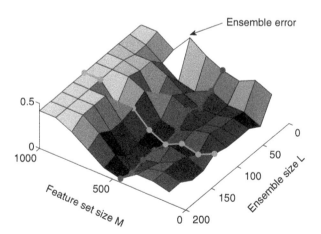

FIGURE 9.12   RS error on the real fMRI data set as a function of the ensemble size L and the feature size M [225].

The figure shows that M has a more profound effect on the error than L, and $M = \frac{n}{2}$ works well for most values of L. This demonstrates the robustness of RS with respect to L.

9.4.3   Genetic Algorithms

We can apply various heuristic search techniques for feature subset selection, for example, evolutionary algorithms, tabu search, and simulated annealing. Genetic algorithms (GAs) appeared to be the preferred heuristic search tool for *ensemble* feature selection [162, 238, 242, 294, 390, 437].

9.4.3.1 Basics of Genetic Algorithms   A sketch of a loop in a typical genetic algorithm is shown in Figure 9.13. Genetic algorithms evolve a population of "chromosomes" (genotypes). Applied to feature selection, a chromosome represents a feature subset. Suppose that there are n features. In the typical encoding, a feature subset is represented as a binary vector with n elements. If bit i is 1, the ith feature is included in the subset. A "generation" contains k chromosomes, where k is called the population size. Each chromosome is evaluated to determine its fitness. Any criterion that evaluates a feature subset can be applied as a fitness function. For example, a *filter* approach uses some measure of separability between the classes, whereas a *wrapper* approach trains a classifier (the phenotype) with the features in

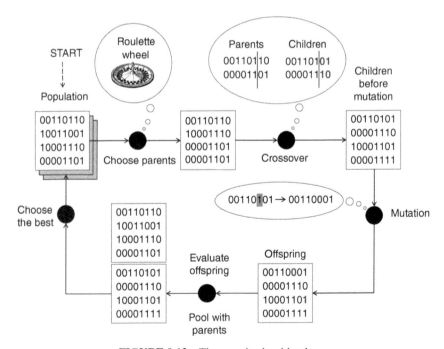

FIGURE 9.13   The genetic algorithm loop.

the chromosome and returns the classification accuracy as the fitness value. A set of offspring chromosomes is created from the current population taken to be the parents. Usually, the parents with higher fitness values are given more chance to reproduce. This selection approach is known as the "roulette wheel" principle.

The two standard genetic operators that create the offspring chromosomes are crossover and mutation. The crossover takes two parents and swaps parts of the chromosomes. In a one-point crossover, a random split point between 1 and n is picked, and the "tails" of the two chromosomes are swapped, thereby creating two children. The offspring set goes through mutation where each bit is flipped to the opposite value with a predefined mutation probability P_m. In real life, mutation probability is very small, but in our genetic algorithm world, this probability is usually set between 0.1 and 0.15.

The sets of offspring and parents are pooled, and the fittest k chromosome survive as the next generation. This is known as the "elitist survival strategy." Alternatively, we may allow only a certain proportion of the population to be replaced, called "the generation gap."

The algorithm is run for a given number of generations or until some stopping criterion is met. Such a criterion could be a plateau on the average population fitness or, equivalently, degeneration of the population to nearly identical chromosomes. The output of a GA is usually the top ranked chromosome. Appendix 9.A.2 gives examples of MATLAB functions for the basic genetic operators for binary chromosomes: roulette wheel selection, crossover, and mutation.

By design, a genetic algorithm is meant to converge to a single best solution. The solution is not guaranteed to be the global maximum of the fitness but the hope is that many local maxima will be overcome in favor of a better one. Convergence to a single solution presents a problem if GAs are to be used for feature selection for the member classifiers. If each chromosome is associated with a feature subset, and the GA is run to convergence, there will be only one feature subset or near clones thereof. The diversity in the ensemble will be close to zero. The following two strategies have been applied to avoid this problem.

Ensemble = Population   In this case L feature subsets are evolved individually. The GA operates on a population of individuals. Each feature subset is a member of the population. The GA is aiming at finding a single individual of the highest possible quality measured by its fitness. The aim in creating an ensemble is not finding one best classifier but finding classifiers that will be jointly better than the single best individual. To achieve this, the individual members of the population have to be both accurate and diverse. While accuracy is accounted for by the fitness function, diversity is not. Hence there should be a mechanism for maintaining diversity and stopping before convergence. Diversity of the feature subsets (the genotype) does not guarantee diversity of the classifier outputs (the phenotype). Nonetheless, diversity in the genotype is the easily available option within this approach [78, 390]. The deviation of a feature subset from the remaining members of the population is one possible measure that can be used together with the accuracy of the classifier built on this feature subset. In fact, any diversity measure can be applied here.

⬛ Example 9.7   Diversity in a population of feature subsets

Let $X = \{X_1, \ldots, X_{10}\}$ be a set of features. A population of six chromosomes (individuals), $S_1, \ldots, S_6$, is displayed below:

S_1	0	1	0	1	0	0	0	1	0	0	$\{X_2, X_4, X_8\}$
S_2	1	1	1	0	0	0	0	0	0	1	$\{X_1, X_2, X_3, X_{10}\}$
S_3	0	0	1	1	1	1	0	0	0	0	$\{X_3, X_4, X_5, X_6\}$
S_4	0	0	0	0	0	0	0	1	0	1	$\{X_8, X_{10}\}$
S_5	0	1	1	0	0	0	0	0	0	0	$\{X_2, X_2\}$
S_6	1	1	0	1	1	1	0	0	1	1	$\{X_1, X_2, X_4, X_5, X_6, X_9, X_{10}\}$

The simplest way of calculating how different subset S_i is from the remaining subsets in the population is to use the averaged Hamming distance between S_i and each of the other $M - 1$ subsets. This is equivalent to the disagreement diversity measure despite the fact that the chromosomes are not oracle outputs. The disagreement measure is symmetric, which means that the 0s and the 1s can be swapped with no change of the value. For the above example, the five distances for S_1 are 5, 5, 3, 3, and 6, respectively. Then the "diversity" of S_1 can be calculated as $d_1 = \frac{5+5+3+3+6}{5} = 4.4$. Equivalently, d_is can be calculated as the sum of the absolute difference between bit k of chromosome S_i and the "averaged" chromosome $\bar{S} = \frac{1}{5}\sum S_i$. The individual diversities of the other subsets are $d_2 = 4.4$, $d_3 = 5.2$, $d_4 = 4.8$, $d_5 = 4.0$, and $d_6 = 6.0$. Suitably weighted, these diversity values can be taken together with the accuracies as components of the fitness function.

9.4.3.2   Ensemble = Chromosome   In this approach, each individual in the population represents the entire ensemble. We can use disjoint subsets (Approach B.1) or allow for intersection (Approach B.2).

B.1. To represent disjoint subsets, we can keep the length of the vector (the chromosome) at n and use integers from 0 to L. The value at position i will denote which classifier uses feature X_i; zero will indicate that feature X_i is not used at all. An integer-valued GA can be used to evolve a population of ensembles. The fitness of a chromosome can be directly a measure of the accuracy of the ensemble represented by that chromosome. For example, let X be a set of ten features. Consider an ensemble of four classifiers. A chromosome $[1, 1, 4, 0, 3, 3, 1, 2, 3, 3]$ will denote an ensemble $D = \{D_1, D_2, D_3, D_4\}$ where D_1 uses features X_1, X_2, and X_7; D_2 uses X_8; D_3 uses X_5, X_6, X_9, and X_{10}; and D_4 uses X_3. Feature X_4 is not used by any of the classifiers.

B.2. To allow any subset of features to be picked by any classifier, the ensemble can be represented by a binary chromosome of length $L \times n$. The first n bits will represent the feature subset for classifier D_1, followed by the n bits for classifier D_2, and so on. A standard binary-valued GA can be used with this representation.

Approach B.1 is simpler than B.2 but the requirement that the ensemble members use disjoint feature subsets might be too restrictive.

9.5   NONRANDOM SELECTION

9.5.1   The "Favorite Class" Model

Oza and Tumer [295] suggest a simple algorithm for selecting feature subsets which they call *input decimation*. The ensemble consists of $L = c$ classifiers, where c is the number of classes. Each classifier has a "favorite" class. To find the feature subset for classifier D_i with favorite class ω_i, we start by calculating the correlation between each feature and the class label variable. The class label variable has value 0 for all objects which are not in class ω_i and 1 for all objects which are in class ω_i.

■ **Example 9.8   Correlation between a feature and a class label variable**
Suppose the data set consists of nine objects labeled in three classes. The values of feature X_1 and the class labels for the objects are as follows:

Object	z_1	z_2	z_3	z_4	z_5	z_6	z_7	z_8	z_9
X_1	2	1	0	5	4	8	4	9	3
Class	ω_1	ω_1	ω_1	ω_2	ω_2	ω_2	ω_3	ω_3	ω_3

The class label variable for ω_1 has values $[1, 1, 1, 0, 0, 0, 0, 0, 0]$ for the nine objects. Its correlation with X_1 is -0.75.

The n correlations are sorted by absolute value and the features corresponding to the n_i largest correlations are chosen as the subset for classifier D_i. The value n_i, $n_i < n$, is the desired number of retained features for class ω_i. These c numbers should be chosen in advance. Using the selected features, D_i is trained to recognize all the c classes. Selecting the subsets in this way creates diversity within the ensemble. Even with this simple selection procedure, the ensemble demonstrated better performance than the random subset selection method [295].

There are numerous feature selection methods and techniques which can be used instead of the sorted correlations. Such are the methods from the sequential group (forward and backward selection) [3, 90], the floating selection methods [315], and so on. Given that the number of subsets needed is the same as the number of classes, using a more sophisticated feature selection technique will not be too computationally expensive and will ensure higher quality of the selected feature subset.

9.5.2   The Iterative Model

Another "favorite class" feature selection method is proposed by Puuronen et al. [318]. They devise various criteria instead of the correlation with the class variable, and suggest an iterative procedure by which the selected subsets are updated. The procedure consists of the following general steps:

1. Generate an initial ensemble of c classifiers according to the "favorite class" procedure based on correlation.
2. Identify the classifier whose output differs the least from the outputs of the other classifiers. We shall call this the *median* classifier. The median classifier is identified using some pairwise measure of diversity, which we will denote by $\Delta(D_i, D_j)$. High values of Δ will denote large disagreement between the outputs of classifiers D_i and D_j. $\Delta(D_i, D_j) = 0$ means that D_i and D_j produce identical outputs. The median classifier D_k is found as

$$D_k = \arg\min_i \sum_{j=1}^{L} \Delta(D_i, D_j). \qquad (9.15)$$

3. Take the feature subset for D_k. Altering the present/absent status of each feature, one at a time, produce n classifier candidates to replace the median classifier. For example, let $n = 4$ and let D_k be built on features X_1 and X_3. We can represent this set as the binary mask $[1, 0, 1, 0]$. The classifier candidates to replace D_k will use the following subsets of features: $[0, 0, 1, 0]$ (X_3), $[1, 1, 1, 0]$ (X_1, X_2, X_3), $[1, 0, 0, 0]$ (X_1), and $[1, 0, 1, 1]$ (X_1, X_3, X_4). Calculate the ensemble accuracy with each replacement. If there is an improvement, then keep the replacement with the highest improvement, dismiss the other candidate classifiers and continue from step 2.
4. Else, stop and return the current ensemble.

This greedy algorithm has been shown experimentally to converge quickly and to improve upon the initial ensemble. Numerous variants of this simple iterative procedure can be designed. First, there is no need to select the initial ensemble according to the "favorite class" procedure. Any ensemble size and any initial subset of features might be used. In the iterative algorithm in Ref. [389], the initial ensemble is generated through the random subspace method. Another possible variation of the procedure, as suggested in Ref. [389], is to check all the ensemble members, not only the median classifier. In this case, the task of calculating ensemble accuracy for each classifier candidate for replacement might become computationally prohibitive.

9.5.3   The Incremental Model

Günter and Bunke [163] propose a method for creating classifier ensembles based on feature subsets. Their ensemble is built gradually, one classifier at a time, so that the feature *subsets* selected for each of the previous classifiers are not allowed to be chosen for the subsequent classifiers. However, intersection of the subsets is allowed. The authors suggest that any feature selection method could be used and advocate the floating search for being both robust and computationally reasonable. There are various ways in which the ban on the previous subsets can be implemented. Günter and Bunke use different feature selection algorithms relying on their suboptimality to produce different feature subsets. To estimate a subset of features S, they use the ensemble performance rather than the performance of the individual classifier built on the subset of features. This performance criterion, on its own, will stimulate diversity in the ensemble. It is not clear how successful such heuristics are in the general case, but the idea of building the ensemble incrementally by varying the feature subsets is certainly worth a deeper look.

9.6   A STABILITY INDEX

Why do we need to worry about stability? If an ensemble based on different feature subsets works well, it is likely that these subsets are diverse. In this scenario, to fit within the "feature selection FOR the ensemble" approach (Figure 9.3), stability is actually undesirable. However, if the aim is to return an informative and highly discriminative feature subset to the user, the perspective changes. This is the case of "feature selection BY the ensemble," and we should have confidence that the returned feature subset is meaningful and robust. In other words, if a different data set sampled from the same distribution is used to select features, the resultant subset will not be much different. Stability is a very desirable property in this case.

In spite of the relatively recent start, the research on feature selection stability is gaining momentum [15, 111, 170, 201, 219, 236, 345, 365, 375, 409, 427]. Here we reproduce the stability index proposed in Ref. [236].

9.6.1   Consistency Between a Pair of Subsets

Let A and B be subsets of features, $A, B \subset X$, of the same cardinality k. Let $r = |A \cap B|$ be the cardinality of the intersection of the two subsets. A list of desirable properties of a consistency index for a pair of subsets is given below:

1. *Monotonicity.* For a fixed subset size, k, and number of features, n, the larger the intersection between the subsets, the higher the value of the consistency index.
2. *Limits.* The index should be bound by constants which do not depend on n or k. The maximum value should be attained when the two subsets are identical, that is, for $r = k$.
3. *Correction for chance.* The index should have a constant value for independently drawn subsets of features of the same cardinality, k.

A general form of such an index is

$$\frac{\text{Observed } r - \text{Expected } r}{\text{Maximum } r - \text{Expected } r}. \qquad (9.16)$$

Maximum r equals k, achieved when A and B are identical subsets. To evaluate the expected cardinality of the intersection, consider r to be a random variable obtained from randomly drawn A and B of size k from a set X of size n (without replacement). We can think of subset A as fixed. Suppose that the elements of $X \setminus A$ are colored in white and those in A are colored in black. A set B of size k is selected without replacement from X. The number of objects from A (black) selected also in B is a random variable Y with hypergeometric distribution with probability mass function

$$P(Y = r) = \frac{\binom{k}{r}\binom{n-k}{k-r}}{\binom{n}{k}}. \tag{9.17}$$

The expected value of Y for given k and n is $\frac{k^2}{n}$.

Definition 1. The consistency index for two subsets $A \subset X$ and $B \subset X$, such that $|A| = |B| = k$, where $0 < k < |X| = n$, is

$$I_C(A, B) = \frac{r - \frac{k^2}{n}}{k - \frac{k^2}{n}} = \frac{rn - k^2}{k(n - k)}. \tag{9.18}$$

This index satisfies the three properties above. First, for fixed k and n, $I_C(A, B)$ increases with increasing r. Second, the maximum value of the index, $I_C(A, B) = 1$, is achieved when $r = k$. The minimum value of the index is bound from below by -1. The limit value is attained for $k = \frac{n}{2}$ and $r = 0$. Note that $I_C(A, B)$ is not defined for $k = 0$ and $k = n$. These are the trivial cases where either no feature is selected or all features are selected. They are not interesting from the point of view of comparing feature subsets, so the lack of values for $I_C(A, B)$ in these cases is not important. For completeness, we can assume that $I_C(A, B) = 0$ for both cases. Finally, $I_C(A, B)$ will assume values close to zero for independently drawn A and B because r is expected to be around $\frac{k^2}{n}$.

▣ Example 9.9   Consistency between two feature rankings

Consider the following two hypothetical sequences of features obtained from two runs of SFS on a data set with 10 features.

$$S_1 = \{X_9, X_7, X_2, X_1, X_3, X_{10}, X_8, X_4, X_5, X_6\}$$
$$S_2 = \{X_3, X_7, X_9, X_{10}, X_2, X_4, X_8, X_6, X_1, X_5\}$$

Denote by $S_i(k)$ the subset of the first k features of sequence S_i.
For example, the cardinality of the intersection of $S_1(3)$ and $S_2(3)$ is $|\{X_7, X_9\}| = 2$. Then

$$I_C(S_1(3), S_2(3)) = \frac{2 \times 10 - 3^2}{3(10 - 3)} = \frac{11}{21} \approx 0.5238.$$

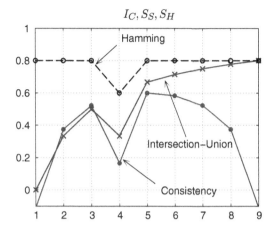

FIGURE 9.14   Consistency index I_C and similarities S_S (intersection union) and S_H (Hamming) for sequences S_1 and S_2 (the example in the text) plotted against the subset size k.

Figure 9.14 shows $I_C(S_1(k), S_2(k))$ against the set size k. By introducing the correction for chance the consistency index I_C differs from two indices proposed previously. Kalousis et al. [201] introduce the similarity index between two subsets of features, A and B, as

$$S_S(A, B) = 1 - \frac{|A| + |B| - 2|A \cap B|}{|A| + |B| - |A \cap B|} = \frac{|A \cap B|}{|A \cup B|}, \tag{9.19}$$

where $|\cdot|$ denotes cardinality, "$\cap$" denotes intersection and "$\cup$" denotes union of sets. Dunne et al. [111] suggest measuring the stability using the relative Hamming distance between the masks corresponding to the two subsets, which in set notation is

$$S_H(A, B) = 1 - \frac{|A \setminus B| + |B \setminus A|}{n}, \tag{9.20}$$

where "$\setminus$" is the set-minus operation and n is the total number of features. The two indices were calculated for $S_1(k), S_2(k)$ from the example above, where k was varied from 1 to n. The results are plotted also in Figure 9.14. While all three indices detect the dip at $k = 4$ features, S_S and S_H have a tendency to increase when the size of the selected set approaches the total number of features n. The point of view advocated here is that consistency should have high value only if it exceeds the consistency by chance or by design.

9.6.2   A Stability Index for K Sequences

Let $S_1, S_2, \ldots, S_K$ be the sequences of features on a given data set.

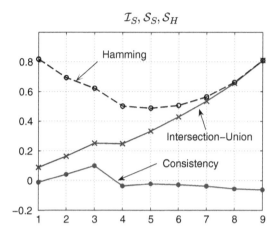

FIGURE 9.15   Consistency index $\mathcal{I}_S$ and similarities $\mathcal{S}_S$ (intersection union) and $\mathcal{S}_H$ (Hamming) for 10 random sequences plotted against the subset size k.

Definition 2. The stability index for a set of sequences of features, $\mathcal{A} = \{S_1, S_2, \dots S_K\}$, for a given set size, k, is the average of all pairwise consistency indices

$$\mathcal{I}_S(\mathcal{A}(k)) = \frac{2}{K(K-1)} \sum_{i=1}^{K-1} \sum_{j=i+1}^{K} I_C(S_i(k), S_j(k)). \qquad (9.21)$$

Averaging the pairwise similarities to arrive at a single index is also the approach adopted for both S_S and S_H [111, 201]. Denote the averaged indices by $\mathcal{S}_S$ and $\mathcal{S}_H$, respectively. To strengthen the argument for correction for chance, Figure 9.15 shows $\mathcal{I}_S$, $\mathcal{S}_S$, and $\mathcal{S}_H$ across all pairs of 10 independently generated random sequences. Only $\mathcal{I}_S$ gives consistency around zero for any number of features k. Similarly $\mathcal{S}_S$ favors large subsets and $\mathcal{S}_H$ favors large and small but not medium-size subsets.

MATLAB code for the stability index is given in Appendix 9.A.3.

9.6.3   An Example of Applying the Stability Index

The AUC ranker was tried on the spam data set. Twenty runs were carried out, where the data was randomly split into a training part with 100 objects and a testing part with 4501 objects. For each run i, $i = 1, \dots, 20$

- The AUC ranker was applied to the training data to create feature ranking S_i. An LDC was trained on subsets of $1, 2, \dots, 57$ features taken in order from S_i. The 57 values of the classification error were stored in vector P_i.

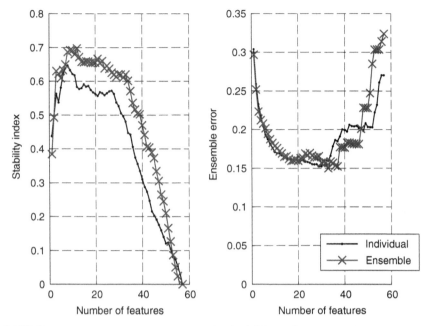

FIGURE 9.16   Stability and ensemble error for the ROC (AUC) ranker on the spam data set.

- An ensemble of $L = 9$ feature rankers was trained on bootstrap samples from the training data. The ensemble feature ranking S'_i was obtained by averaging the nine individual rankings. Again, an LDC was trained on subsets of $1, 2, \dots, 57$ features taken in order from S'_i. The 57 values of the ensemble classification error were stored in vector P'_i.

The stability index was calculated for all feature values (1 to 57) for the individual rankings $S_1, \dots, S_{20}$, and also for the ensemble rankings $S'_1, \dots, S'_{20}$. Figure 9.16a plots the two stability indices.

Finally, the individual and the ensemble error vectors were averaged across the 20 runs. The averaged curves are plotted in Figure 9.16b.

Expectedly, the feature ranking ensemble stability is higher than the stability of the individual ranker [1]. To complete the study, we must pair the stability with the ensemble error. An extremely stable but inaccurate ensemble would be useless. The purpose of the ensemble is to compile a stable yet accurate feature subset. The dip of the error curves is roughly where the highest stability is. This indicates that the data set contains a core set of about 15–20 features, which can be identified and communicated back to the end user. This example demonstrates that by using a bootstrap ensemble of feature rankers, we can gain stability without sacrificing accuracy.

APPENDIX

9.A.1   MATLAB CODE FOR THE NUMERICAL EXAMPLE OF ENSEMBLE RANKING

The code below generates four variants of a random permutation of the integers from 1 to 10 interpreted as feature rankings. Each ranking is a distorted version of the initial permutation. A number of swaps between 1 and 5 is chosen, and then the swaps are carried out. The version is stored as the ith ranking in array b. Lines 14 and 15 calculate the ranks for the features, the average ranks, and the final ranking.

```
1  %----------------------------------------------------------------%
2  a = randperm(10); % the "true" ranking
3  b = zeros(4,10); % array for rankings R_1 ... R_4
4  for i = 1:4 % create 4 distorted versions
5      k = randi(5); % choose the # of swaps
6      t = a; % restore the original permutation
7      for j = 1:k
8          r = randperm(10); % choose the swap
9          tt = t([r(1),r(2)]); % take the two values
10         t([r(2),r(1)]) = tt; % assign to swapped places
11     end
12     b(i,:) = t; % store ranking R_i
13 end
14 [~,fr] = sort(b,2); % calculate the feature ranks
15 [~,FinalRanking] = sort(mean(fr)); % sort by average rank
16 disp(FinalRanking) % show the final ranking
17 %----------------------------------------------------------------%
```

9.A.2   MATLAB GA NUGGETS

The code below shows a self-contained example of a genetic algorithm. The roulette wheel selection, crossover, and mutation are written as functions within the same MATLAB file but can be saved and used separately. The toy GA has the objective to create a chromosome whose binary entries alternate so that the ideal solution is 101010101... . The fitness function created as a function handle in lines 6 and 7 measures the proportion of 0s and 1s in correct positions. The code will produce a figure showing the fitness of the best chromosome and the average population fitness, as illustrated in Figure 9.A.1.

```
1  %----------------------------------------------------------------%
2  function GA_nuggets
3  n = 50; % chromosome size
4  ps = 20; % population size
```

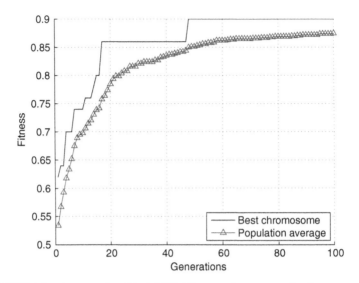

FIGURE 9.A.1   An example of the MATLAB output from a run of code GA_nuggets.

```
 5  Pm = 0.15; % mutation probability
 6  fitness = @(x) (sum(x(:,1:2:end),2) +...
 7      sum(1-x(:,2:2:end),2))/n;
 8
 9  figure, xlabel('Generations'), ylabel('Fitness')
10  P = rand(ps,n) > 0.5; % population
11  f = fitness(P);
12
13  for i = 1:100 % 100 generations
14      O = roulette_wheel(P,f);
15      O = crossover(O,ps/2);
16      O = mutate(O,Pm);
17      [f,ix] = sort([f;fitness(O)],'descend');
18      G = [P;O]; P = G(ix(1:ps),:); % new population
19      f = f(1:ps); F(i) = f(1); MF(i) = mean(f); % new f
20      cla, grid on, hold on, plot(1:i,F,'k-',1:i,MF,'r^-')
21      drawnow
22  end
23  fprintf('Final fitness = %.2f\n',f(1))
24  legend('Best chromosome','Population average',...
25      'location','SouthEast')
26  end
27
28  function O = roulette_wheel(P,f)
29  ps = size(P,1); % population size and chromosome length
```

```
30  s = cumsum(f) / sum(f); % roulette probability scale
31  O = []; % offspring
32  for i = 1:ps % assuming that the desired #parents is ps
33      idx = find(rand < s); O = [O;P(idx(1),:)];
34  end
35  end
36
37  function C = crossover(O,k)
38  % k = numer of desired cross-overs (usually ps/2)
39  C = []; % set with the 2*k children
40  for i = 1:k % assuming that the desired #parents is ps
41      rp = randperm(size(O,1));
42      c1 = O(rp(1),:); c2 = O(rp(2),:); % the 2 parents
43      cp = randi(size(O,2)-1); % the crosover point
44      C = [C;c1(1:cp) c2(cp+1:end);c2(1:cp) c1(cp+1:end)];
45  end
46  end
47
48  function M = mutate(C,Pm)
49  % Pm = mutation probability
50  mi = rand(size(C)) < Pm; % mutation index
51  M = C; M(mi) = 1-M(mi); % mutated C
52  end
53  %-----------------------------------------------------------%
```

9.A.3   MATLAB CODE FOR THE STABILITY INDEX

The code below calculates the stability index applicable for complete rankings. The output S contains the stability index for feature sizes $1, 2, \ldots, n$. The input A is a matrix of size $n \times m$, containing the m feature rankings ($m > 1$). Each column is a permutation of the integers $1 \ldots n$. Element $a(i,j)$ is the number of the feature (out of n) ranked ith in the jth ranking.

```
1   %-----------------------------------------------------------%
2   function S = stability_index(A)
3   [n,m] = size(A); S = [];
4   for i = 1:n
5       Sm = 0;
6       for r1 = 1 : m - 1
7           for r2 = r1 + 1 : m
8               Sm = Sm + consistency_index(A(1:i,r1),...
9                   A(1:i,r2), n);
10          end
11      end
```

```
12      S = [S; Sm * 2 /(m * (m - 1))];
13 end
14 %....................................................
15 function ind = consistency_index(A,B,n)
16 k = length(A); r = numel(intersect(A,B));
17 if k == 0 || k == n, ind = 0;
18 else ind = (r*n-k*k)/(k*n-k*k); end
19 %--------------------------------------------------------%
```

10

A FINAL THOUGHT

One of my friends who kindly read the manuscript and returned comments to me said: "Chapter 10. There isn't one? No conclusion or retrospective on the past decade?"

Well, wise and prophetic words do not come easily to me. Only that famous Russian aphorism springs to mind "Нельзя объять необъятное," which loosely means "you can't embrace immensity," or rather "you can't embrace the unembraceable." To me, the classifier ensemble field is starting to resemble the unembraceable. The lively interest in the area has led to significant advancements both in the structuring of the field and in populating this structure with exciting discoveries, ideas, methods, and results. However, it is impossible to compile a list of the most significant achievements in the past decade without offending or alienating half of my readership. I will leave this task to the main judge – time.

What is clear, though, is that classifier ensembles are here to stay. Solving the challenges of modern pattern recognition and machine learning, surfacing in application areas such as computer vision, medical research, geosciences, and many more, will require bespoke research arsenals. It is likely that the field will branch out, and the relevant fundamental research will be compartmentalized. I look forward to seeing this happen. I just hope we do not suffer the fate of those building the Tower of Babel, and we will still be able to understand one another.

Combining Pattern Classifiers: Methods and Algorithms, Second Edition. Ludmila I. Kuncheva.
© 2014 John Wiley & Sons, Inc. Published 2014 by John Wiley & Sons, Inc.

REFERENCES

1. T. Abeel, T. Helleputte, Y. Van de Peer, P. Dupont, and Y. Saeys. Robust biomarker identification for cancer diagnosis with ensemble feature selection methods. *Bioinformatics (Oxford, England)*, 26(3):392–398, 2010.

2. M. Abouelenien and X. Yuan. SampleBoost: improving boosting performance by destabilizing weak learners based on weighted error analysis. In: *Proceedings of the 21st International Conference on Pattern Recognition (ICPR)*, pp. 585–588. IEEE, 2012.

3. D. W. Aha and R. L Bankert. A comparative evaluation of sequential feature selection algorithms. In: *Proceedings of the 5th International Workshop on AI and Statistics*, Ft Lauderdale, FL, 1995, pp. 1–7.

4. D. W. Aha and R. L. Blankert. Cloud classification using error-correcting output codes. *Artificial Intelligence Applications: Natural Resources, Agriculture and Environmental Science*, 11(1):13–28, 1997.

5. W. I. Ai and P. Langley. Induction of one-level decision trees. In: *Proceedings of the Ninth International Conference on Machine Learning*, pp. 233–240. Morgan Kaufmann, 1992.

6. M. Aksela. Comparison of classifier selection methods for improving committee performance. In: T. Windeatt and F. Roli, editors, *Proc. 4rd Int. Workshop on Multiple Classifier Systems (MCS 2003)*, Guildford, UK. Volume 2709: Lecture Notes in Computer Science, pp. 84–93. Springer, 2003.

7. I. Aleksander and H. Morton. *Introduction to Neural Computing*, 2nd edition. International Thomson Computer Press, London, 1995.

8. L. A. Alexandre, A. C. Campilho, and M. Kamel. On combining classifiers using sum and product rules. *Pattern Recognition Letters*, 22(12):1283–1289, 2001.

Combining Pattern Classifiers: Methods and Algorithms, Second Edition. Ludmila I. Kuncheva.
© 2014 John Wiley & Sons, Inc. Published 2014 by John Wiley & Sons, Inc.

9. F. M. Alkoot and J. Kittler. Experimental evaluation of expert fusion strategies. *Pattern Recognition Letters*, 20:1361–1369, 1999.

10. E. L. Allwein, R. E. Schapire, and Y. Singer. Reducing multiclass to binary: a unifying approach for margin classifiers. *Journal of Machine Learning Research*, 1:113–141, 2000.

11. E. Alpaydin. Comparison of statistical and neural classifiers and their applications to optical character recognition and speech classification. In: C. T. Leondes, editor, *Image Processing and Pattern Recognition*, Volume 5: Neural Network Systems, pp. 61–88. Academic Press, 1998.

12. E. Alpaydin. *Introduction to Machine Learning*. MIT Press, Cambridge, MA, 2004.

13. E. Alpaydin and C. Kaynak. Cascading classifiers. *KYBERNETIKA*, 34(4):369–374, 1998.

14. E. Alpaydin and E. Mayoraz. Learning error-correcting output codes from data. In: *Proceedings of ICANN'99*, pp. 743–748. MIT AI Laboratory, 1999.

15. W. Altidor, T. M. Khoshgoftaar, and A. Napolitano. A noise-based stability evaluation of threshold-based feature selection techniques. In: *IEEE International Conference on Information Reuse and Integration (IRI)*, pp. 240–245. IEEE Systems, Man, and Cybernetics Society, 2011.

16. H. Altinçay. On naive Bayesian fusion of dependent classifiers. *Pattern Recognition Letters*, 26(15):2463–2473, 2005.

17. J. A. Anderson and E. Rosenfeld. *Neurocomputing. Foundations of Research*. The MIT Press, Cambridge, MA, 1988.

18. P. Arabie, L. J. Hubert, and G. De Soete. *Clustering and Classification*. World Scientific, Singapore, 1996.

19. G. Armano, C. Chira, and N. Hatami. A new gene selection method based on random subspace ensemble for microarray cancer classification. In: M. Loog, L. Wessels, M. J. T. Reinders, and D. Ridder, editors, *Pattern Recognition in Bioinformatics*, Volume 7036: Lecture Notes in Computer Science, pp. 191–201. Springer, Berlin, Heidelberg, 2011.

20. R. Avnimelech and N. Intrator. Boosted mixture of experts: an ensemble learning scheme. *Neural Computation*, 11(2):483–497, 1999.

21. B. Ayerdi, J. Maiora, and M. Grana. Active learning of hybrid extreme rotation forests for CTA image segmentation. In: *Proceedings of the 12th International Conference on Hybrid Intelligent Systems (HIS)*, pp. 543–548. IEEE, 2012.

22. K. Bache and M. Lichman. UCI Machine Learning Repository, 17 December 2013. http://archive.ics.uci.edu/ml.

23. M. A. Bagheri, Q. Gao, and S. Escalera. A genetic-based subspace analysis method for improving error-correcting output coding. *Pattern Recognition*, 46(10):2830–2839, 2013.

24. T. Bailey and A. K. Jain. A note on distance-weighted *k*-nearest neighbor rules. *IEEE Transactions on Systems, Man, and Cybernetics*, 8(4):311–313, 1978.

25. R. E. Banfield, L. O. Hall, K. W. Bowyer, and W. P. Kegelmeyer. A new ensemble diversity measure applied to thinning ensembles. In: T. Windeatt and F. Roli, editors, *Proc 4rd Int. Workshop on Multiple Classifier Systems (MCS 2003)*, Guildford, UK. Volume 2709: Lecture Notes in Computer Science, pp. 306–316. Springer, 2003.

26. Y. L. Barabash. *Collective Statistical Decisions in Recognition*. Radio i Sviaz, Moscow, 1983 (In Russian).

27. Z. Barutcuoglu, P. Long, and R. Servedio. One-pass boosting. In: J. C. Platt, D. Koller, Y. Singer, and S. Roweis, editors, *Advances in Neural Information Processing Systems (NIPS)*, Volume 20, pp. 73–80. MIT Press, Cambridge, MA, 2007.

28. R. Battiti and A. M. Colla. Democracy in neural nets: voting schemes for classification. *Neural Networks*, 7:691–707, 1994.

29. E. Bauer and R. Kohavi. An empirical comparison of voting classification algorithms: bagging, boosting, and variants. *Machine Learning*, 36:105–142, 1999.

30. M. A. Bautista, S. Escalera, X. Baró, P. Radeva, J. Vitriá, and O. Pujol. Minimal design of error-correcting output codes. *Pattern Recognition Letters*, 33(6):693–702, 2012.

31. A. Bella, C. Ferri, J. Hernández-Orallo, and M. J. Ramírez-Quintana. On the effect of calibration in classifier combination. *Applied Intelligence*, 38(4):566–585, 2013.

32. S. Ben-David, P. M. Long, and Y. Mansour. Agnostic boosting. In: David Helmbold and Bob Williamson, editors, *Computational Learning Theory*. Volume 2111: Lecture Notes in Computer Science, pp. 507–516. Springer, Berlin, Heidelberg, 2001.

33. A. Ben-Hur and J. Weston. A user's guide to support vector machines. In: *Data Mining Techniques for the Life Sciences*. Volume 609: Methods in Molecular Biology, pp. 223–239. Springer, 2008.

34. J. A. Benediktsson and P. H. Swain. Consensus theoretic classification methods. *IEEE Transactions on Systems, Man, and Cybernetics*, 22:688–704, 1992.

35. C. Berenstein, L. N. Kanal, and D. Lavine. Consensus rules. In: L. N. Kanal and J. F. Lemmer, editors, *Uncertainty in Artificial Intelligence*, pp. 27–32. Elsevier Science Publishers B.V., 1986.

36. A. Bertoni, R. Folgieri, and G. Valentini. Random subspace ensembles for the biomolecular diagnosis of tumors. In: *Models and Metaphors from Biology to Bioinformatics Tools, NETTAB*, 2004.

37. A. Bertoni, R. Folgieri, and G. Valentini. Feature selection combined with random subspace ensemble for gene expression based diagnosis of malignancies. In: B. Apolloni, M. Marinaro, and R. Tagliaferri, editors, *Biological and Artificial Intelligence Environments*, pp. 29–36. Springer, 2005.

38. J. C. Bezdek. *Pattern Recognition with Fuzzy Objective Function Algorithms*. Plenum Press, New York, 1981.

39. J. C. Bezdek, J. M. Keller, R. Krishnapuram, and N. R. Pal. *Fuzzy Models and Algorithms for Pattern Recognition and Image Processing*. Kluwer Academic Publishers, 1999.

40. C. M. Bishop. *Neural Networks for Pattern Recognition*. Clarendon Press, Oxford, 1995.

41. C. M. Bishop. *Pattern Recognition and Machine Learning*. Springer, New York, 2006.

42. A. Blum and P. Langley. Selection of relevant features and examples in machine learning. *Artificial Intelligence*, 97(1–2):245–271, 1997.

43. P. Bonissone, J. M. Cadenas, M. C. Garrido, and R. A. Díaz-Valladares. A fuzzy random forest. *International Journal of Approximate Reasoning*, 51(7):729–747, 2010.

44. R. F. Bordley. A multiplicative formula for aggregating probability assessments. *Management Science*, 28:1137–1148, 1982.

45. J. K. Bradley and R. Schapire. FilterBoost: regression and classification on large datasets. In: *Advances in Neural Information Processing Systems (NIPS)*, Volume 20. MIT Press, Cambridge, MA, 2008.

46. L. Breiman. Bagging predictors. Technical Report 421, Department of Statistics, University of California, Berkeley, CA, 1994.

47. L. Breiman. Bagging predictors. *Machine Learning*, 26(2):123–140, 1996.

48. L. Breiman. Arcing classifiers. *The Annals of Statistics*, 26(3):801–849, 1998.

49. L. Breiman. Pasting small votes for classification in large databases and on-line. *Machine Learning*, 36:85–103, 1999.

50. L. Breiman. Random forests. *Machine Learning*, 45:5–32, 2001.

51. L. Breiman, J. Friedman, R. Olshen, and C. Stone. *Classification and Regression Trees*. Wadsworth International, Belmont, CA, 1984.

52. G. Brown. An information theoretic perspective on multiple classifier systems. In: *Proceedings of the 8th International Workshop on Multiple Classifier Systems*, LNCS, pp. 344–353. Springer, 2009.

53. G. Brown. Ensemble learning. In: C. Sammut and G. Webb, editors, *Encyclopedia of Machine Learning*. Springer, 2010.

54. G. Brown and L. I. Kuncheva. GOOD and BAD diversity in majority vote ensembles. In: *Proc. 9th International Workshop on Multiple Classifier Systems (MCS'10)*, Cairo, Egypt. Volume 5997: Lecture Notes in Computer Science, pp. 124–133. Springer, 2010.

55. G. Brown, A. Pocock, M. Zhao, and M. Lujan. Conditional likelihood maximisation: a unifying framework for information theoretic feature selection. *Journal of Machine Learning Research*, 13:27–66, 2012.

56. G. Brown, J. Wyatt, R. Harris, and X. Yao. Diversity creation methods: a survey and categorisation. *Information Fusion*, 6(1):5–20, 2005.

57. L. Bruzzone and R. Cossu. A multiple-cascade-classifier system for a robust and partially unsupervised updating of land-cover maps. *IEEE Transactions on Geoscience and Remote Sensing*, 40(9):1984–1996, 2002.

58. R. Bryll, R. Gutierrez-Osuna, and F. Quek. Attribute bagging: improving accuracy of classifier ensembles by using random feature subsets. *Pattern Recognition*, 36(6):1291–1302, 2003.

59. P. Bühlmann and B. Yu. Boosting with the L2 loss: regression and classification. *Journal of the American Statistical Association*, 98:324–339, 2003.

60. P. Bühlmann and T. Hothorn. Twin boosting: improved feature selection and prediction. *Statistics and Computing*, 20(2):119–138, 2010.

61. C. J. C. Burges. A tutorial on support vector machines for pattern recognition. *Data Mining and Knowledge Discovery*, 2(2):121–167, 1998.

62. J. B. D. Cabrera. On the impact of fusion strategies on classification errors for large ensembles of classifiers. *Pattern Recognition*, 39:1963–1978, 2006.

63. R. A. Calvo and S. D'Mello. Affect detection: an interdisciplinary review of models, methods, and their applications. *IEEE Transactions on Affective Computing*, 1(1):18–37, 2010.

64. A. M. P. Canuto, M. C. C. Abreu, L. M. Oliveira, J. C. Xavier Jr., and A. de M. Santos. Investigating the influence of the choice of the ensemble members in accuracy and diversity of selection-based and fusion-based methods for ensembles. *Pattern Recognition Letters*, 28(4):472–486, 2007.

65. S.-H. Cha. Comprehensive survey on distance/similarity measures between probability density functions. *International Journal of Mathematical Models and Methods in Applied Sciences*, 1(4):300–307, 2007.

66. P. K. Chan, Wei Fan, A. L. Prodromidis, and S. J. Stolfo. Distributed data mining in credit card fraud detection. *IEEE Intelligent Systems and their Applications*, 14(6):67–74, 1999.

67. O. Chapelle, V. Vapnik, O. Bousquet, and S. Mukherjee. Choosing multiple parameters for support vector machines. *Machine Learning*, 46(1):131–159, 2002.

68. N. V. Chawla, L. O. Hall, K. W. Bowyer, T. E. Moore Jr., and W. P. Kegelmeyer. Distributed pasting of small votes. In F. Roli and J. Kittler, editors, *Proc. 3d International Workshop on Multiple Classifier Systems, MCS'02*, Cagliari, Italy. Volume 2364: Lecture Notes in Computer Science, pp. 51–62. Springer, 2002.

69. N. V. Chawla, A. Lazarevic, L. O. Hall, and K. W. Bowyer. SMOTEBoost: improving prediction of the minority class in boosting. In: *Knowledge Discovery in Databases: PKDD 2003*. Volume 2838: Lecture Notes in Computer Science, pp. 107–119. Springer, Berlin, Heidelberg, 2003.

70. N. V. Chawla and K. W. Bowyer. Random subspaces and subsampling for 2-d face recognition. In: *Proceedings of the IEEE Computer Society Conference on Computer Vision and Pattern Recognition, (CVPR)*, Volume 2, pp. 582–589, 2005.

71. D. Chen and X. Cheng. An asymptotic analysis of some expert fusion methods. *Pattern Recognition Letters*, 22:901–904, 2001.

72. K. Chen, L. Wang, and H. Chi. Methods of combining multiple classifiers with different features and their applications to text-independent speaker identification. *International Journal on Pattern Recognition and Artificial Intelligence*, 11(3):417–445, 1997.

73. S. B. Cho. Pattern recognition with neural networks combined by genetic algorithm. *Fuzzy Sets and Systems*, 103:339–347, 1999.

74. S. S. Choi, S. H. Cha, and C. Tappert. A survey of binary similarity and distance measures. *Journal on Systemics, Cybernetics and Informatics*, 8(1):43–48, 2010.

75. Dan Claudiu Ciresan, Ueli Meier, Luca Maria Gambardella, and Jürgen Schmidhuber. Deep big simple neural nets excel on handwritten digit recognition. *Neural Computation*, 22:3207–3220, 2010.

76. W. W. Cohen. Fast effective rule induction. In: *Twelfth International Conference on Machine Learning*, pp. 115–123. Morgan Kaufmann, 1995.

77. T. M. Cover. The best two independent measurements are not the two best. *IEEE Transactions on Systems, Man, and Cybernetics*, 4:116–117, 1974.

78. P. Cunningham and J. Carney. Diversity versus quality in classification ensembles based on feature selection. Technical Report TCD-CS-2000-02, Department of Computer Science, Trinity College, Dublin, 2000.

79. D. Richard Cutler, Thomas C. Edwards Jr., Karen H. Beard, Adele Cutler, and Kyle T. Hess. Random forests for classification in ecology. *Ecology*, 88(11):2783–2792, 2007.

80. F. Cutzu. Polychotomous classification with pairwise classifiers: a new voting principle. In: *Proc. 4th International Workshop on Multiple Classifier Systems (MCS 2003)*, Guildford, UK. Volume 2709: Lecture Notes in Computer Science, pp. 115–124. Springer, 2003.

81. B. V. Dasarathy. *Nearest Neighbor (NN) Norms: NN Pattern Classification Techniques*. IEEE Computer Society Press, Los Alamitos, CA, 1990.

82. B. V. Dasarathy and B. V. Sheela. A composite classifier system design: concepts and methodology. *Proceedings of IEEE*, 67:708–713, 1979.

83. M. Dash and H. Liu. Feature selection for classification. *Intelligent Data Analysis*, 1:131–156, 1997.

84. W. H. E. Day. Consensus methods as tools for data analysis. In: H. H. Bock, editor, *Classification and Related Methods for Data Analysis*, pp. 317–324. Elsevier Science Publishers B.V. (North Holland), 1988.

85. K. W. De Bock and D. Van den Poel. An empirical evaluation of rotation-based ensemble classifiers for customer churn prediction. *Expert Systems with Applications*, 38(10):12293–12301, 2011.

86. F. De Martino, G. Valente, N. Staeren, J. Ashburner, and R. Goebel a E. Formisano. Combining multivariate voxel selection and support vector machines for mapping and classification of fMRI spatial patterns. *NeuroImage*, 43(1):44–58, 2008.

87. M. Demirekler and H. Altincay. Plurality voting-based multiple classifier systems: statistically independent with respect to dependent classifiers sets. *Pattern Recognition*, 35:2365–2379, 2002.

88. A. Demiriz, K. P. Bennett, and J. Shawe-Taylor. Linear programming boosting via column generation. *Machine Learning*, 46:225–254, 2002.

89. J. Demšar. Statistical comparison of classifiers over multiple data sets. *Journal of Machine Learning Research*, 7:1–30, 2006.

90. P. A. Devijver and J. Kittler. *Pattern Recognition: A Statistical Approach*. Prentice-Hall, Inc., Englewood Cliffs, NJ, 1982.

91. L. Didaci and G. Giacinto. Dynamic classifier selection by adaptive k-nearest-neighbourhood rule. In: *Proceedings of the 5th International Workshop on Multiple Classifier Systems. (MCS'04)*, Cambridge, UK. Volume 3077: Lecture Notes in Computer Science, pp. 174–183. Springer, 2004.

92. L. Didaci, G. Giacinto, F. Roli, and G. L. Marcialis. A study on the performances of dynamic classifier selection based on local accuracy estimation. *Pattern Recognition*, 38(11):2188–2191, 2005.

93. C. Dietrich, G. Palm, and F. Schwenker. Decision templates for the classification of bioacoustic time series. *Information Fusion*, 4:101–109, 2003.

94. T. Dietterich. An experimental comparison of three methods for constructing ensembles of decision trees: bagging, boosting and randomization. *Machine Learning*, 40(2):139–157, 2000.

95. T. G. Dietterich. Ensemble methods in machine learning. In: J. Kittler and F. Roli, editors, *Multiple Classifier Systems*, Cagliari, Italy. Volume 1857: Lecture Notes in Computer Science, pp. 1–15. Springer, 2000.

96. T. G. Dietterich. Bias-variance analysis of ensemble learning. In: *7th Course of the International School on Neural Networks "E.R. Caianiello", Ensemble Methods for Learning Machines*, Vietri-sul-Mare, Salerno-Italy, 2002.

97. T. G. Dietterich and G. Bakiri. Error-correcting output codes: a general method for improving multiclass inductive learning programs. In: *Proc. 9th National Conference on Artificial Intelligence, AAAI-91*, pp. 572–577. AAAI Press, 1991.

98. T. G. Dietterich and G. Bakiri. Solving multiclass learning problems via error-correcting output codes. *Journal of Artificial Intelligence Research*, 2:263–286, 1995.

99. T. G. Dietterich. Approximate statistical tests for comparing supervised classification learning algorithms. *Neural Computation*, 7(10):1895–1924, 1998.

100. C. Domingo and O. Watanabe. Madaboost: a modification of AdaBoost. In: *Proceedings of the Thirteenth Annual Conference on Computational Learning Theory, COLT '00*, San Francisco, CA, USA. pp. 180–189. Morgan Kaufmann, 2000.

101. P. Domingos. Why does bagging work? A Bayesian account and its implications. In: *Knowledge Discovery and Data Mining*, pp. 155–158. AAAI Press, 1997.

102. P. Domingos. A unified bias-variance decomposition and its applications. In: *Proc. 7th International Conference on Machine Learning*, Stanford, CA, pp. 231–238. Morgan Kaufmann, 2000.

103. P. Domingos and M. Pazzani. On the optimality of the simple Bayesian classifier under zero-one loss. *Machine Learning*, 29:103–130, 1997.

104. H. Drucker, C. Cortes, L. D. Jackel, Y. LeCun, and V. Vapnik. Boosting and other ensemble methods. *Neural Computation*, 6:1289–1301, 1994.

105. D. Dubois and H. Prade. A review of fuzzy set aggregation connectives. *Information Sciences*, 36:85–121, 1985.

106. R. O. Duda and P. E. Hart. *Pattern Classification and Scene Analysis*. John Wiley & Sons, New York, 1973.

107. R. O. Duda, P. E. Hart, and D. G. Stork. *Pattern Classification*, 2nd edition. John Wiley & Sons, New York, 2001.

108. R. P. W. Duin. The combining classifier: to train or not to train? In: *Proc. 16th International Conference on Pattern Recognition, ICPR'02, Canada*, pp. 765–770. IEEE Computer Society, 2002.

109. R. P. W. Duin and D. M. J. Tax. Experiments with classifier combination rules. In: J. Kittler and F. Roli, editors, *Multiple Classifier Systems*, Cagliari, Italy. Volume 1857: Lecture Notes in Computer Science, pp. 16–29, Springer, 2000.

110. R. P. W. Duin. A note on comparing classifiers. *Pattern Recognition Letters*, 17:529–536, 1996.

111. K. Dunne, P. Cunningham, and F. Azuaje. Solution to instability problems with sequential wrapper-based approaches to feature selection. Technical Report TCD-CS-2002-28, Department of Computer Science, Trinity College, Dublin, 2002.

112. S. Džeroski and B. Ženko. Is combining classifiers with stacking better than selecting the best one? *Machine Learning*, 54(3):255–273, 2004.

113. D. E. Eckhardt and L. D. Lee. A theoretical basis for the analysis of multiversion software subject to coincident errors. *IEEE Transactions on Software Engineering*, 11(12):1511–1517, 1985.

114. N. Edakunni, G. Brown, and T. Kovacs. Boosting as a product of experts. In: *Proceedings of the 27th Conference on Uncertainty in Artificial Intelligence*, pp. 187–194. Association for Uncertainty in Artificial Intelligence Press, 2011.

115. B. Efron and R. Tibshirani. *An Introduction to the Bootstrap*. Chapman & Hall, New York, 1993.

116. M. Egmont-Petersen, W. R. M. Dassen, and J. H. C. Reiber. Sequential selection of discrete features for neural networks—a Bayesian approach to building a cascade. *Pattern Recognition Letters*, 20(11–13):1439–1448, 1999.

117. G. Eibl and K. P. Pfeiffer. How to make AdaBoost.M1 work for weak base classifiers by changing only one line of the code. In: *Proceedings of the 13th European Conference on Machine Learning, ECML '02*, pp. 72–83. Springer, Lecture Notes in Computer Science, 2002.

118. G. Eibl and K.-P. Pfeiffer. Multiclass boosting for weak classifiers. *Journal of Machine Learning Research*, 6:189–210, 2005.

119. R. Elwell and R. Polikar. Incremental learning of concept drift in nonstationary environments. *IEEE Transactions on Neural Networks*, 22(10):1517–1531, 2011.

120. H. Erdogan and M. U. Sen. A unifying framework for learning the linear combiners for classifier ensembles. In: *Proceedings of the 20th International Conference on Pattern Recognition (ICPR)*, pp. 2985–2988. IEEE, 2010.

121. S. Escalera, O. Pujol, and P. Radeva. Error-correcting output codes library. *Journal of Machine Learning Research*, 11:661–664, 2010.

122. S. Escalera, O. Pujol, and P. Radeva. On the decoding process in ternary error-correcting output codes. *IEEE Transactions on Pattern Analysis and Machine Intelligence*, 32(1):120–134, 2010.

123. S. Escalera, D. M. J. Tax, O. Pujol, P. Radeva, and R. P. W. Duin. Subclass problem-dependent design for error-correcting output codes. *IEEE Transactions on Pattern Analysis and Machine Intelligence*, 30(6):1041–1054, 2008.

124. F. Esposito, D. Malerba, and G. Semeraro. A comparative analysis of methods for pruning decision trees. *IEEE Transactions on Pattern Analysis and Machine Intelligence*, 19(5):476–491, 1997.

125. A. Esuli, T. Fagni, and F. Sebastiani. MP-Boost: a multiple-pivot boosting algorithm and its application to text categorization. In: *String Processing and Information Retrieval*. Volume 4209: Lecture Notes in Computer Science, pp. 1–12. Springer, Berlin, Heidelberg, 2006.

126. B. Everitt. *Cluster Analysis*. John Wiley and Sons, New York, 1993.

127. R. A. Fisher. The use of multiple measurements in taxonomic problems. *Annals of Eugenics*, 7:179–188, 1936.

128. J. L. Fleiss. *Statistical Methods for Rates and Proportions*. John Wiley & Sons, 1981.

129. G. Forman. An extensive empirical study of feature selection metrics for text classification. *Journal of Machine Learning Research*, 3:1289–1305, 2003.

130. E. Frank and S. Kramer. Ensembles of nested dichotomies for multi-class problems. In: *Proceedings of the 21st International Conference of Machine Learning (ICML)*, pp. 305–312. ACM Press, 2004.

131. Y. Freund. An adaptive version of the boost by majority algorithm. *Machine Learning*, 43(3):293–318, 2001.

132. Y. Freund. A more robust boosting algorithm, eprint arXiv:0905.2138, Statistics – Machine Learning, 2009.

133. Y. Freund and R. E. Schapire. Experiments with a new boosting algorithm. In: *Thirteenth International Conference on Machine Learning*, San Francisco, pp. 148–156, Morgan Kaufmann, 1996.

134. Y. Freund and R. E. Schapire. A decision-theoretic generalization of on-line learning and an application to boosting. *Journal of Computer and System Sciences*, 55(1):119–139, 1997.

135. Y. Freund and R. E. Schapire. Discussion of the paper "Arcing Classifiers" by Leo Breiman. *The Annals of Statistics*, 26(3):824–832, 1998.

136. J. Friedman, T. Hastie, and R. Tibshirani. Additive logistic regression: a statistical view of boosting. *Annals of Statistics*, 28(2):337–374, 2000.

137. J. H. Friedman. Regularized discriminant analysis. *Journal of the American Statistical Association*, 84(405):165–175, 1989.

138. J. H. Friedman. Stochastic gradient boosting. *Computational Statistics and Data Analysis*, 38:367–378, 2002.

139. N. Friedman, D. Geiger, and M. Goldszmid. Bayesian network classifiers. *Machine Learning*, 29(2):131–163, 1997.

140. K. S. Fu. *Syntactic Pattern Recognition and Applications*. Prentice Hall, Englewood Cliffs, NJ, 1982.

141. K. Fukunaga. *Introduction to Statistical Pattern Recognition*. Academic Press, Orlando, FL, 1972.

142. K. Fukushima. Cognitron: a self-organizing multilayered neural network. *Biological Cybernetics*, 20:121–136, 1975.

143. G. Fumera and F. Roli. Performance analysis and comparison of linear combiners for classifier fusion. In: *Proc. 16th International Conference on Pattern Recognition*. IEEE Computer Society, 2002.

144. G. Fumera and F. Roli. Linear combiners for classifier fusion: some theoretical and experimental results. In: T. Windeatt and F. Roli, editors, *Proc. 4rd Int. Workshop on Multiple Classifier Systems (MCS 2003)*, Guildford, UK. Volume 2709: Lecture Notes in Computer Science, pp. 74–83. Springer, 2003.

145. G. Fumera and F. Roli. A theoretical and experimental analysis of linear combiners for multiple classifier systems. *IEEE Transactions on Pattern Analysis and Machine Intelligence*, 27:942–956, 2005.

146. J. Gamma and P. Brazdil. Cascade generalization. *Machine Learning*, 41(3):315–343, 2000.

147. S. García and F. Herrera. An extension on "Statistical Comparisons of Classifiers over Multiple Data Sets" for all pairwise comparisons. *Journal of Machine Learning Research*, 9:2677–2694, 2008.

148. N. García-Pedrajas, J. Maudes-Raedo, C. García-Osorio, and J. J. Rodríguez-Díez. Supervised subspace projections for constructing ensembles of classifiers. *Information Sciences*, 193:1–21, 2012.

149. P. Geurts, D. deSeny, M. Fillet, M.-A. Meuwis, M. Malaise, M.-P. Merville, and L. Wehenkel. Proteomic mass spectra classification using decision tree based ensemble methods. *Bioinformatics*, 21(14):3138–3145, 2005.

150. S. Ghahramani. *Fundamentals of Probability*, 2nd edition. Prentice Hall, Englewood Cliffs, NJ, 2000.

151. J. Ghosh. Multiclassifier systems: back to the future. In: F. Roli and J. Kittler, editors, *Proc. 3d International Workshop on Multiple Classifier Systems, MCS'02*, Cagliari, Italy. Volume 2364: Lecture Notes in Computer Science, pp. 1–15. Springer, 2002.

152. K. Ghosh, Y. S. Ng, and R. Srinivasan. Evaluation of decision fusion strategies for effective collaboration among heterogeneous fault diagnostic methods. *Computers & Chemical Engineering*, 35(2):342–355, 2011.

153. G. Giacinto and F. Roli. Dynamic classifier selection. In: *Proceedings of the First International Workshop on Multiple Classifier Systems*, Lecture Notes in Computer Science, pp. 177–189. Springer, 2000.

154. G. Giacinto and F. Roli. An approach to the automatic design of multiple classifier systems. *Pattern Recognition Letters*, 22:25–33, 2001.

155. G. Giacinto and F. Roli. Design of effective neural network ensembles for image classification processes. *Image Vision and Computing Journal*, 19(9-10):699–707, 2001.

156. G. Giacinto and F. Roli. Dynamic classifier selection based on multiple classifier behaviour. *Pattern Recognition*, 34(9):1879–1881, 2001.

157. P. O. Gislason, J. A. Benediktsson, and J. R. Sveinsson. Random forests for land cover classification. *Pattern Recognition Letters*, 27(4):294–300, 2006.

158. A. D. Gordon. *Classification*. Chapman & Hall /CRC, Boca Raton, FL, 1999.

159. B. Grofman, G. Owen, and S. L. Feld. Thirteen theorems in search of the truth. *Theory and Decisions*, 15:261–278, 1983.

160. V. Grossi and F. Turini. Stream mining: a novel architecture for ensemble-based classification. *Knowledge and Information Systems*, 30:247–281, 2012.

161. Y. Guermeur. Combining discriminant models with new multi-class SVMs. *Pattern Analysis & Applications*, 5(2):168–179, 2002.

162. C. Guerra-Salcedo and D. Whitley. Genetic approach to feature selection for ensemble creation. In: *Proc. Genetic and Evolutionary Computation Conference*, pp. 236–243. Morgan Kaufmann, 1999.

163. S. Günter and H. Bunke. Creation of classifier ensembles for handwritten word recognition using feature selection algorithms. In: *Proc. 8th International Workshop on Frontiers in Handwriting Recognition*, Canada, 2002.

164. I. Guyon and A. Elisseeff. An introduction to variable and feature selection. *Journal of Machine Learning Research*, 3:1157–1182, 2003.

165. I. Guyon, J. Weston, S. Barnhill, and V. Vapnik. Gene selection for cancer classification using support vector machines. *Machine Learning*, 46:389–422, 2002.

166. M. S. Haghighi, A. Vahedian, and H. S. Yazdi. Creating and measuring diversity in multiple classifier systems using support vector data description. *Applied Soft Computing*, 11(8):4931–4942, 2011.

167. M. Hall, E. Frank, G. Holmes, B. Pfahringer, P. Reutemann, and I. H. Witten. The WEKA data mining software: an update. *SIGKDD Explorations*, 11, 2009.

168. J. Ham, Y. C. Chen, M. M. Crawford, and J. Ghosh. Investigation of the random forest framework for classification of hyperspectral data. *IEEE Transactions on Geoscience and Remote Sensing*, 43(3):492–501, 2005.

169. M. Han, X. Zhu, and W. Yao. Remote sensing image classification based on neural network ensemble algorithm. *Neurocomputing*, 78:133–138, 2012.

170. Y. Han and L. Yu. A variance reduction framework for stable feature selection. In: *2010 IEEE International Conference on Data Mining*, pp. 206–215, 2010.

171. B. Hanczar, J. Hua, C. Sima, J. Weinstein, M. Bittner, and E. R. Dougherty. Small-sample precision of ROC-related estimates. *Bioinformatics (Oxford, England)*, 26(6):822–830, 2010.

172. D. J. Hand and K. Yu. Idiot's Bayes—not so stupid after all? *International Statistical Review*, 69:385–398, 2001.

173. D. J. Hand. Classifier technology and the illusion of progress (with discussion). *Statistical Science*, 21:1–34, 2006.

174. L. K. Hansen and P. Salamon. Neural network ensembles. *IEEE Transactions on Pattern Analysis and Machine Intelligence*, 12(10):993–1001, 1990.

175. S. Hashem. Optimal linear combinations of neural networks. *Neural Networks*, 10(4):599–614, 1997.

176. S. Hashem. Treating harmful collinearity in neural network ensembles. In: A. J. C. Sharkey, editor, *Combining Artificial Neural Nets*, pp. 101–125. Springer, London, 1999.

177. S. Hashem, B. Schmeiser, and Y. Yih. Optimal linear combinations of neural networks: an overview. In: *IEEE International Conference on Neural Networks*, Orlando, FL, pp. 1507–1512. IEEE, 1994.

178. T. Hastie, R. Tibshirani, and J. Friedman. *The Elements of Statistical Learning*. Springer, New York, 2001.

179. T. Hastie, R. Tibshirani, and J. Friedman. *The Elements of Statistical Learning. Data Mining, Inference, and Prediction*, 2nd edition. Springer, 2009.

180. S. Haykin. *Neural Networks. A Comprehensive Foundation*. Macmillan College Publishing Company, New York, 1994.

181. D. Hernandez-Lobato, G. Martinez-Mu noz, and A. Suarez. How large should ensembles of classifiers be? *Pattern Recognition*, 46(5):1323–1336, 2013.

182. T. K. Ho. The random space method for constructing decision forests. *IEEE Transactions on Pattern Analysis and Machine Intelligence*, 20(8):832–844, 1998.

183. T. K. Ho. Multiple classifier combination: lessons and the next steps. In: A. Kandel and H. Bunke, editors, *Hybrid Methods in Pattern Recognition*, pp. 171–198. World Scientific, 2002.

184. T. K. Ho, J. J. Hull, and S. N. Srihari. Decision combination in multiple classifier systems. *IEEE Transactions on Pattern Analysis and Machine Intelligence*, 16:66–75, 1994.

185. R. C. Holte. Very simple classification rules perform well on most commonly used datasets. *Machine Learning*, 11(1):63–91, 1993.

186. T. Hothorn and B. Lausen. Double-bagging: combining classifiers by bootstrap aggregation. *Pattern Recognition*, 36(6):1303–1309, 2003.

187. R. Hu and R. I. Damper. A 'No Panacea Theorem' for multiple classifier combination. In: *The 18th International Conference on Pattern Recognition (ICPR'06)*, Hong Kong, IEEE Computer Society, 2006.

188. Y.-M. Huang, C.-M. Hung, and H. C. Jiau. Evaluation of neural networks and data mining methods on a credit assessment task for class imbalance problem. *Nonlinear Analysis: Real World Applications*, 7:720–747, 2006.

189. Y. S. Huang and C. Y. Suen. A method of combining multiple classifiers—a neural network approach. In: *12th International Conference on Pattern Recognition*, Jerusalem, Israel, pp. 473–475. IEEE Computer Society, 1994.

190. Y. S. Huang and C. Y. Suen. A method of combining multiple experts for the recognition of unconstrained handwritten numerals. *IEEE Transactions on Pattern Analysis and Machine Intelligence*, 17:90–93, 1995.

191. H. Ishwaran, U. B. Kogalur, E. H. Blackstone, and M. S. Lauer. Random survival forests. *Annals of Applied Statistics*, 2(3):841–860, 2008.

192. K. Jackowski and M. Wozniak. Method of classifier selection using the genetic approach. *Expert Systems*, 27(2):114–128, 2010.

193. R. A. Jacobs. Methods for combining experts' probability assessments. *Neural Computation*, 7:867–888, 1995.

194. R. A. Jacobs, M. I. Jordan, S. J. Nowlan, and G. E. Hinton. Adaptive mixtures of local experts. *Neural Computation*, 3:79–87, 1991.

195. A. K. Jain and R. C. Dubes. *Algorithms for Clustering Data*. Prentice Hall, Englewood Cliffs, NJ, 1988.

196. A. K. Jain and J. Mao. Guest editorial: special issue on artificial neural networks and statistical pattern recognition. *IEEE Transactions on Neural Networks*, 8(1):1–3, 1997.

197. B. Jin, A. Strasburger, S. Laken, F. A. Kozel, K. Johnson, M. George, and X. Lu. Feature selection for fMRI-based deception detection. *BMC Bioinformatics*, 10(Suppl 9):S15, 2009.

198. M. I. Jordan and R. A. Jacobs. Hierarchies of adaptive experts. In: *International Conference on Neural Networks, San Diego, CA*, pp. 192–196. IEEE, 1992.

199. M. I. Jordan and L. Xu. Convergence results for the EM approach to mixtures of experts architectures. *Neural Networks*, 8:1409–1431, 1995.

200. A. Jóźwik and G. Vernazza. Recognition of leucocytes by a parallel k-nn classifier. In: *Proc. International Conference on Computer-Aided Medical Diagnosis, Warsaw*, pp. 138–153, 1987.

201. A. Kalousis, J. Prados, and M. Hilario. Stability of feature selection algorithms. In: *Proc. 5th IEEE International Conference on Data Mining (ICDM'05)*, pp. 218–225. IEEE, 2005.

202. E. M. Karabulut and T. Ibrikci. Effective diagnosis of coronary artery disease using the rotation forest ensemble method. *Journal of Medical systems*, 36(5):3011–3018, 2012.

203. T. Kavzoglu and I. Colkesen. An assessment of the effectiveness of a rotation forest ensemble for land-use and land-cover mapping. *International Journal of Remote Sensing*, 34(12):4224–4241, 2013.

204. H. Kim, H. Kim, H. Moon, and H. Ahn. A weight-adjusted voting algorithm for ensembles of classifiers. *Journal of the Korean Statistical Society*, 40(4):437–449, 2011.

205. K. Kira and L. A. Rendell. A practical approach to feature selection. In: *Proceedings of the Ninth International Workshop on Machine Learning*, pp. 249–256. Morgan Kaufmann, 1992.

206. J. Kittler and F. M. Alkoot. Relationship of sum and vote fusion strategies. In: F. Roli and J. Kittler, editors, *Proc. Second International Workshop on Multiple Classifier Systems, MCS'01*, Cambridge, UK. Volume 2096: Lecture Notes in Computer Science, pp. 339–348. Springer, 2001.

207. J. Kittler, M. Hatef, R. P. W. Duin, and J. Matas. On combining classifiers. *IEEE Transactions on Pattern Analysis and Machine Intelligence*, 20(3):226–239, 1998.

208. J. Kittler, A. Hojjatoleslami, and T. Windeatt. Strategies for combining classifiers employing shared and distinct representations. *Pattern Recognition Letters*, 18:1373–1377, 1997.

209. A. Klautau, N. Jevtić, and A. Orlitsky. On nearest-neighbor error-correcting output codes with application to all-pairs multiclass support vector machines. *Journal of Machine Learning Research*, 4:1–15, 2003.

210. E. M. Kleinberg. On the algorithmic implementation of stochastic discrimination. *IEEE Transactions on Pattern Analysis and Machine Intelligence*, 22(5):473–490, 2000.

211. A. H. R. Ko, R. Sabourin, A. de Souza Britto Jr., and L. Oliveira. Pairwise fusion matrix for combining classifiers. *Pattern Recognition*, 40(8):2198–2210, 2007.

212. R. Kohavi. Scaling up the accuracy of Naive-Bayes classifiers: a decision-tree hybrid. In: *Proceedings of the 2nd International Conference on Knowledge Discovery and Data Mining*, 1996.

213. R. Kohavi, B. Becker, and D. Sommerfield. Improving simple Bayes. Technical report, Data Mining and Visualization Group, Silicon Graphics Inc, California, 1997.

214. R. Kohavi and G. John. Wrappers for feature subset selection. *Artificial Intelligence Journal*, 97:273–324, 1997.

215. R. Kohavi and D. H. Wolpert. Bias plus variance decomposition for zero-one loss functions. In: L. Saitta, editor, *Machine Learning: Proc. 13th International Conference*, pp. 275–283. Morgan Kaufmann, 1996.

216. R. Kohavi. A study of cross-validation and bootstrap for accuracy estimation and model selection. In: *Proceedings of the 14th International Joint Conference on Artificial Intelligence - Volume 2*, IJCAI'95, pp. 1137–1143. Morgan Kaufmann Publishers Inc. San Francisco, CA, USA, 1995.

217. E. B. Kong and T. G. Dietterich. Error-correcting output coding corrects bias and variance. In: *Proc. 12th International Conference on Machine Learning*, California, pp. 313–321. Morgan Kaufmann, 1995.

218. I. Kononenko. Inductive and Bayesian learning in medical diagnosis. *Applied Artificial Intelligence*, 7:317–337, 1993.

219. P. Krizek, J. Kittler, and V. Hlavac. Improving stability of feature selection methods. In: *Proceedings of the 12th International Conference on Computer Analysis of Images and Patterns*, Berlin, Heidelberg. Volume 4673: CAIP'07, pages 929–936. Springer-Verlag, 2007.

220. A. Krogh and J. Vedelsby. Neural network ensembles, cross validation and active learning. In: G. Tesauro, D. S. Touretzky, and T. K. Leen, editors, *Advances in Neural Information Processing Systems*. Volume 7, pp. 231–238. MIT Press, Cambridge, MA, 1995.

221. S. Kumar, J. Ghosh, and M. M. Crawford. Hierarchical fusion of multiple classifiers for hyperspectral data analysis. *Pattern Analysis & Applications*, 5(2):210–220, 2002.

222. L. I. Kuncheva. 'fuzzy' vs 'non-fuzzy' in combining classifiers designed by boosting. *IEEE Transactions on Fuzzy Systems*, 11(6):729–741, 2003.

223. L. I. Kuncheva. *Combining Pattern Classifiers. Methods and Algorithms*. John Wiley and Sons, New York, 2004.

224. L. I. Kuncheva. Using diversity measures for generating error-correcting output codes in classifier ensembles. *Pattern Recognition Letters*, 26:83–90, 2005.

225. L. I. Kuncheva and C. O. Plumpton. Choosing parameters for random subspace ensembles for fMRI classification. In: *Proc. Proc. Multiple Classifier Systems (MCS'10)*, Cairo, Egypt, pp. 124–133. Springer, 2010.

226. L. I. Kuncheva and J. J. Rodríguez. An experimental study on rotation forest ensembles. In: *Proceedings of the 7th International Workshop on Multiple Classifier Systems (MCS'07)*, Prague, Czech Republic. Volume 4472: Lecture Notes in Computer Science, pp. 459–468. Springer, 2007.

227. L. I. Kuncheva and J. J. Rodríguez. Classifier ensembles for fMRI data analysis: an experiment. *Magnetic Resonance Imaging*, 28(4):583–593, 2010.

228. L. I. Kuncheva, J. J. Rodríguez, C. O. Plumpton, D. E. J. Linden, and S. J. Johnston. Random subspace ensembles for fMRI classification. *IEEE Transactions on Medical Imaging*, 29(2):531–542, 2010.

229. L. I. Kuncheva, C. J. Smith, S. Yasir, C. O. Phillips, and K. E. Lewis. Evaluation of feature ranking ensembles for high-dimensional biomedical data: a case study. In: *Proceedings of the Workshop on Biological Data Mining and its Applications in Healthcare (BioDM), IEEE 12th International Conference on Data Mining*, Brussels, Belgium, pp. 49–56. IEEE, 2012.

230. L. I. Kuncheva. Change-glasses approach in pattern recognition. *Pattern Recognition Letters*, 14:619–623, 1993.

231. L. I. Kuncheva. On combining multiple classifiers. In: *Proc. 7th International Conference on Information Processing and Management of Uncertainty (IPMU'98)*, Paris, France, pp. 1890–1891, 1998. ISBN 2-9516453-2-5.

232. L. I. Kuncheva. Clustering-and-selection model for classifier combination. In: *Proc. Knowledge-Based Intelligent Engineering Systems and Applied Technologies*, Brighton, UK, pp. 185–188, 2000.

233. L. I. Kuncheva. Using measures of similarity and inclusion for multiple classifier fusion by decision templates. *Fuzzy Sets and Systems*, 122(3):401–407, 2001.

234. L. I. Kuncheva. Switching between selection and fusion in combining classifiers: an experiment. *IEEE Transactions on Systems, Man, and Cybernetics*, 32(2):146–156, 2002.

235. L. I. Kuncheva. A theoretical study on six classifier fusion strategies. *IEEE Transactions on Pattern Analysis and Machine Intelligence*, 24(2):281–286, 2002.

236. L. I. Kuncheva. A stability index for feature selection. In: *Proc. IASTED, Artificial Intelligence and Applications*, Innsbruck, Austria, pp. 390–395, 2007.

237. L. I. Kuncheva, J. C. Bezdek, and R. P. W. Duin. Decision templates for multiple classifier fusion: an experimental comparison. *Pattern Recognition*, 34(2):299–314, 2001.

238. L. I. Kuncheva and L. C. Jain. Designing classifier fusion systems by genetic algorithms. *IEEE Transactions on Evolutionary Computation*, 4(4):327–336, 2000.

239. L. I. Kuncheva and J. J. Rodríguez. Classifier ensembles with a random linear oracle. *IEEE Transactions on Knowledge and Data Engineering*, 19(4):500–508, 2007.

240. L. I. Kuncheva, C. J. Whitaker, C. A. Shipp, and R. P. W. Duin. Is independence good for combining classifiers? In: *Proc. 15th International Conference on Pattern Recognition*, Barcelona, Spain, Volume 2, pp. 169–171. IEEE Computer Society, 2000.

241. L. I. Kuncheva, C. J. Whitaker, C. A. Shipp, and R. P. W. Duin. Limits on the majority vote accuracy in classifier fusion. *Pattern Analysis and Applications*, 6:22–31, 2003.

242. L. I. Kuncheva. Genetic algorithm for feature selection for parallel classifiers. *Information Processing Letters*, 46(4):163–168, 1993.

243. L. I. Kuncheva. A bound on kappa-error diagrams for analysis of classifier ensembles. *IEEE Transactions on Knowledge and Data Engineering*, 25(3):494–501, 2013.

244. L. I. Kuncheva and J. J. Rodríguez. A weighted voting framework for classifiers ensembles. *Knowledge and Information Systems* 38, 259–275, 2014.

245. P. A. Lachenbruch. Multiple reading procedures: the performance of diagnostic tests. *Statistics in Medicine*, 7:549–557, 1988.

246. C. Lai, M. J. T. Reinders, and L. Wessels. Random subspace method for multivariate feature selection. *Pattern Recognition Letters*, 27(10):1067–1076, 2006.

247. L. Lam. Classifier combinations: implementations and theoretical issues. In: J. Kittler and F. Roli, editors, *Multiple Classifier Systems*, Cagliari, Italy. Volume 1857: Lecture Notes in Computer Science, pp. 78–86. Springer, 2000.

248. L. Lam and A. Krzyzak. A theoretical analysis of the application of majority voting to pattern recognition. In: *12th International Conference on Pattern Recognition*, Jerusalem, Israel, pp. 418–420. IEEE Computer Society, 1994.

249. L. Lam and C. Y. Suen. Optimal combination of pattern classifiers. *Pattern Recognition Letters*, 16:945–954, 1995.

250. L. Lam and C. Y. Suen. Application of majority voting to pattern recognition: an analysis of its behavior and performance. *IEEE Transactions on Systems, Man, and Cybernetics*, 27(5):553–568, 1997.

251. P. Langley. Selection of relevant features in machine learning. In: *Proc. AAAI Fall Symposium on Relevance*, pp. 140–144, 1994.

252. P. Langley. The changing science of machine learning. *Machine Learning*, 82:275–279, 2011.

253. P. Langley, W. Iba, and K. Thompson. An analysis of Bayesian classifiers. In: *Proceedings of the 10th National Conference on Artificial Intelligence*, pages 399–406. AAAI Press, Menlo Park, California, 1992.

254. P. Latinne, O. Debeir, and C. Decaestecker. Different ways of weakening decision trees and their impact on classification accuracy of DT combination. In: J. Kittler and F. Roli, editors, *Multiple Classifier Systems*, Cagliari, Italy. Volume 1857: Lecture Notes in Computer Science, pp. 200–209. Springer, 2000.

255. A. Lazarevic and Z. Obradovic. Adaptive boosting techniques in heterogeneous and spatial databases. *Intelligent Data Analysis*, 5:1–24, 2001.

256. T. Lee, J. A. Richards, and P. H. Swain. Probabilistic and evidential approaches for multi-source data analysis. *IEEE Transactions on Geoscience and Remote Sensing*, 25(3):283–293, 1987.

257. Li, S., Hou, X., Zhang, H. & Cheng, Q. Learning Spatially Localized, Parts-Based Representation, Proceedings of the 2001 IEEE Computer Society Conference on Computer Vision and Pattern Recognition (CVPR), IEEE, 2001, 1, I-207-I-212.

258. Y. Li, M. Dong, and R. Kothari. Classifiability-based omnivariate decision trees. *IEEE Transactions on Neural Networks*, 16(6):1547–1560, 2005.

259. A. Liaw and M. Wiener. Classification and regression by randomForest. *R News*, 2(3):18–22, 2002.

260. X. Lin, S. Yacoub, J. Burns, and S. Simske. Performance analysis of pattern classifier combination by plurality voting. *Pattern Recognition Letters*, 24(12):1795–1969, 2003.

261. F. Lingenfelser, J. Wagner, and E. André. A systematic discussion of fusion techniques for multi-modal affect recognition tasks. In: *Proceedings of the 13th International Conference on Multimodal Interfaces*, ICMI'11, pp. 19–26. ACM New York, NY, USA, 2011.

262. B. Littlewood and D. R. Miller. Conceptual modeling of coincident failures in multiversion software. *IEEE Transactions on Software Engineering*, 15(12):1596–1614, 1989.

263. H. Liu, R. Stine, and L. Auslender, editors. *Proceedings of the Workshop on Feature Selection for Data Mining*, Newport Beach, CA, 2005.

264. K.-H. Liu and D.-S. Huang. Cancer classification using Rotation Forest. *Computers in Biology and Medicine*, 38(5):601–610, 2008.

265. T. Löfström. Estimating, optimizing and combining diversity and performance measures in ensemble creation. PhD thesis, Department of Technology, Orebro University, 2008.

266. P. Long and R. Servedio. Martingale boosting. In *Proceedings of the 18th Annual Conference on Computational Learning Theory (COLT)*, pp. 79–94. Springer, Lecture Notes in Computer Science, 2005.

267. P. M. Long and V. B. Vega. Boosting and microarray data. *Machine Learning*, 52:31–44, 2003.

268. C. G. Looney. *Pattern Recognition Using Neural Networks. Theory and Algorithms for Engineers and Scientists*. Oxford University Press, Oxford, 1997.

269. W. S. MacCulloch and W. Pitts. A logical calculus of the ideas immanent in nervous activity. *Bulletin of Mathematical Biophysics*, 5:115–133, 1943.

270. S. Mao, L. C. Jiao, L. Xiong, and S. Gou. Greedy optimization classifiers ensemble based on diversity. *Pattern Recognition*, 44(6):1245–1261, 2011.

271. D. D. Margineantu and T. G. Dietterich. Pruning adaptive boosting. In: *Proc. 14th International Conference on Machine Learning*, San Francisco, CA, pp. 378–387. Morgan Kaufmann, 1997.

272. A. I. Marques, V. Garcia, and J. S. Sanchez. Two-level classifier ensembles for credit risk assessment. *Expert Systems with Applications*, 39(12):10916–10922, 2012.

273. G. Martinez-Muñoz, D. Hernandez-Lobato, and A. Suarez. An analysis of ensemble pruning techniques based on ordered aggregation. *IEEE Transactions on Pattern Analysis and Machine Intelligence*, 31(2):245–259, 2009.

274. L. Mason, P. L. Bartlet, and J. Baxter. Improved generalization through explicit optimization of margins. *Machine Learning*, 38(3):243–255, 2000.

275. F. Masulli and G. Valentini. Comparing decomposition methods for classification. In: *Proc. International Conference on Knowledge-Based Intelligent Engineering Systems and Applied Technologies (Kes 2000)*, Brighton, UK, pp. 788–792, 2000.

276. F. Masulli and G. Valentini. Effectiveness of error-correcting output codes in multiclass learning problems. In: J. Kittler and F. Roli, editors, *Multiple Classifier Systems* Cagliari, Italy. Volume 1857: Lecture Notes in Computer Science, pp. 107–116. Springer, 2000.

277. O. Matan. On voting ensembles of classifiers (extended abstract). In: *Proceedings of AAAI-96 Workshop on Integrating Multiple Learned Models*, pp. 84–88. AAAI Press, 1996.

278. J. Maudes, J. J. Rodríguez, C. Garcia-Osorio, and N. Garcia-Pedrajas. Random feature weights for decision tree ensemble construction. *Information Fusion*, 13(1):20–30, 2012.

279. J. Meynet and J.-P. Thiran. Information theoretic combination of pattern classifiers. *Pattern Recognition*, 43(10):3412–3421, 2010.

280. L. Miclet and A. Cornuéjols. What is the place of Machine Learning between Pattern Recognition and Optimization? In: *Proceedings of the TML-2008 Conference (Teaching Machine Learning)*, Saint-Etienne, France, 2008.

281. D. J. Miller and L. Yan. Critic-driven ensemble classification. *IEEE Transactions on Signal Processing*, 47(10):2833–2844, 1999.

282. M. Minsky and S. Papert. *Perceptrons: An Introduction to Computational Geometry*. MIT Press, Cambridge, MA, 1969.

283. P. Moerland and E. Mayoraz. Dynaboost: combining boosted hypotheses in a dynamic way. Technical Report IDIAP-RR99-09, IDIAP (Dalle Molle Institute for Perceptual Artificial Intelligence), 1999.

284. L. C. Molina, L. Belanche, and A. Nebot. Feature selection algorithms: a survey and experimental evaluation. In: *Proc. the IEEE International Conference on Data Mining (ICDM'02)*, Japan. IEEE Computer Society, 2002.

285. A. M. Mood, F. A. Graybill, and D. C. Boes. *Introduction to the Theory of Statistics*, 3rd edition. McGraw-Hill Series in Probability and Statistics. McGraw-Hill, 1974.

286. C. Nadeau and Y. Bengio. Inference for the generalization error. *Machine Learning*, 62:239–281, 2003.

287. G. Nagy. Candide's practical principles of experimental pattern recognition. *IEEE Transactions on Pattern Analysis and Machine Intelligence*, 5(2):199–200, 1983.

288. L. Nanni and A. Lumini. An experimental comparison of ensemble of classifiers for bankruptcy prediction and credit scoring. *Expert Systems with Applications*, 36(2, Part 2): 3028–3033, 2009.

289. K. C. Ng and B. Abramson. Consensus diagnosis: a simulation study. *IEEE Transactions on Systems, Man, and Cybernetics*, 22:916–928, 1992.

290. N. J. Nilsson. *Learning Machines*. McGraw-Hill, New York, 1962.

291. S. J. Nowlan and G. E. Hinton. Evaluation of adaptive mixtures of competing experts. In: R. P. Lippmann, J. E. Moody, and D. S. Touretzky, editors, *Advances in Neural Information Processing Systems 3*, pp. 774–780. Morgan Kaufmann, 1991.

292. S.-B. Oh. On the relationship between majority vote accuracy and dependency in multiple classifier systems. *Pattern Recognition Letters*, 24:359–363, 2003.

293. D. Opitz and J. Shavlik. A genetic algorithm approach for creating neural network ensembles. In: A. J. C. Sharkey, editor, *Combining Artificial Neural Nets*, pp. 79–99. Springer, London, 1999.

294. D. W. Opitz. Feature selection for ensembles. In: *Proc. 16th National Conference on Artificial Intelligence, (AAAI)*, Orlando, FL, pp. 379–384. AAAI Press, 1999.

295. N. Oza and K. Tumer. Input decimation ensembles: decorrelation through dimensionality reduction. In: *Proc. 12nd International Workshop on Multiple Classifier Systems, MCS'01*, Cambridge, UK. Volume 2096: Lecture Notes in Computer Science, pp. 238–247. Springer, 2001.

296. N. C. Oza. Boosting with averaged weight vectors. In: T. Windeatt and F. Roli, editors, *Proc. 4rd Int. Workshop on Multiple Classifier Systems (MCS 2003)*, Guildford, UK. Volume 2709: Lecture Notes in Computer Science, pp. 15–24. Springer, 2003.

297. N. C. Oza and K. Tumer. Classifier ensembles: select real-world applications. *Information Fusion*, 9(1):4–20, 2008.

298. A. Ozcift. SVM feature selection based rotation forest ensemble classifiers to improve computer-aided diagnosis of Parkinson disease. *Journal of Medical Systems*, 36(4):2141–2147, 2012.

299. M. Pal. Random forest classifier for remote sensing classification. *International Journal of Remote Sensing*, 26(1):217–222, 2005.

300. I. Partalas, G. Tsoumakas, and I. Vlahavas. Pruning an ensemble of classifiers via reinforcement learning. *Neurocomputing*, 72:1900–1909, 2009.

301. I. Partalas, G. Tsoumakas, and I. Vlahavas. An ensemble uncertainty aware measure for directed hill climbing ensemble pruning. *Machine Learning*, 81(3):257–282, 2010.

302. D. Partridge and W. Krzanowski. Refining multiple classifier system diversity. Technical Report 348, Computer Science Department, University of Exeter, UK, 2003.

303. D. Partridge and W. J. Krzanowski. Software diversity: practical statistics for its measurement and exploitation. *Information & Software Technology*, 39:707–717, 1997.

304. A. Passerini, M. Pontil, and P. Frasconi. New results on error correcting output codes of kernel machines. *IEEE Transactions on Neural Networks*, 15(1):45–54, 2004.

305. E. A. Patrick. *Fundamentals of Pattern Recognition*. Prentice-Hall, Inc., Englewood Cliffs, NJ, 1972.

306. E. Pękalska, R. P. W. Duin, and M. Skurichina. A discussion on the classifier projection space for classifier combining. In: F. Roli and J. Kittler, editors, *Proc. 3rd International Workshop on Multiple Classifier Systems, MCS'02*, Cagliari, Italy. Volume 2364: Lecture Notes in Computer Science, pp. 137–148. Springer, 2002.

307. E. Pękalska, M. Skurichina, and R. P. W. Duin. Combining Fisher linear discriminant for dissimilarity representations. In: J. Kittler and F. Roli, editors, *Multiple Classifier*

Systems, Cagliari, Italy. Volume 1857: Lecture Notes in Computer Science, pp. 230–239. Springer, 2000.

308. F. Pereira, T. Mitchell, and M. Botvinick. Machine learning classifiers and fMRI: a tutorial overview. *NeuroImage*, 45(1, Supplement 1):S199–S209, 2009.

309. W. Pierce. Improving reliability of digital systems by redundancy and adaptation. PhD thesis, Electrical Engineering, Stanford University, 1961.

310. J. Platt. Probabilistic outputs for support vector machines and comparison to regularized likelihood methods. In: A. J. Smola, P. Bartlett, B. Schoelkopf, and D. Schuurmans, editors, *Advances in Large Margin Classifiers*, pp. 61–74. MIT Press, 2000.

311. R. Polikar. Ensemble based systems in decision making. *IEEE Circuits and Systems Magazine*, 6:21–45, 2006.

312. S. Prabhakar and A. K. Jain. Decision-level fusion in fingerprint verification. *Pattern Recognition*, 35(4):861–874, 2002.

313. L. Prechelt. PROBEN1—A set of neural network benchmark problems and benchmarking rules. Technical Report 21/94, University of Karlsruhe, Karlsruhe, Germany, 1994.

314. F. Provost and P. Domingos. Tree induction for probability-based ranking. *Machine Learning*, 52(3):199–215, 2003.

315. P. Pudil, J. Novovičová, and J. Kittler. Floating search methods in feature selection. *Pattern Recognition Letters*, 15:1119–1125, 1994.

316. O. Pujol, S. Escalera, and P. Radeva. An incremental node embedding technique for error correcting output codes. *Pattern Recognition*, 41:713–725, 2008.

317. O. Pujol, P. Radeva, and J. Vitria. Discriminant ECOC: a heuristic method for application dependent design of error correcting output codes. *IEEE Transactions on Pattern Analysis and Machine Intelligence*, 28(6):1007–1012, 2006.

318. S. Puuronen, A. Tsymbal, and I. Skrypnyk. Correlation-based and contextual merit-based ensemble feature selection. In: *Proc. 4th International Conference on Advances in Intelligent Data Analysis, IDA'01, Cascais, Portugal*. Volume 2189: Lecture Notes in Computer Science, pp. 135–144. Springer, 2001.

319. J. R. Quinlan. Bagging, boosting and C4.5. In: *Proc 13rd Int. Conference on Artificial Intelligence AAAI-96*, Cambridge, MA, pp. 725–730. AAAI Press, 1996.

320. J. R. Quinlan. Induction of decision trees. *Machine Learning*, 1(1):81–106, 1986.

321. R. Ranawana and V. Palade. Multi-classifier systems: review and a roadmap for developers. *International Journal of Hybrid Intelligent Systems*, 3(1):35–61, 2006.

322. L. A. Rastrigin and R. H. Erenstein. *Method of Collective Recognition*. Energoizdat, Moscow, 1981. (In Russian).

323. G. Rätsch, T. Onoda, and K.-R. Müller. Soft margins for AdaBoost. *Machine Learning*, 42(3):287–320, 2001.

324. Š. Raudys. Trainable fusion rules. I. Large sample size case. *Neural Networks*, 19(10): 1506–1516, 2006.

325. Š. Raudys. Trainable fusion rules. II. Small sample-size effects. *Neural Networks*, 19(10): 1517–1527, 2006.

326. J. Read, A. Bifet, G. Holmes, and B. Pfahringer. Scalable and efficient multi-label classification for evolving data streams. *Machine Learning*, 88(1–2):243–272, 2012.

327. G. Reid, S.and Grudic. Regularized linear models in stacked generalization. In: J. A. Benediktsson, J. Kittler, and F. Roli, editors, *Multiple Classifier Systems*.

Volume 5519: Lecture Notes in Computer Science, pp. 112–121. Springer, Berlin, Heidelberg, 2009.

328. M. D. Richard and R. P. Lippmann. Neural network classifiers estimate Bayesian: a posteriori probabilities. *Neural Computation*, 3:461–483, 1991.

329. R. Rifkin and A. Klautau. In defense of one-vs-all classification. *Journal of Machine Learning Research*, 5:101–141, 2004.

330. B. D. Ripley. *Pattern Recognition and Neural Networks*. University Press, Cambridge, 1996.

331. J. J. Rodríguez, L. I. Kuncheva, and C. J. Alonso. Rotation forest: a new classifier ensemble method. *IEEE Transactions on Pattern Analysis and Machine Intelligence*, 28(10):1619–1630, 2006.

332. J. J. Rodríguez, C. García-Osorio, and J. Maudes. Forests of nested dichotomies. *Pattern Recognition Letters*, 31(2):125–132, 2010.

333. L. Rokach. Collective-agreement-based pruning of ensembles. *Computational Statistics & Data Analysis*, 53(4):1015–1026, 2009.

334. L. Rokach. Taxonomy for characterizing ensemble methods in classification tasks: a review and annotated bibliography. *Computational Statistics and Data Analysis*, 53:4046–4072, 2009.

335. L. Rokach. Ensemble-based classifiers. *Artificial Intelligence Review*, 33:1–39, 2010.

336. F. Roli, G. Giacinto, and G. Vernazza. Methods for designing multiple classifier systems. In: J. Kittler and F. Roli, editors, *Proc. Second International Workshop on Multiple Classifier Systems*, Cambridge, UK. Volume 2096: Lecture Notes in Computer Science, pp. 78–87. Springer, 2001.

337. F. Roli, J. Kittler, G. Fumera, and D. Muntoni. An experimental comparison of classifier fusion rules for multimodal personal identity verification systems. In: F. Roli and J. Kittler, editors, *Multiple Classifier Systems*. Volume 2364: Lecture Notes in Computer Science, pp. 325–335. Springer, Berlin, Heidelberg, 2002.

338. F. Roli, J. Kittler, T. Windeatt, N. Oza, R. Polikar, M. Haindl, J. A. Benediktsson, N. El-Gayar, C. Sansone, and Z. H. Zhou, editors. *Proc. of the international Workshops on Multiple Classifier Systems*, Lecture Notes in Computer Science (LNCS). Springer, 2000–2013.

339. L. Rosasco, E. De, Vito A. Caponnetto, M. Piana, and A. Verri. Are loss functions all the same? *Neural Computation*, 15:1063–1076, 2004.

340. F. Rosenblatt. *Principles of Neurodynamics: Perceptrons and the Theory of Brain Mechanisms*. Spartan Books, Washington DC, 1962.

341. D. W. Ruck, S. K. Rojers, M. Kabrisky, M. E. Oxley, and B. W. Suter. The multilayer perceptron as an approximation to a Bayes optimal discriminant function. *IEEE Transactions on Neural Networks*, 1(4):296–298, 1990.

342. D. Ruta. Classifier diversity in combined pattern recognition systems. PhD thesis, University of Paisley, Scotland, UK, 2003.

343. D. Ruta and B. Gabrys. A theoretical analysis of the limits of majority voting errors for multiple classifier systems. Technical Report 11, ISSN 1461-6122, Department of Computing and Information Systems, University of Paisley, December 2000.

344. D. Ruta and B. Gabrys. Analysis of the correlation between majority voting error and the diversity measures in multiple classifier systems. In: *Proc. SOCO 2001*, Paisley, Scotland, 2001.

345. Y. Saeys, T. Abeel, and Y. Peer. Robust feature selection using ensemble feature selection techniques. In: *Proceedings of the European conference on Machine Learning and Knowledge Discovery in Databases—Part II*, pp. 313–325. Springer, 2008.

346. Y. Saeys, I. Inza, and P. Larrañaga. A review of feature selection techniques in bioinformatics. *Bioinformatics*, 23(19):2507–2517, 2007.

347. S. Salzberg. On comparing classifiers: pitfalls to avoid and a recommended approach. *Data Mining and Knowledge Discovery*, 1:317–328, 1997.

348. S. Salzberg. On comparing classifiers: a critique of current research and methods. *Data Mining and Knowledge Discovery*, 1:1–12, 1999.

349. F. Scarselli and A. C. Tsoi. Universal approximation using feedforward neural networks: a survey of some existing methods, and some new results. *Neural Networks*, 11(1):15–37, 1998.

350. R. E. Schapire. Theoretical views of boosting. In: *Proc. 4th European Conference on Computational Learning Theory*, pp. 1–10. Springer, 1999.

351. R. E. Schapire and Y. Freund. *Boosting: Foundations and Algorithms*. MIT Press, 2012.

352. R. E. Schapire, Y. Freund, P. Bartlett, and W. S. Lee. Boosting the margin: a new explanation for the effectiveness of voting methods. *The Annals of Statistics*, 26(5):1651–1686, 1998.

353. G. S. Sebestyen. *Decision-Making Process in Pattern Recognition*. The Macmillan Company, New York, 1962.

354. C. Seiffert, T. M. Khoshgoftaar, J. Van Hulse, and A. Napolitano. RUSBoost: a hybrid approach to alleviating class imbalance. *IEEE Transactions on Systems, Man, and Cybernetics Part A*, 40(1):185–197, 2010.

355. G. Seni and J. F. Elder. *Ensemble Methods in Data Mining: Improving Accuracy Through Combining Predictions*. Synthesis Lectures on Data Mining and Knowledge Discovery. Morgan & Claypool Publishers, 2010.

356. R. A. Servedio. Smooth boosting and learning with malicious noise. *Journal of Machine Learning Research*, 4:633–648, 2003.

357. M. Sewell. Ensemble learning. Technical Report RN/11/02, Department of Computer Science, UCL, London, 2011.

358. R. E. Shapire. Using output codes to boost multiclass learning problems. In: *Proc. 14th International Conference on Machine Learning*. Morgan Kaufmann, 1997.

359. L. Shapley and B. Grofman. Optimizing group judgemental accuracy in the presence of interdependencies. *Public Choice*, 43:329–343, 1984.

360. A. J. C. Sharkey, editor. *Combining Artificial Neural Nets. Ensemble and Modular Multi-Net Systems*. Springer, London, 1999.

361. A. J. C. Sharkey, N. E. Sharkey, U. Gerecke, and G. O. Chandroth. The test-and-select approach to ensemble combination. In: J. Kittler and F. Roli, editors, *Multiple Classifier Systems*, Cagliari, Italy. Volume 1857: Lecture Notes in Computer Science, pp. 30–44. Springer, 2000.

362. L. Shen and L. Bai. MutualBoost learning for selecting Gabor features for face recognition. *Pattern Recognition Letters*, 27(15):1758–1767, 2006.

363. H. W. Shin and S. Y. Sohn. Selected tree classifier combination based on both accuracy and error diversity. *Pattern Recognition*, 38(2):191–197, 2005.

364. J. Shotton, A. Fitzgibbon, M. Cook, T. Sharp, M. Finocchio, R. Moore, A. Kipman, and A. Blake. Real-time human pose recognition in parts from single depth images. In: *Proceedings of the 2011 IEEE Conference on Computer Vision and Pattern Recognition*, pp. 1297–1304. IEEE, 2011.

365. C. Sima, S. Attor, U. Brag-Neto, J. Lowey, E. Suh, and E. R. Dougherty. Impact of error estimation on feature selection. *Pattern Recognition*, 38:2472–2482, 2005.

366. P. Simeone, C. Marrocco, and F. Tortorella. Design of reject rules for ECOC classification systems. *Pattern Recognition*, 45(2):863–875, 2012.

367. S. Singh and M. Singh. A dynamic classifier selection and combination approach to image region labelling. *Signal Processing—Image Communication*, 20(3):219–231, 2005.

368. D. B. Skalak. The sources of increased accuracy for two proposed boosting algorithms. In: *Proc. American Association for Artificial Intelligence, AAAI-96, Integrating Multiple Learned Models Workshop*. AAAI Press, 1996.

369. M. Skurichina. Stabilizing weak classifiers. PhD thesis, Delft University of Technology, Delft, The Netherlands, 2001.

370. P. Smialowski, D. Frishman, and S. Kramer. Pitfalls of supervised feature selection. *Bioinformatics*, 26(3):440–443, 2010.

371. F. Smieja. The pandemonium system of reflective agents. *IEEE Transactions on Neural Networks*, 7:97–106, 1996.

372. P. C. Smits. Multiple classifier systems for supervised remote sensing image classification based on dynamic classifier selection. *IEEE Transactions on Geoscience and Remote Sensing*, 40(4):801–813, 2002.

373. P. H. A. Sneath and R. R. Sokal. *Numerical Taxonomy*. W.H. Freeman & Co, 1973.

374. P. Somol, J. Grim, J. Novovičová, and P. Pudil. Improving feature selection process resistance to failures caused by curse-of-dimensionality effects. *Kybernetika*, 47(3):401–425, 2011.

375. P. Somol and J. Novovičová. Evaluating stability and comparing output of feature selectors that optimize feature subset cardinality. *IEEE Transactions on Pattern Analysis and Machine Intelligence*, 32(11):1921–1939, 2010.

376. R. J. Stapenhurst. Diversity, margins and non-stationary learning. PhD thesis, School of Computer Science, The University of Manchester, UK, 2012.

377. G. Stiglic, J. J. Rodríguez, and P. Kokol. Rotation of random forests for genomic and proteomic classification problems. In H. R. Arabnia and Q. N. Tran, editors, *Software Tools and Algorithms for Biological Systems*. Volume 696: Advances in Experimental Medicine and Biology, pp. 211–221. Springer, 2011.

378. V. Svetnik, A. Liaw, C. Tong, J. C. Culberson, R. P. Sheridan, and B. P. Feuston. Random forest: a classification and regression tool for compound classification and QSAR modeling. *Journal of Chemical Information and Computer Sciences*, 43(6):1947–1958, 2003.

379. E. K. Tang, P. N. Suganthan, and X. Yao. An analysis of diversity measures. *Machine Learning*, 65(1):247–271, 2006.

380. E. K. Tang, P. N. Suganthan, and X. Yao. Gene selection algorithms for microarray data based on least squares support vector machine. *BMC Bioinformatics*, 7(1):95, 2006.

381. D. Tao, X. Tang, X. Li, and X. Wu. Asymmetric bagging and random subspace for support vector machines-based relevance feedback in image retrieval. *IEEE Transactions on Pattern Analysis and Machine Intelligence*, 28(7):1088–1099, 2006.

382. D. M. J. Tax, R. P. W. Duin, and M. van Breukelen. Comparison between product and mean classifier combination rules. In: *Proc. Workshop on Statistical Pattern Recognition*, Prague, Czech, 1997.

383. D. M. J. Tax, M. van Breukelen, R. P. W. Duin, and J. Kittler. Combining multiple classifier by averaging or multiplying? *Pattern Recognition*, 33:1475–1485, 2000.

384. K. M. Ting and I. H. Witten. Issues in stacked generalization. *Journal of Artificial Intelligence Research*, 10:271–289, 1999.

385. D. M. Titterington, G. D. Murray, L. S. Murray, D. J. Spiegelhalter, A. M. Skene, J. D. F. Habbema, and G. J. Gelpke. Comparison of discrimination techniques applied to a complex data set of head injured patients. *Journal of the Royal Statistical Society. Series A (General)*, 144:145–175, 1981.

386. J. T. Tou and R. C. Gonzalez. *Pattern Recognition Principles*. Reading, MA: Addison-Wesley, 1974.

387. G. T. Toussaint. Note on optimal selection of independent binary-valued features for pattern recognition. *IEEE Transactions on Information Theory*, 17:618, 1971.

388. V. Tresp and M. Taniguchi. Combining estimators using non-constant weighting functions. In: G. Tesauro, D. S. Touretzky, and T. K. Leen, editors, *Advances in Neural Information Processing Systems 7*. MIT Press, Cambridge, MA, 1995.

389. A. Tsymbal, P. Cunnigham, M. Pechenizkiy, and S. Puuronen. Search strategies for ensemble feature selection in medical diagnostics. Technical report, Trinity College Dublin, Ireland, 2003.

390. A. Tsymbal, M. Pechenizkiy, and P. Cunningham. Diversity in search strategies for ensemble feature selection. *Information Fusion*, 6(1):83–98, 2005.

391. J. D. Tubbs and W. O. Alltop. Measures of confidence associated with combining classification rules. *IEEE Transactions on Systems, Man, and Cybernetics*, 21:690–692, 1991.

392. S. Tulyakov, S. Jaeger, V. Govindaraju, and D. Doermann. Review of classifier combination methods. In: Simone Marinai and Hiromichi Fujisawa, editors, *Machine Learning in Document Analysis and Recognition*. Volume 90: *Studies in Computational Intelligence*, pp. 361–386. Springer, Berlin, Heidelberg, 2008.

393. K. Tumer and J. Ghosh. Analysis of decision boundaries in linearly combined neural classifiers. *Pattern Recognition*, 29(2):341–348, 1996.

394. K. Tumer and J. Ghosh. Linear and order statistics combiners for pattern classification. In: A. J. C. Sharkey, editor, *Combining Artificial Neural Nets*, pp. 127–161. Springer, London, 1999.

395. N. Ueda. Optimal linear combination of neural networks for improving classification performance. *IEEE Transactions on Pattern Analysis and Machine Intelligence*, 22(2):207–215, 2000.

396. G. Valentini. Ensemble methods based on bias-variance analysis. PhD thesis, University of Genova, Genova, Italy, 2003.

397. G. Valentini and M. Re. Ensemble methods: a review. In: *Advances in Machine Learning and Data Mining for Astronomy*. Chapman & Hall, 2012.

398. V. Valev and A. Asaithambi. Multidimensional pattern recognition problems and combining classifiers. *Pattern Recognition Letters*, 22(12):1291–1297, 2001.

399. M. van Breukelen, R. P. W Duin, D. M. J. Tax, and J. E. den Hartog. Combining classifiers for the recognition of handwritten digits. In: *I-st IAPR TC1 Workshop on Statistical Techniques in Pattern Recognition*, Prague, Czech Republic, pp. 13–18, 1997.

400. M. Van Erp, L. Vuurpijl, and L. Schomaker. An overview and comparison of voting methods for pattern recognition. In: *Proceedings of the 8th IEEE International Workshop on Frontiers in Handwriting Recognition (WFHR02)*, pp. 195–200. IEEE, 2002.

401. S. Van Landeghem, T. Abeel, Y. Saeys, and Y. Van de Peer. Discriminative and informative features for biomolecular text mining with ensemble feature selection. *Bioinformatics (Oxford, England)*, 26(18):i554–i560, 2010.

402. V. N. Vapnik. *The Nature of Statistical Learning Theory*. Springer, 1995.

403. V. N. Vapnik. An overview of statistical learning theory. *IEEE Transactions on Neural Networks*, 10(5):988–999, 1999.

404. S. B. Vardeman and M. D. Morris. Majority voting by independent classifiers can increase error rates. *The American Statistician*, 67(2):94–96, 2013.

405. P. Viola and M. J. Jones. Robust real-time face detection. *International Journal of Computer Vision*, 57(2):137–154, 2004.

406. E. A. Wan. Neural network classification: a Bayesian interpretation. *IEEE Transactions on Neural Networks*, 1(4):303–305, 1990.

407. G. Wang, J. Ma, L. Huang, and K. Xu. Two credit scoring models based on dual strategy ensemble trees. *Knowledge-Based Systems*, 26:61–68, 2012.

408. H. Wang, T. M. Khoshgoftaar, and A. Napolitano. A comparative study of ensemble feature selection techniques for software defect prediction. In: *2010 Ninth International Conference on Machine Learning and Applications (ICMLA)*, pp. 135–140. IEEE, 2010.

409. H. Wang, T. M. Khoshgoftaar, and R. Wald. Measuring robustness of feature selection techniques on software engineering datasets. In: *2011 IEEE International Conference on Information Reuse and Integration (IRI)*, pp. 309–314. IEEE, 2011.

410. X. Wang and X. Tang. Random sampling for subspace face recognition. *International Journal of Computer Vision*, 70:91–104, 2006.

411. M. Warmuth, J. Liao, and G. Ratsch. Totally corrective boosting algorithms that maximize the margin. In: *Proceedings of the 23rd International Conference on Machine Learning*, New York, pp. 1001–1008. ACM, ACM International Conference Proceeding Series, ISBN 1-59593-383-2, 2006.

412. G. I. Webb. MultiBoosting: a technique for combining boosting and wagging. *Machine Learning*, 40(2):159–196, 2000.

413. G. I. Webb, J. Boughton, and Z. Wang. Not so naive Bayes: aggregating one-dependence estimators. *Machine Learning*, 58(1):5–24, 2005.

414. W. Wei, T. K. Leen, and E. Barnard. A fast histogram-based postprocessor that improves posterior probability estimates. *Neural Computation*, 11(5):1235–1248, 1999.

415. T. Windeatt. Diversity measures for multiple classifier system analysis and design. *Information Fusion*, 6(1):21–36, 2005.

416. T. Windeatt and G. Ardeshir. Boosted ECOC ensembles for face recognition. In: *International Conference on Visual Information Engineering VIE 2003*, pp. 165–168. IEEE, 2003.

417. T. Windeatt and R. Ghaderi. Coding and decoding strategies for multi-class learning problems. *Information fusion*, 4:11–21, 2003.

418. T. Woloszynski and M. Kurzynski. A probabilistic model of classifier competence for dynamic ensemble selection. *Pattern Recognition*, 44(10–11):2656–2668, 2011.

419. T. Woloszynski, M. Kurzynski, P. Podsiadlo, and G. W. Stachowiak. A measure of competence based on random classification for dynamic ensemble selection. *Information Fusion*, 13(3):207–213, 2012.

420. D. H. Wolpert. Stacked generalization. *Neural Networks*, 5(2):241–260, 1992.

421. K. Woods, W. P. Kegelmeyer, and K. Bowyer. Combination of multiple classifiers using local accuracy estimates. *IEEE Transactions on Pattern Analysis and Machine Intelligence*, 19:405–410, 1997.

422. X. Wu, V. Kumar, J. R. Quinlan, J. Ghosh, Q. Yang, H. Motoda, G. J. McLachlan, A. Ng, B. Liu, P. S. Yu, Z.-H. Zhou, M. Steinbach, D. J. Hand, and D. Steinberg. Top 10 algorithms in data mining. *Knowledge and Information Systems*, 14(1):1–37, 2008.

423. J.-F. Xia, K. Han, and D.-S. Huang. Sequence-based prediction of protein-protein interactions by means of Rotation Forest and autocorrelation descriptor. *Protein and Peptide Letters*, 17(1):137–145, 2010.

424. J. Xiao, C. He, X. Jiang, and D. Liu. A dynamic classifier ensemble selection approach for noise data. *Information Sciences*, 180(18):3402–3421, 2010.

425. L. Xu, A. Krzyzak, and C. Y. Suen. Methods of combining multiple classifiers and their application to handwriting recognition. *IEEE Transactions on Systems, Man, and Cybernetics*, 22:418–435, 1992.

426. O. T. Yildiz and E. Alpaydin. Omnivariate decision trees. *IEEE Transactions on Neural Networks*, 12(6):1539–1546, 2001.

427. L. Yu, Y. Han, and M. E. Berens. Stable gene selection from microarray data via sample weighting. *IEEE/ACM Transactions on Computational Biology and Bioinformatics*, 9(1):262–272, 2012.

428. L. Yu and H. Liu. Feature selection for high-dimensional data: a fast correlation-based filter solution. In: *Proceedings of the 20th International Conference on Machine Learning (ICML2003)*, Washington, DC, AAAI Press, 2003.

429. G. U. Yule. On the association of attributes in statistics. *Phil. Trans., A*, 194:257–319, 1900.

430. B. Zadrozny and C. Elkan. Obtaining calibrated probability estimates from decision trees and naive Bayesian classifiers. In: *Proceedings of the Eighteenth International Conference on Machine Learning (ICML'01)*, pp. 609–616. Morgan Kaufmann, 2001.

431. G. Zenobi and P. Cunningham. Using diversity in preparing ensembles of classifiers based on different feature subsets to minimize generalization error. In: Lecture Notes in Computer Science, pp. 576–587. Springer, 2001.

432. B. Zhang. Reliable classification of vehicle types based on cascade classifier ensembles. *IEEE Transactions on Intelligent Transportation Systems*, 14(1):322–332, 2013.

433. B. Zhang and T. D. Pham. Phenotype recognition with combined features and random subspace classifier ensemble. *BMC Bioinformatics*, 12(128), 2011.

434. C.-X. Zhang and R. P. W. Duin. An experimental study of one- and two-level classifier fusion for different sample sizes. *Pattern Recognition Letters*, 32(14):1756–1767, 2011.

435. C.-X. Zhang and J.-S. Zhang. Rotboost: a technique for combining rotation forest and adaboost. *Pattern Recognition Letters*, 29(10):1524–1536, 2008.

436. L. Zhang and W.-D. Zhou. Sparse ensembles using weighted combination methods based on linear programming. *Pattern Recognition*, 44(1):97–106, 2011.

437. Z. Zhang and P. Yang. An ensemble of classifiers with genetic algorithm based feature selection. *IEEE Intelligent Informatics Bulletin*, 9(1):18–24, 2008.

438. G. Zhong and C.-L. Liu. Error-correcting output codes based ensemble feature extraction. *Pattern Recognition*, 46(4):1091–1100, 2013.

439. Z.-H. Zhou. *Ensemble Methods: Foundations and Algorithm*. CRC Press – Business & Economics, 2012.

440. Z.-H. Zhou and N. Li. Multi-information ensemble diversity. In: N. Gayar, J. Kittler, and F. Roli, editors, *Multiple Classifier Systems*. Volume 5997: Lecture Notes in Computer Science, pp. 134–144. Springer, Berlin, Heidelberg, 2010.

441. J. Zhu, S. Rosset, H. Zou, and T. Hastie. Multiclass AdaBoost. Technical Report 430, Department of Statistics, University of Michigan, 2006.

442. Y. Zhu, J. Liu, and S. Chen. Semi-random subspace method for face recognition. *Image and Vision Computing*, 27(9):1358–1370, 2009.

443. Yu. A. Zuev. A probability model of a committee of classifiers. *USSR Comput. Math. Math. Phys.*, 26(1):170–179, 1987.

INDEX

Combining Pattern Classifiers: Methods and Algorithms, Second Edition. Ludmila I. Kuncheva.
© 2014 John Wiley & Sons, Inc. Published 2014 by John Wiley & Sons, Inc.